Introduction to Fourier Analysis on Euclidean Spaces

PRINCETON MATHEMATICAL SERIES

Editors: WU-CHUNG HSIANG, JOHN MILNOR, and ELIAS M. STEIN

1. The Classical Groups, By HERMANN WEYL
3. An Introduction to Differential Geometry, By LUTHER PFAHLER EISENHART
4. Dimension Theory, By W. HUREWICZ and H. WALLMAN
6. The Laplace Transform, By D. V. WIDDER
7. Integration, By EDWARD J. MCSHANE
8. Theory of Lie Groups: 1, By C. CHEVALLEY
9. Mathematical Methods of Statistics, By HARALD CRAMÉR
10. Several Complex Variables, By S. BOCHNER and W. T. MARTIN
11. Introduction to Topology, By S. LEFSHETZ
12. Algebraic Geometry and Topology, edited By R. H. FOX, D. C. SPENCER, and A. W. TUCKER
14. The Topology of Fibre Bundles, By NORMAN STEENROD
15. Foundations of Algebraic Topology, By SAMUEL EILENBERG and NORMAN STEENROD
16. Functionals of Finite Riemann Surfaces, By MENAHEM SCHIFFER and DONALD C. SPENCER.
17. Introduction to Mathematical Logic, Vol. 1, By ALONZO CHURCH
19. Homological Algebra, By H. CARTAN and S. EILENBERG
20. The Convolution Transform, By I. I. HIRSCHMAN and D. V. WIDDER
21. Geometric Integration Theory, By H. WHITNEY
22. Qualitative Theory of Differential Equations, By V. V. NEMYTSKII and V. V. STEPANOV
23. Topological Analysis, By GORDON T. WHYBURN (revised 1964)
24. Analytic Functions, By AHLFORS, BEHNKE and GRAUERT, BERS, et al.
25. Continuous Geometry, By JOHN VON NEUMANN
26. Riemann Surfaces, by L. AHLFORS and L. SARIO
27. Differential and Combinatorial Topology, edited By S. S. CAIRNS
28. Convex Analysis, By R. T. ROCKAFELLAR
29. Global Analysis, edited By D. C. SPENCER and S. IYANAGA
30. Singular Integrals and Differentiability Properties of Functions, By E. M. STEIN
31. Problems in Analysis, edited By R. C. GUNNING
32. Introduction to Fourier Analysis on Euclidean Spaces, By E. M. STEIN and G. WEISS

Introduction to

FOURIER ANALYSIS ON EUCLIDEAN SPACES

By Elias M. Stein
& Guido Weiss

PRINCETON, NEW JERSEY
PRINCETON UNIVERSITY PRESS

L.C. Card: 73-106394

ISBN: 0-691-08078-X

AMS 1970: 31B05, 31B25, 32A07, 33A45, 42A02, 42A18,
42A40, 42A68, 44A25, 46E30, 46E35

Second Printing, with corrections, 1975

Sixth Printing, 1990

9 8

Princeton University Press books are printed on acid-free paper and meet the guidelines for permanence and durability of the Committee on Production Guidelines for Book Longevity of the Council on Library Resources

This book has been composed in Times New Roman type

Printed in the United States of America

THIS BOOK IS DEDICATED TO

Antoni Zygmund

IN APPRECIATION FOR HIS FRIENDSHIP,
HIS TEACHING, AND THE INSPIRATION
HE GAVE US

Preface

This book is designed to be an introduction to harmonic analysis in Euclidean spaces. The subject has seen a considerable flowering during the past twenty years. We have not tried to cover all phases of this development. Rather, our chief concern was to illustrate various methods used in this aspect of Fourier analysis that exploit the structure of Euclidean spaces. In particular, we try to show the role played by the action of translations, dilations, and rotations. Another concern, not independent of this chief one, is to motivate the study of harmonic analysis on more general spaces having an analogous structure (such as arises in symmetric spaces). It is our feeling that the study of Fourier analysis in that context and, also, in other general settings, is more meaningful once the special Euclidean case is understood.

Because of these concerns we have not included several topics that are usually presented in more general treatments of harmonic analysis. For example, results centering around the Wiener Tauberian theorem, which hold in the case of general locally compact Abelian groups, do not involve the special features of Euclidean spaces. Stated very briefly, our selection of topics was motivated by showing how real variable and complex variable methods extend from the one-dimensional to the many-dimensional case.

We require that the readers have mastered the material that is usually covered in a course in the theory of integration and in the theory of functions of a complex variable. We also believe that this book is much more meaningful to someone who has some previous acquaintance with harmonic analysis. If the reader has no such previous experience, he might find it profitable first to look at the expository article "Harmonic Analysis" (see Guido Weiss [3]).

The idea of writing this book first occurred to us when we presented some of these topics in graduate courses given during the academic year 1958–59. Since that time each of us has lectured on this subject at various times and places. We are thankful to our colleagues and students who helped us clarify our ideas and organize this material.

More specifically, one of us (Weiss) taught a course on this subject at Washington University jointly with Mitchell Taibleson during the academic year 1963–64. We are grateful for his contribution to the organization of this course and his continued interest in our effort. The other author gave courses at the University of Chicago (1961–62) and Princeton (1963–65) that dealt with the topics of this book. Our task of writing the book was considerably facilitated by having the lecture notes, which were prepared by R. Askey, N. J. Weiss, and D. Levine. It is a pleasure to express our appreciation to them. Ronald R. Coifman was a constant help throughout the writing of this book. His many suggestions are incorporated in much of the presentation. Miguel de Guzmán, J. R. Hattemer, Stephen Wainger, and N. J. Weiss read the entire manuscript. We are grateful for their corrections and comments. We are also indebted to I. I. Hirschman Jr., who read part of the manuscript and made several useful suggestions, and to Mrs. A. Bonami, J. L. Clerc, L. J. Dickson, S. S. Gelbart, and S. Zucker who helped us correct the proof sheets.

Princeton, N.J.
St. Louis, Mo.

Contents

Introduction to Fourier Analysis on Euclidean Spaces

CHAPTER I

The Fourier Transform

In this chapter we introduce the Fourier transform and study its more elementary properties. Since most of the material of this chapter is rather standard our treatment here will be brief. We begin by considering the behavior of the Fourier transform on the spaces $L^1(E_n)$ and $L^2(E_n)$. This will be done in the first two sections. The Fourier transform's formal aspects are more easily described in the context of distributions; therefore, in the third section we extend its definition to the space of tempered distributions. The reader will note that in this chapter we are mainly exploiting the translation structure of Euclidean spaces. In the following chapters (and specifically in Chapter IV), however, the action of rotations on these spaces plays an important role.

1. The Basic L^1 Theory of the Fourier Transform

We begin by introducing some notation that will be used throughout this work. E_n denotes n-dimensional (real) Euclidean space. We consistently write $x = (x_1, x_2, \ldots, x_n)$, $y = (y_1, y_2, \ldots, y_n), \ldots$ for the elements of E_n. The *inner product* of $x, y \in E_n$ is the number $x \cdot y = \sum_{j=1}^{n} x_j y_j$; the *norm* of $x \in E_n$ is the (nonnegative) number $|x| = \sqrt{x \cdot x}$; furthermore, $dx = dx_1\, dx_2 \cdots dx_n$ denotes the element of ordinary Lebesgue measure.

We will deal with various spaces of functions defined on E_n. The simplest of these are the $L^p = L^p(E_n)$ spaces, $1 \leqq p < \infty$, of all measurable functions f such that $\|f\|_p = (\int_{E_n} |f(x)|^p\, dx)^{1/p} < \infty$. The number $\|f\|_p$ is called the *L^p norm* of f. The space $L^\infty(E_n)$ consists of all essentially bounded functions on E_n and, for $f \in L^\infty(E_n)$, we let $\|f\|_\infty$ be the essential supremum of $|f(x)|$, $x \in E_n$.[1] Often, the space C_0 of all continuous functions vanishing at infinity, with the L^∞ norm just described, arises more naturally than $L^\infty = L^\infty(E_n)$. Unless otherwise specified, all functions are assumed to be

[1] We say f is *equivalent* to g if $f(x) = g(x)$ for almost every $x \in E_n$, whenever f $g \in L^p(E_n)$, $1 \leqq p \leqq \infty$. If we consider the equivalence classes obtained from this relation and define the *norm* of a class to be the norm of any one of its representatives (clearly $\|f\|_p = \|g\|_p$ if f is equivalent to g) we obtain a Banach space. We shall denote this space by $L^p(E_n)$ as well. It will be obvious from the context which of these spaces $L^p(E_n)$ is under discussion.

complex valued; it will be assumed, throughout the book, that all functions are (Borel) measurable.

If $f \in L^1(E_n)$ the *Fourier transform* of f is the function $\hat{f}$ defined by letting

$$\hat{f}(x) = \int_{E_n} f(t) e^{-2\pi i x \cdot t}\, dt$$

for all $x \in E_n$. It is easy to establish the following results:

THEOREM 1.1. (a) *The mapping $f \to \hat{f}$ is a bounded linear transformation from $L^1(E_n)$ into $L^\infty(E_n)$. In fact $\|\hat{f}\|_\infty \leqq \|f\|_1$;*

(b) *If $f \in L^1(E_n)$ then $\hat{f}$ is uniformly continuous.*

THEOREM 1.2 (Riemann–Lebesgue). *If $f \in L^1(E_n)$ then $\hat{f}(x) \to 0$ as $|x| \to \infty$; thus, in view of the last result, we can conclude that $\hat{f}$ belongs to the class C_0.*

Theorem 1.1 is obvious; moreover, so is Theorem 1.2 when f is the characteristic function of the n-dimensional interval $I = \{x \in E_n;\, a_1 \leqq x_1 \leqq b_1, \ldots, a_n \leqq x_n \leqq b_n\}$ (for we can calculate $\hat{f}$ explicitly as an iterated integral). The same is therefore true for a finite linear combination of such characteristic functions. The result for a general $f \in L^1(E_n)$ follows easily by approximating f in the L^1 norm by such a linear combination g; for then $f = g + (f - g)$, where $\hat{f} - \hat{g}$ is uniformly small (by Theorem 1.1, part (a)) while $\hat{g}(x) \to 0$ as $|x| \to \infty$.

Theorem 1.2 gives a necessary condition for a function to be a Fourier transform. Belonging to the class C_0, however, is far from being sufficient (see 4.1). There seems to be no simple satisfactory condition characterizing Fourier transforms of functions in $L^1(E_n)$.

The above definition of the Fourier transform extends immediately to finite Borel measures: if μ is such a measure on E_n we define $\hat{\mu}$ by letting

$$\hat{\mu}(x) = \int_{E_n} e^{-2\pi i x \cdot t}\, d\mu(t).$$

Theorem 1.1 is valid for these Fourier transforms if we replace the L^1-norm by the total variation of μ.

In addition to the vector-space operations, $L^1(E_n)$ is endowed with a "multiplication" making this space a Banach algebra. This operation, called *convolution*, is defined in the following way: If f and g belong to $L^1(E_n)$ their convolution $h = f * g$ is the function whose value at $x \in E_n$ is

$$h(x) = \int_{E_n} f(x - y) g(y)\, dy.$$

One can show by an elementary argument that $f(x - y)g(y)$ is a measurable function of the two variables x and y. It then follows immediately from Fubini's theorem on the interchange of the order of integration that $h \in L^1(E_n)$ and $\|h\|_1 \leqq \|f\|_1\|g\|_1$. Furthermore, this operation is commutative and associative. More generally, $h = f * g$ is defined whenever $f \in L^p(E_n)$, $1 \leqq p \leqq \infty$, and $g \in L^1(E_n)$. In fact we have the following result:

THEOREM 1.3. *If $f \in L^p(E_n)$, $1 \leqq p \leqq \infty$, and $g \in L^1(E_n)$ then $h = f * g$ is well defined and belongs to $L^p(E_n)$. Moreover,*

$$\|h\|_p \leqq \|f\|_p\|g\|_1.$$

It is clear that $|h(x)| \leqq \int_{E_n} |f(x - y)|\,|g(y)|\,dy$. Thus the desired result is an easy consequence of Minkowski's integral inequality:

$$\left(\int_{E_n} |h(x)|^p\,dx\right)^{1/p} \leqq \int_{E_n}\left(\int_{E_n} |f(x - y)|^p\,dx\right)^{1/p} |g(y)|\,dy$$
$$= \|f\|_p\|g\|_1.$$

As was the case for the Fourier transform, we can extend this operation to include finite Borel measures: if μ is such a measure on E_n we define $h = f * d\mu$ by letting

$$h(x) = \int_{E_n} f(x - y)\,d\mu(y),$$

for $x \in E_n$ and $f \in L^p$. Theorem 1.3 is valid for these convolutions if we replace the L^1-norm of g by the total variation of μ.

An essential feature of harmonic analysis is the fact that the Fourier transform of the convolution of two functions is the (pointwise) product of their Fourier transforms. More precisely, the following result is an easy consequence of the definitions:

THEOREM 1.4. *If f and g belong to $L^1(E_n)$ then*

$$(f * g)\hat{} = \hat{f}\hat{g}.$$ [2]

Many other important operations of analysis have particularly simple relations with the Fourier transform. For example, if we let τ_h denote *translation* by $h \in E_n$ (by this we mean the operator mapping the function $g(x)$ into the function $g(x - h)$) we have

$$\text{(1.5)} \qquad \begin{aligned} &\text{(i)} \quad (\tau_h f)\hat{}(x) = e^{-2\pi i h\cdot x}\hat{f}(x), \\ &\text{(ii)} \quad (e^{2\pi i t\cdot h} f(t))\hat{}(x) = (\tau_h \hat{f})(x), \end{aligned}$$

whenever $f \in L^1(E_n)$.

[2] We shall use this notation consistently: $(\ldots)\hat{}$ denotes the Fourier transform of $(\ldots)$.

If $a > 0$ we let δ_a denote *dilation by a*; that is, δ_a is the operator mapping the function $g(x)$ into the function $g(ax)$. Whenever $f \in L^1(E_n)$ we then have

$$a^n(\delta_a f)\hat{}\,(x) = \hat{f}(a^{-1}x). \tag{1.6}$$

Differentiation and Fourier transformation are related in the following way:

THEOREM 1.7. *Suppose $f \in L^1(E_n)$ and $x_k f(x) \in L^1(E_n)$, where x_k is the k-th coordinate function. Then $\hat{f}$ is differentiable with respect to x_k and*

$$\frac{\partial \hat{f}}{\partial x_k}(x) = (-2\pi i t_k f(t))\hat{}\,(x).$$

PROOF. Letting $h = (0, \ldots, h_k, \ldots, 0)$ be a nonzero vector along the k-th coordinate axis, we have, by part (ii) of (1.5) and the Lebesgue dominated convergence theorem,

$$\frac{\hat{f}(x+h) - \hat{f}(x)}{h_k} = \left\{\left(\frac{e^{-2\pi i t \cdot h} - 1}{h_k}\right) f(t)\right\}\hat{}\,(x) \to (-2\pi i t_k f(t))\hat{}\,(x)$$

as $h_k \to 0$.

Theorem 1.7 asserts that applying the Fourier transform after multiplying by the k-th coordinate function is equivalent (up to a multiplicative constant) to taking the partial derivative with respect to the k-th variable of the Fourier transform. It is also true that the Fourier transforms of such partial derivatives are obtainable (again, up to a multiplicative constant) by multiplying the Fourier transform by the corresponding coordinate functions. We shall encounter many versions of this result. In order to make a precise statement of one of these versions (perhaps the simplest to prove) we introduce the following concept: We say that f is *differentiable in the L^p norm with respect to x_k* whenever $f \in L^p(E_n)$ and there exists a g in $L^p(E_n)$ such that

$$\left(\int_{E_n} \left|\frac{f(x+h) - f(x)}{h_k} - g(x)\right|^p dx\right)^{1/p} \to 0,$$

as $h_k \to 0$ (we are using the notation established in the proof of Theorem 1.7). The function g is called the *partial derivative of f (with respect to x_k) in the L^p norm.*

Applying part (i) of (1.5) and part (a) of Theorem 1.1 to

$$|\hat{g}(x) - \hat{f}(x)(e^{2\pi i (h \cdot x)} - 1)/h_k|,$$

and then letting $h_k \to 0$ we obtain:

THEOREM 1.8. *If $f \in L^1(E_n)$ and g is the partial derivative of f with respect to x_k in the L^1 norm then*

$$\hat{g}(x) = 2\pi i x_k \hat{f}(x).$$

Both Theorems 1.7 and 1.8 can be extended to higher derivatives. Without going into details, we note the following formulas:

(1.9)
$$\text{(i)}\quad P(D)\hat{f}(x) = (P(-2\pi i t)f(t))\hat{\ }(x),$$
$$\text{(ii)}\quad (P(D)f)\hat{\ }(x) = P(2\pi i x)\hat{f}(x),$$

where, for an n-tuple $\alpha = (\alpha_1, \ldots, \alpha_n)$ of nonnegative integers we let $x^\alpha = x_1^{\alpha_1} x_2^{\alpha_2} \cdots x_n^{\alpha_n}$, $D^\alpha = \partial^{\alpha_1+\alpha_2+\cdots+\alpha_n}/\partial x_1^{\alpha_1}\, \partial x_2^{\alpha_2} \cdots \partial x_n^{\alpha_n}$, P is a polynomial in the n variables $x_1, x_2, \ldots, x_n$ and $P(D)$ is the associated differential operator (i.e., we replace x^α by D^α in $P(x)$).

We now turn to the problem of inverting the Fourier transform. That is, we shall consider the question: *Given the Fourier transform $\hat{f}$ of an integrable function f, how do we obtain f back again from $\hat{f}$?* The reader familiar with the elementary theory of Fourier series and integrals would expect $f(t)$ to be equal to the integral

$$\int_{E_n} \hat{f}(x)e^{2\pi i t\cdot x}\, dx. \tag{1.10}$$

Unfortunately, $\hat{f}$ need not be integrable (for example, let $n = 1$ and f be the characteristic function of a finite interval). In order to get around this difficulty we shall use certain summability methods for integrals. We first introduce the *Abel* method of summability, whose analog for series is very well-known. For each $\varepsilon > 0$ we define the *Abel mean* $A_\varepsilon = A_\varepsilon(f)$ to be the integral

$$A_\varepsilon(f) = A_\varepsilon = \int_{E_n} f(x)e^{-\varepsilon|x|}\, dx.$$

It is clear that if $f \in L^1(E_n)$ then $\lim_{\varepsilon\to 0} A_\varepsilon(f) = \int_{E_n} f(x)\, dx$. On the other hand, these Abel means are well-defined even when f is not integrable (if we only assume, for example, that f is bounded, then $A_\varepsilon(f)$ is defined for all $\varepsilon > 0$). Moreover, their limit

$$\lim_{\varepsilon\to 0} A_\varepsilon(f) = \lim_{\varepsilon\to 0} \int_{E_n} f(x)e^{-\varepsilon|x|}\, dx \tag{1.11}$$

may exist even when f is not integrable. A classical example of such a case is obtained by letting $f(x) = \sin x/x$ when $n = 1$.[3]

Whenever the limit in (1.11) exists and is finite we say that $\int_{E_n} f$ is *Abel summable* to this limit.

A somewhat similar method of summability is *Gauss summability*. This method is defined by the *Gauss* (sometimes called *Gauss–Weierstrass*)

[3] As is well known, in this case $\lim_{p\to\infty} \int_0^p f(x)\, dx$ exists. It is an easy exercise to show that whenever f is locally integrable and such a limit, l, exists the Abel means $A_\varepsilon = \int_0^\infty e^{-\varepsilon x} f(x)\, dx$ converge to l.

means

$$G_\varepsilon(f) = G_\varepsilon = \int_{E_n} f(x)e^{-\varepsilon|x|^2}\,dx.$$

We say that $\int_{E_n} f$ is *Gauss summable* (to l) if

$$\lim_{\varepsilon\to 0} G_\varepsilon(f) = \lim_{\varepsilon\to 0} \int_{E_n} f(x)e^{-\varepsilon|x|^2}\,dx \tag{1.11'}$$

exists and equals the number l.

We see that both (1.11) and (1.11′) can be put in the form

$$M_{\varepsilon,\Phi}(f) = M_\varepsilon(f) = \int_{E_n} \Phi(\varepsilon x)f(x)\,dx, \tag{1.12}$$

where $\Phi \in C_0$ and $\Phi(0) = 1$. Then $\int_{E_n} f$ is summable to l if $\lim_{\varepsilon\to 0} M_\varepsilon(f) = l$. We shall call $M_\varepsilon(f)$ the Φ *means* of this integral.

We shall need the Fourier transforms of the functions $e^{-\varepsilon|x|^2}$ and $e^{-\varepsilon|x|}$. The calculation of the first one is easy and is reducible to the one dimensional case: Since

$$\int_{E_n} e^{-4\pi^2|x|^2}e^{-2\pi i x\cdot t}\,dx = \prod_{j=1}^{n}\int_{-\infty}^{\infty} e^{-4\pi^2 x_j^2}e^{-2\pi i x_j t_j}\,dx_j$$

$$= \prod_{j=1}^{n}\frac{1}{2\sqrt{\pi}}e^{-t_j^2/4} = 2^{-n}\pi^{-n/2}e^{-|t|^2/4},$$

the change of variables $(\sqrt{\alpha}/2\sqrt{\pi})y = x$ gives us

THEOREM 1.13. *For all $\alpha > 0$ we have*

$$\int_{E_n} e^{-\pi\alpha|y|^2}e^{-2\pi i t\cdot y}\,dy = \alpha^{-n/2}e^{-\pi|t|^2/\alpha}.\text{[4]}$$

The second Fourier transform is somewhat harder to obtain:

THEOREM 1.14. *For all $\alpha > 0$ we have*

$$\int_{E_n} e^{-2\pi|y|\alpha}e^{-2\pi i t\cdot y}\,dy = c_n\frac{\alpha}{(\alpha^2 + |t|^2)^{(n+1)/2}},$$

where $c_n = \Gamma[(n+1)/2]/(\pi^{(n+1)/2})$.

PROOF. By a change of variables we see that it suffices to show this result when $\alpha = 1$. In order to show this let us momentarily assume that

$$e^{-\beta} = \frac{1}{\sqrt{\pi}}\int_0^\infty \frac{e^{-u}}{\sqrt{u}}e^{-\beta^2/4u}\,du \tag{i}$$

[4] We note that this theorem shows that $e^{-\pi|x|^2}$ is its own Fourier transform.

whenever $\beta > 0$. Then, using Theorem 1.13 to establish the third equality,

$$\int_{E_n} e^{-2\pi|y|}e^{-2\pi i t\cdot y}\,dy = \int_{E_n}\left\{\frac{1}{\sqrt{\pi}}\int_0^\infty \frac{e^{-u}}{\sqrt{u}}e^{-4\pi^2|y|^2/4u}\,du\right\}e^{-2\pi i t\cdot y}\,dy$$

$$= \frac{1}{\sqrt{\pi}}\int_0^\infty \frac{e^{-u}}{\sqrt{u}}\left\{\int_{E_n} e^{-4\pi^2|y|^2/4u}e^{-2\pi i t\cdot y}\,dy\right\}du$$

$$= \frac{1}{\sqrt{\pi}}\int_0^\infty \frac{e^{-u}}{\sqrt{u}}\left\{\left(\sqrt{\frac{u}{\pi}}\right)^{n/2} e^{-u|t|^2}\right\}du$$

$$= \frac{1}{\pi^{(n+1)/2}}\int_0^\infty e^{-u}u^{(n-1)/2}e^{-u|t|^2}\,du$$

$$= \frac{1}{\pi^{(n+1)/2}}\frac{1}{(1+|t|^2)^{(n+1)/2}}\int_0^\infty e^{-s}s^{(n-1)/2}\,ds$$

$$= \frac{\Gamma[(n+1)/2]}{\pi^{(n+1)/2}}\frac{1}{(1+|t|^2)^{(n+1)/2}}.$$

Consequently the theorem will be established once we show (i). We do this by using the identities

(ii) $$e^{-\beta} = \frac{2}{\pi}\int_0^\infty \frac{\cos\beta x}{1+x^2}\,dx, \qquad \beta > 0 \qquad \text{and}$$

(iii) $$\frac{1}{1+x^2} = \int_0^\infty e^{-(1+x^2)u}\,du.$$

The second of these equalities is obvious, while the first is an easy application of the theory of residues to the function $e^{i\beta z}/(1+z^2)$. Thus,

$$e^{-\beta} = \frac{2}{\pi}\int_0^\infty \frac{\cos\beta x}{1+x^2}\,dx = \frac{2}{\pi}\int_0^\infty \cos\beta x\left\{\int_0^\infty e^{-u}e^{-ux^2}\,du\right\}dx$$

$$= \frac{2}{\pi}\int_0^\infty e^{-u}\left\{\int_0^\infty e^{-ux^2}\cos\beta x\,dx\right\}du = \frac{2}{\pi}\int_0^\infty e^{-u}\left\{\frac{1}{2}\int_{-\infty}^\infty e^{-ux^2}e^{i\beta x}\,dx\right\}du$$

$$= \frac{2}{\pi}\int_0^\infty e^{-u}\left\{\pi\int_{-\infty}^\infty e^{-4\pi^2uy^2}e^{-2\pi i\beta y}\,dy\right\}du = \frac{2}{\pi}\int_0^\infty e^{-u}\left\{\frac{1}{2}\sqrt{\frac{\pi}{u}}\,e^{-\beta^2/4u}\right\}du.$$

and the theorem is proved.

We shall denote the Fourier transforms of $e^{-4\pi^2\alpha|y|^2}$ and $e^{-2\pi\alpha|y|}$, $\alpha > 0$, by W and P, respectively. That is, $W(t,\alpha) = (4\pi\alpha)^{-(n/2)}e^{-|t|^2/4\alpha}$ and $P(t,\alpha) = c_n[\alpha/(\alpha^2+|t|^2)^{(n+1)/2}]$. The first of these two functions is called the *Weierstrass* (or *Gauss–Weierstrass*) *kernel* while the second is called the *Poisson kernel.*[5]

We shall show, in particular, that the Abel and Gauss means of the integral (1.10) converge to f (both in the norm and a.e.). This will be done by deriving a general formula from which we will be able to express the Abel and Gauss means of this integral in terms of convolutions of f with the Poisson and Weierstrass kernels. In order to obtain these expressions we will use the following important result.

THEOREM 1.15 (The Multiplication Formula). *If f and g belong to $L^1(E_n)$ then*

$$\int_{E_n} \hat{f}(x)g(x)\,dx = \int_{E_n} f(x)\hat{g}(x)\,dx.$$

PROOF. Applying Fubini's theorem, we have

$$\begin{aligned}\int_{E_n} \hat{f}(x)g(x)\,dx &= \int_{E_n}\left\{\int_{E_n} f(t)e^{-2\pi i t\cdot x}\,dt\right\}g(x)\,dx\\ &= \int_{E_n}\left\{\int_{E_n} g(x)e^{-2\pi i t\cdot x}\,dx\right\}f(t)\,dt\\ &= \int_{E_n} f(t)\hat{g}(t)\,dt.\end{aligned}$$

Suppose now that the function Φ in (1.12) is integrable and its Fourier transform is $\hat{\Phi} = \varphi$. If we let $\varphi_\varepsilon(x) = \varepsilon^{-n}\varphi(x/\varepsilon)$ for $\varepsilon > 0$, then (1.6) implies $(\delta_\varepsilon\Phi)\hat{}\,(x) = \varepsilon^{-n}\varphi(x/\varepsilon) = \varphi_\varepsilon(x)$. For example, Theorem 1.13 shows us that when $\Phi(x) = e^{-4\pi^2|x|^2}$ we have $\varphi_\varepsilon(x) = W(x, \varepsilon^2)$ and, similarly from Theorem 1.14 we see that $\varphi_\varepsilon(x) = P(x, \varepsilon)$ when $\Phi(x) = e^{-2\pi|x|}$. Then, applying the multiplication formula (Theorem 1.15) to $f(x)$ and $e^{2\pi i t\cdot x}\delta_\varepsilon\Phi(x)$ and using (1.5), part (ii), in order to evaluate the Fourier transform of the latter function we obtain

THEOREM 1.16. *If f and Φ belong to $L^1(E_n)$ and $\varphi = \hat{\Phi}$ then*

$$\int_{E_n} \hat{f}(x)e^{2\pi i t\cdot x}\Phi(\varepsilon x)\,dx = \int_{E_n} f(x)\varphi_\varepsilon(x-t)\,dx$$

for all $\varepsilon > 0$. In particular,

$$\int_{E_n} \hat{f}(x)e^{2\pi i t\cdot x}e^{-2\pi\varepsilon|x|}\,dx = \int_{E_n} f(x)P(x-t, \varepsilon)\,dx$$

[5] In later sections we shall encounter other Poisson kernels; these will be associated with certain domains. For this reason we shall, later on, call the kernel we just introduced the *Poisson kernel associated with the upper half-space*

$$E^+_{n+1} = \{x = (x_1, \ldots, x_n, x_{n+1}) \in E_{n+1};\, x_{n+1} > 0\}.$$

and

$$\int_{E_n} \hat{f}(x)e^{2\pi i t\cdot x}e^{-4\pi^2\alpha|x|^2}\,dx = \int_{E_n} f(x)W(x-t,\alpha)\,dx$$

for all $\alpha > 0$.

We shall now show that the integral (1.10) is summable to f for a large class of methods of summability that includes both Abel and Gauss summability. We shall first show that the means $\int_{E_n}\hat{f}(x)e^{2\pi i t\cdot x}\Phi(\varepsilon x)\,dx$ (see (1.12)) converge to f in the L^1 norm under the following very general conditions: Φ and $\hat{\Phi} = \varphi$ are both integrable and $\int_{E_n}\varphi(x)\,dx = 1$. The following lemma will show that the Poisson and Gauss–Weierstrass kernels satisfy these conditions.

LEMMA 1.17. (a) $\int_{E_n} W(x,\alpha)\,dx = 1$ *for all* $\alpha > 0$;
(b) $\int_{E_n} P(x,\varepsilon)\,dx = 1$ *for all* $\varepsilon > 0$.

PROOF. By a change of variable we first note that $\int_{E_n} W(x,\alpha)\,dx = \int_{E_n} W(x,1)\,dx$ and $\int_{E_n} P(x,\varepsilon)\,dx = \int_{E_n} P(x,1)\,dx$. Thus, it suffices to prove the lemma when $\alpha = 1 = \varepsilon$. In this case, part (a) is an immediate consequence of the one dimensional result $\int_{-\infty}^{\infty} e^{-x^2/4}\,dx = 2\sqrt{\pi}$ (for the n-dimensional integral can be written as the product of n such integrals). In order to establish (b) we first note that $1/c_n = \pi^{(n+1)/2}/\Gamma[(n+1)/2]$ is half the surface area of the unit sphere Σ_n of E_{n+1}. Denoting this surface area by ω_n, we see that (b) is then equivalent to

$$\int_{E_n} \frac{dx}{(1+|x|^2)^{(n+1)/2}} = \frac{\omega_n}{2}.$$

But, letting $r = |x|$, $x' = x/r$ (when $x \neq 0$), $\Sigma_{n-1} = \{x \in E_n;\ |x| = 1\}$, dx' the element of surface area on Σ_{n-1} and, finally, putting $r = \tan\theta$, we have

$$\begin{aligned}\int_{E_n} \frac{dx}{(1+|x|^2)^{(n+1)/2}} &= \int_0^{\infty}\left(\int_{\Sigma_{n-1}} \frac{1}{(1+r^2)^{(n+1)/2}}\,dx'\right)r^{n-1}\,dr \\ &= \omega_{n-1}\int_0^{\infty} \frac{r^{n-1}}{(1+r^2)^{(n+1)/2}}\,dr \\ &= \omega_{n-1}\int_0^{\pi/2} \sin^{n-1}\theta\,d\theta.\ [6]\end{aligned}$$

[6] We are, of course, expressing our original integral in "polar coordinates" (r, x'). We remind the reader that $dx = r^{n-1}\,dr\,dx'$ (see (4.14) and the bibliographical notes for further details).

But $\omega_{n-1} \sin^{n-1} \theta$ is clearly the surface area of the sphere of radius $\sin \theta$ obtained by intersecting Σ_n with the hyperplane $x_n = \cos \theta$. Thus, the area of the upper half of Σ_n is obtained by summing these $(n - 1)$ dimensional areas as θ ranges from 0 to $\pi/2$; that is,

$$\omega_{n-1} \int_0^{\pi/2} \sin^{n-1} \theta \, d\theta = \frac{\omega_n}{2},$$

which is the desired result.

THEOREM 1.18. *Suppose $\varphi \in L^1(E_n)$ with $\int_{E_n} \varphi(x)\, dx = 1$ and for $\varepsilon > 0$ let $\varphi_\varepsilon(x) = \varepsilon^{-n}\varphi(x/\varepsilon)$. If $f \in L^p(E_n)$, $1 \leqq p < \infty$, or $f \in C_0 \subset L^\infty(E_n)$, then $\|f * \varphi_\varepsilon - f\|_p \to 0$ as $\varepsilon \to 0$. In particular, $u(x, \varepsilon) = \int_{E_n} f(t)P(x - t, \varepsilon)\, dt$ and $s(x, \varepsilon) = \int_{E_n} f(t)W(x - t, \varepsilon)\, dt$ converge to f in the L^p norm as $\varepsilon \to 0$.*[7]

PROOF. We have, by a change of variables,

$$\int_{E_n} \varphi_\varepsilon(t)\, dt = \int_{E_n} \varepsilon^{-n} \varphi(t/\varepsilon)\, dt = \int_{E_n} \varphi(t)\, dt = 1.$$

Hence,

$$\begin{aligned}(f * \varphi_\varepsilon)(x) - f(x) &= \int_{E_n} f(x - t)\varphi_\varepsilon(t)\, dt - \int_{E_n} f(x)\varphi_\varepsilon(t)\, dt \\ &= \int_{E_n} [f(x - t) - f(x)]\varphi_\varepsilon(t)\, dt.\end{aligned}$$

Therefore, by Minkowski's inequality for integrals

$$\begin{aligned}\|f * \varphi_\varepsilon - f\|_p &\leqq \int_{E_n} \left(\int_{E_n} |f(x - t) - f(x)|^p\, dx \right)^{1/p} \varepsilon^{-n}|\varphi(t/\varepsilon)|\, dt \\ &= \int_{E_n} \left(\int_{E_n} |f(x - \varepsilon t) - f(x)|^p\, dx \right)^{1/p} |\varphi(t)|\, dt.\end{aligned}$$

The expression $(\int_{E_n} |f(x + h) - f(x)|^p\, dx)^{1/p} = \omega_{p,f}(h) = \omega(h)$ is known as the *L^p modulus of continuity of f*. It is obviously bounded as a function of h since $\omega(h) \leqq 2\|f\|_p$. Moreover, $\omega(h) \to 0$ as $|h| \to 0$. This is clearly the case when f is continuous and has compact support. The general case is easily obtained by approximating f (in the L^p norm) by such functions.

[7] The function $u(x, \varepsilon)$, defined for $x \in E_n$ and $\varepsilon > 0$, is called the *Poisson integral of f*; $s(x, \varepsilon)$ is known as the *Gauss–Weierstrass integral of f*.

We have shown, therefore, that

$$\|f * \varphi_\varepsilon - f\|_p \leqq \int_{E_n} \omega(-\varepsilon t)|\varphi(t)|\, dt.$$

But, by the Lebesgue dominated convergence theorem, the integral on the right tends to 0 with ε since the integrand tends to 0 and is majorized by the integrable function $2\|f\|_p|\varphi(t)|$. This proves the theorem.

Though it is not related to our study of the inversion of the Fourier integral we observe that essentially the same argument gives us the following result, which will be useful to us in the future:

COROLLARY 1.19. *Suppose $\varphi \in L^1(E_n)$ and $\int_{E_n} \varphi(t)\, dt = 0$ then $\|f * \varphi_\varepsilon\|_p \to 0$ as $\varepsilon \to 0$ whenever $f \in L^p(E_n)$, $1 \leqq p < \infty$, or $f \in C_0 \subset L^\infty(E_n)$.*

PROOF. Once we observe that $(f * \varphi_\varepsilon)(x) = (f * \varphi_\varepsilon)(x) - f(x)\cdot 0 = (f * \varphi_\varepsilon)(x) - f(x)\int_{E_n} \varphi_\varepsilon(t)\, dt = \int_{E_n} [f(x - t) - f(x)]\varphi_\varepsilon(t)\, dt$, the rest of the argument is precisely that used in the last proof.

From Theorems 1.16 and 1.18 we obtain the following solution to the Fourier inversion problem:

THEOREM 1.20. *If Φ and its Fourier transform $\varphi = \hat{\Phi}$ are integrable and $\int_{E_n} \varphi(x)\, dx = 1$, then the Φ means of the integral $\int_{E_n} \hat{f}(t)e^{2\pi i t\cdot x}\, dt$ converge to $f(x)$ in the L^1 norm. In particular, the Abel and Gauss means of this integral converge to $f(x)$ in the L^1 norm.*

Since $s(x, \alpha) = \int_{E_n} \hat{f}(t)e^{2\pi i x\cdot t}e^{-4\pi^2\alpha|t|^2}\, dt$ converges in L^1 to $f(x)$ as $\alpha > 0$ tends to 0, we can find a sequence $\alpha_k \to 0$ such that $s(x, \alpha_k) \to f(x)$ for a.e. x. If we further assume that $\hat{f} \in L^1(E_n)$ the Lebesgue dominated convergence theorem gives us the following pointwise equality:

COROLLARY 1.21. *If both f and $\hat{f}$ are integrable then*

$$f(x) = \int_{E_n} \hat{f}(t)e^{2\pi i x\cdot t}\, dt$$

for almost every x. [8]

We have singled out the Gauss–Weierstrass and the Abel methods of summability. The former is probably the simplest; we shall see in the second chapter that the latter is intimately connected with harmonic functions and provides us with very powerful tools in Fourier analysis.[9]

[8] We know from Theorem 1.1 that $\hat{f}$ is continuous. If $\hat{f}$ is integrable, the integral $\int_{E_n} \hat{f}(t)e^{2\pi i x\cdot t}\, dt$ also defines a continuous function (in fact, it equals $(\hat{f})\hat{\ }(-x)$). Thus, by changing f on a set of measure 0 we can obtain equality in Corollary 1.21 for all x.

[9] In the same way, the Gauss–Weierstrass method is intimately connected with the solution of the heat equation.

A typical example of the relative simplicity of the Gauss–Weierstrass method is Theorem 1.13 whose proof is much simpler than the corresponding result for the Abel method (Theorem 1.14).

We also have seen that part (a) of Lemma 1.17 was much easier to prove than part (b). It is of interest to remark that Corollary 1.21 enables us to derive the latter without making the calculation occurring in the proof of Lemma 1.17. In fact, applying Corollary 1.21 to $f(x) = e^{-2\pi\varepsilon|x|}$ and using theorem 1.14 we obtain $\int_{E_n} P(t, \varepsilon)e^{2\pi i x \cdot t}\, dt = e^{-2\pi\varepsilon|x|}$. Putting $x = 0$ we obtain $\int_{E_n} P(t, \varepsilon)\, dt = 1$.

It is clear from THEOREM 1.20 that if $\hat{f}(x) = 0$ for all x then $f(t) = 0$ for almost every t. Applying this to $f = f_1 - f_2$ we obtain the following uniqueness result for the Fourier transform:

COROLLARY 1.22. *If f_1 and f_2 belong to $L^1(E_n)$ and $\hat{f}_1(x) = \hat{f}_2(x)$ for $x \in E_n$ then $f_1(t) = f_2(t)$ for almost every $t \in E_n$.*

The Fourier inversion problem also has a solution in the pointwise sense. We shall show, in particular, that the Abel and Gauss means of the integral (1.10) converge almost everywhere to $f(t)$. In order to do this we need to introduce some new concepts and notation. We first state an important result in the theory of differentiation of integrals of functions defined on E_n. If f is locally integrable on E_n then for almost every $x \in E_n$

$$\frac{1}{r^n} \int_{|t|<r} [f(x - t) - f(x)]\, dt \to 0 \tag{1.23}$$

as $r \to 0$.[10] In particular, this is true for $f \in L^p(E_n)$. We shall call the set of points x for which (1.23) holds *the set of points where the integral of f is differentiable*. This set is naturally associated with each locally integrable function f. There exists, a smaller but closely related set that (as we shall see) is also naturally associated with f. This is the set of all x such that

$$\frac{1}{r^n} \int_{|t|<r} |f(x - t) - f(x)|\, dt \to 0 \tag{1.24}$$

as $r \to 0$. This set is called the *Lebesgue set* of f and it includes almost all points of E_n. To see this we note that since the function $|f(t) - \rho|$ is

[10] If we let $S(x;r)$ denote the sphere $\{t \in E_n;\, |t - x| < r\}$ and $|S(x;r)|$ its measure, then (1.23) is equivalent to the equality $f(x) = \lim_{r\to 0} |S(x;r)|^{-1} \int_{S(x;r)} f(t)\, dt$. When $n = 1$ this becomes $f(x) = \lim_{r\to 0} (1/2r) \int_{x-r}^{x+r} f(t)\, dt$ and it is a classical theorem of Lebesgue that this equality holds a.e. We defer the proof of the fact that (1.23) holds a.e. to the second chapter where we shall see that it is a consequence of a general technique involving the Hardy–Littlewood maximal function (see Corollary 3.14 in that chapter).

locally integrable it follows from (1.23) that the set F_ρ of all $x \in E_n$ for which it is not true that

$$\lim_{r\to 0} \frac{1}{r^n} \int_{|t|<r} \{|f(x-t)-\rho| - |f(x)-\rho|\}\, dt = 0$$

has measure 0. Thus, the union, F, over all rational numbers ρ, of the sets F_ρ has measure 0. We claim that if $x \in E_n - F$ then (1.24) holds. In fact, given such an x and an $\varepsilon > 0$ let ρ be a rational number such that $|f(x) - \rho| < \varepsilon/2$ and let Ω_n denote the volume of the (solid) unit sphere in E_n. Then,

$$\frac{1}{\Omega_n r^n} \int_{|t|<r} |f(x-t) - f(x)|\, dt \leqq \frac{1}{\Omega_n r^n} \int_{|t|<r} |f(x-t)-\rho|\, dt + \frac{1}{\Omega_n r^n} \int_{|t|<r} |\rho - f(x)|\, dt.$$

The first term on the right tends to $|f(x) - \rho| < \varepsilon/2$ as $r \to 0$, while the second term equals $|\rho - f(x)| < \varepsilon/2$. Thus, for r close to 0,

$$(\Omega_n r^n)^{-1} \int_{|t|<r} |f(x-t) - f(x)|\, dt < \varepsilon$$

and it follows that almost all points x of E_n belong to the Lebesgue set of f.

We shall show that under suitable conditions the integral (1.10) is Φ-summable to $f(t)$ whenever t belongs to the Lebesgue set of f. More precisely, the following general result, together with Theorem 1.16, gives us conditions on the Fourier transform, φ, of Φ for this to be the case:

THEOREM 1.25. *Suppose* $\varphi \in L^1(E_n)$. Let $\psi(x) = \operatorname{ess.\,sup.}_{|t|\geqq|x|} |\varphi(t)|$ *and, for* $\varepsilon > 0$, *let* $\varphi_\varepsilon(x) = \varepsilon^{-n}\varphi(x/\varepsilon)$. *If* $\psi \in L^1(E_n)$ *and* $f \in L^p(E_n)$, $1 \leqq p \leqq \infty$, *then* $\lim_{\varepsilon\to 0} (f * \varphi_\varepsilon)(x) = f(x) \int_{E_n} \varphi(t)\, dt$ *whenever x belongs to the Lebesgue set of f. In particular, the Poisson and Gauss–Weierstrass integrals of f,* $\int_{E_n} f(t)P(x-t, \varepsilon)\, dt$ *and* $\int_{E_n} f(t)W(x-t, \varepsilon)\, dt$, *converge to $f(x)$ as $\varepsilon \to 0$ for almost every $x \in E_n$.*

PROOF. We fix a point x of the Lebesgue set of f and choose $\delta > 0$. Then we can find an $\eta > 0$ such that

$$\text{(i)} \qquad r^{-n} \int_{|t|<r} |f(x-t) - f(x)|\, dt < \delta$$

provided $r \leqq \eta$.

We have, since $\int_{E_n} \varphi_\varepsilon(t)\,dt = \int_{E_n} \varphi(t)\,dt = a$ for all $\varepsilon > 0$,

$$\begin{aligned}
|(f * \varphi_\varepsilon)(x) - af(x)| &= \left| \int_{E_n} \{f(x-t) - f(x)\}\varphi_\varepsilon(t)\,dt \right| \\
&\leqq \left| \int_{|t|<\eta} \{f(x-t) - f(x)\}\varphi_\varepsilon(t)\,dt \right| \\
&\qquad + \left| \int_{|t|\geqq\eta} \{f(x-t) - f(x)\}\varphi_\varepsilon(t)\,dt \right| \\
&= I_1 + I_2.
\end{aligned}$$

In order to estimate I_1 we first observe that the function ψ is *radial* (i.e., $\psi(x_1) = \psi(x_2)$ if $|x_1| = |x_2|$) and, if we put $\psi_0(r) = \psi(x)$ when $|x| = r$, then ψ_0 is a decreasing function of r. Thus, if Ω_n denotes the volume of the solid unit sphere of E_n, we have

$$\frac{\Omega_n(2^n - 1)}{2^n} r^n \psi_0(r) \leqq \int_{r/2 \leqq |x| \leqq r} \psi(x)\,dx \to 0$$

as $r \to 0$ or as $r \to \infty$. That is, $\lim r^n\psi_0(r) = 0$ as r tends to 0 or ∞. In particular, there exists a constant A such that $r^n\psi_0(r) \leqq A$ for $0 < r < \infty$. If Σ_{n-1} denotes the surface of the unit sphere $\{t' \in E_n;\ |t'| = 1\}$, we let $g(r) = \int_{\Sigma_{n-1}} |f(x - rt') - f(x)|\,dt'$, where dt' is the element of surface area on Σ_{n-1}. Condition (i) is therefore equivalent to

$$G(r) = \int_0^r s^{n-1} g(s)\,ds \leqq \delta r^n$$

provided $r \leqq \eta$. Using this notation and these observations, therefore, we have

$$\begin{aligned}
I_1 &\leqq \int_{|t|<\eta} |f(x-t) - f(x)|\varepsilon^{-n}\psi(t/\varepsilon)\,dt = \int_0^\eta r^{n-1} g(r)\varepsilon^{-n}\psi_0(r/\varepsilon)\,dr \\
&= G(r)\varepsilon^{-n}\psi_0(r/\varepsilon)]_0^\eta - \int_0^\eta G(r)\,d(\varepsilon^{-n}\psi_0(r/\varepsilon)) \\
&\leqq \delta r^n \varepsilon^{-n}\psi_0(r/\varepsilon)]_0^\eta - \int_0^{\eta/\varepsilon} G(\varepsilon s)\varepsilon^{-n}\,d\psi_0(s) \\
&\leqq \delta A - \int_0^{\eta/\varepsilon} \delta s^n\,d\psi_0(s) \leqq \delta\left(A - \int_0^\infty s^n\,d\psi_0(s)\right).
\end{aligned}$$

But $-\int_0^\infty s^n\, d\psi_0(s) = n\int_0^\infty s^{n-1}\psi_0(s)\, ds = (n/\omega_{n-1})\int_{E_n}\psi(x)\, dx$. This shows that there exists a constant B, depending only on ψ, such that $I_1 \leqq B\delta$.

In order to estimate I_2 we let $\psi_\varepsilon(x) = \varepsilon^{-n}\psi(x/\varepsilon)$ and we denote the characteristic function of the set of all $x \in E_n$ such that $|x| \geqq \eta$ by χ_η. Then, if $(1/p) + (1/p') = 1$,

$$I_2 \leqq \|f\|_p\|\chi_\eta\psi_\varepsilon\|_{p'} + |f(x)|\cdot\|\chi_\eta\psi_\varepsilon\|_1.$$

Since

$$\|\chi_\eta\psi_\varepsilon\|_1 = \int\limits_{|x|\geqq\eta}\psi_\varepsilon(x)\, dx = \int\limits_{|x|\geqq\eta/\varepsilon}\psi(x)\, dx,$$

the second summand tends to 0 with ε. We can show that the same is true of the first summand: Since $p' = 1 + (p'/p)$, an application of Hölder's inequality gives us

$$\|\chi_\eta\psi_\varepsilon\|_{p'} = \left(\int\limits_{|x|\geqq\eta}[\psi_\varepsilon(x)]^{p'}\, dx\right)^{1/p'} = \left(\int\limits_{|x|\geqq\eta}\psi_\varepsilon(x)[\psi_\varepsilon(x)]^{p'/p}\, dx\right)^{1/p'}$$
$$\leqq \|\chi_\eta\psi_\varepsilon\|_\infty^{1/p}\|\chi_\eta\psi_\varepsilon\|_1^{1/p'}.$$

But, as we have already noted, $\|\chi_\eta\psi_\varepsilon\|_\infty = \sup\limits_{|x|\geqq\eta}\psi_\varepsilon(x) = \eta^{-n}(\eta/\varepsilon)^n\psi_0(\eta/\varepsilon) \to 0$ as $\varepsilon \to 0$.

We have shown, therefore, that for ε sufficiently small

$$|(f * \varphi_\varepsilon)(x) - af(x)|$$

is bounded by a constant multiple of δ, this constant depending only on ψ. This clearly proves the theorem.

If x is a point of continuity of f then it certainly belongs to the Lebesgue set of f. Consequently, Theorem 1.25 holds for any such point. If f is continuous at 0, Theorems 1.25 and 1.16 imply

$$\lim_{\varepsilon\to 0}\int\limits_{E_n}\hat{f}(x)e^{-2\pi\varepsilon|x|}\, dx = f(0).$$

If we further assume that $\hat{f} \geqq 0$ it follows from Fatou's lemma that $\hat{f} \in L^1(E_n)$. Thus (see Theorem 1.1, part (b)), the integral (1.10) defines a continuous function of t that equals $f(t)$ almost everywhere by Corollary 1.21. We therefore obtain the following very useful result:

COROLLARY 1.26. *Suppose $f \in L^1(E_n)$ and $\hat{f} \geqq 0$. If f is continuous at 0 then $\hat{f}$ belongs to $L^1(E_n)$ and*

$$f(t) = \int\limits_{E_n}\hat{f}(x)e^{2\pi i t\cdot x}\, dx$$

for almost every x. In particular,

$$f(0) = \int_{E_n} \hat{f}(x)\, dx.$$

An immediate consequence of this corollary and the formulas in Theorems 1.13 and 1.14 is

COROLLARY 1.27. (a) $\int_{E_n} W(x, \alpha) e^{2\pi i t\cdot x}\, dx = e^{-4\pi^2\alpha|t|^2}$

(b) $\int_{E_n} P(x, \alpha) e^{2\pi i t\cdot x}\, dx = e^{-2\pi\alpha|t|}$ *for all* $\alpha > 0$.

From Theorem 1.4 and Corollaries 1.22 and 1.27 we immediately obtain the *semigroup properties* of the Weierstrass and Poisson kernels.

COROLLARY 1.28. *If* α_1 *and* α_2 *are positive real numbers then*

$$\text{(a)} \quad W(x, \alpha_1 + \alpha_2) = \int_{E_n} W(x - t, \alpha_1) W(t, \alpha_2)\, dt$$

and

$$\text{(b)} \quad P(x, \alpha_1 + \alpha_2) = \int_{E_n} P(x - t, \alpha_1) P(t, \alpha_2)\, dt$$

2. The L^2 Theory and the Plancherel Theorem

The integral defining the Fourier transform is not defined in the Lebesgue sense for the general function in $L^2(E_n)$; nevertheless, the Fourier transform has a natural definition on this space and a particularly elegant theory.

If, in addition to being integrable, we assume f to be square-integrable then $\hat{f}$ will also be square-integrable. In fact, we have the following basic result:

THEOREM 2.1. *If* $f \in L^1 \cap L^2$ *then* $\|\hat{f}\|_2 = \|f\|_2$.

PROOF. Let $g(x) = \overline{f(-x)}$. Then, by Theorem 1.3, $h = f * g \in L^1(E_n)$ and, by Theorem 1.4, $\hat{h} = \hat{f}\hat{g}$. But $\hat{g} = \overline{\hat{f}}$; thus, $\hat{h} = |\hat{f}|^2$. Now, an application of Corollary 1.26 shows that $\hat{h} \in L^1(E_n)$ and $h(0) = \int_{E_n} \hat{h}(x)\, dx$ (it is an immediate consequence of Schwarz's inequality and the fact that the L^2 modulus of continuity $\omega_2(f; \delta)$ tends to 0 as $\delta \to 0$ that h, the convolution of the two L^2 functions f and g, is uniformly continuous). We thus have

$$\int_{E_n} |\hat{f}|^2\, dx = \int_{E_n} \hat{h}\, dx = h(0) = \int_{E_n} f(x) g(0 - x)\, dx$$

$$= \int_{E_n} f(x)\overline{f(x)}\, dx = \int_{E_n} |f|^2\, dx.$$

This theorem asserts that the Fourier transform is a bounded linear operator defined on the dense subset $L^1 \cap L^2$ of $L^2(E_n)$ (in fact, it is an isometry). Therefore, there exists a unique bounded extension, $\mathcal{F}$, of this operator to all of L^2. $\mathcal{F}$ will be called the *Fourier transform on* L^2; we shall also use the notation $\hat{f} = \mathcal{F}f$ whenever $f \in L^2(E_n)$.

In general, if $f \in L^2(E_n)$ this definition of the Fourier transform gives us $\hat{f}$ as the L^2 limit of the sequence $\{\hat{h}_k\}$, where $\{h_k\}$ is any sequence in $L^1 \cap L^2$ converging to f in the L^2 norm. It is convenient to choose the sequence $\{h_k\}$, where $h_k(t)$ equals $f(t)$ when $|t| \leqq k$ and is zero elsewhere. Thus, $\hat{f}$ is the L^2 limit of the sequence of functions $\hat{h}_k$ defined by

$$\hat{h}_k(x) = \int_{|t| \leqq k} f(t)e^{-2\pi i x \cdot t}\, dt = \int_{E_n} h_k(t)e^{-2\pi i x \cdot t}\, dt. \tag{2.2}$$

A linear operator on $L^2(E_n)$ that is an isometry and maps onto $L^2(E_n)$ is called a *unitary* operator. It is an immediate consequence of Theorem 2.1 that $\mathcal{F}$ is an isometry. Moreover, we have the additional property that $\mathcal{F}$ is onto:

THEOREM 2.3. *The Fourier transform is a unitary operator on* $L^2(E_n)$.

PROOF. Since $\mathcal{F}$ is an isometry its range is a closed subspace of $L^2(E_n)$. If this subspace were not all of $L^2(E_n)$ we could find a function g such that $\int_{E_n} \hat{f}g\, dx = 0$ for all $f \in L^2$ and $\|g\|_2 \neq 0$. The multiplication formula (Theorem 1.15) obviously extends to L^2; consequently, $\int_{E_n} f\hat{g}\, dx = \int_{E_n} \hat{f}g\, dx = 0$ for all $f \in L^2$. But this implies that $\hat{g}(x) = 0$ for almost every x, contradicting the fact that $\|g\|_2 = \|\hat{g}\|_2 \neq 0$.

Theorem 2.3 is a major part of the basic theorem in the L^2 theory of the Fourier transform:

THEOREM 2.4. *The inverse of the Fourier transform,* $\mathcal{F}^{-1}$, *can be obtained by letting* $(\mathcal{F}^{-1}g)(x) = (\mathcal{F}g)(-x)$ *for all* $g \in L^2(E_n)$.

Theorems 2.3 and 2.4 are usually referred to as the *Plancherel theorem.* Theorem 2.4 follows from the fact that $\mathcal{F}^{-1}\hat{f}$ can be expressed as the L^2 limit of the sequence defined by

$$f_k(t) = \int_{|x| \leqq k} \hat{f}(x)e^{2\pi i t \cdot x}\, dx. \tag{2.5}$$

We first claim that this is clear when $\hat{f} \in L^1 \cap L^2$ (a dense subset of the range). To do this we only need to check that this expression agrees with

$\mathscr{F}^*\hat{f}$.[11] But this is immediate, for letting $\tilde{f}(t) = \int_{E_n} \hat{f}(x)e^{2\pi i t\cdot x}\,dx = \lim_{k\to\infty} f_k(t)$ (in L^2), we have

$$(g,\tilde{f}) = \int_{E_n} g(t)\left(\overline{\int_{E_n} \hat{f}(x)e^{2\pi i t\cdot x}\,dx}\right)dt$$
$$= \int_{E_n}\left(\int_{E_n} g(t)e^{-2\pi i x\cdot t}\,dt\right)\overline{\hat{f}(x)}\,dx = (\mathscr{F}g,\hat{f})$$

for all $g \in L^1 \cap L^2$. That is, $(g,\tilde{f}) = (\mathscr{F}g, \mathscr{F}f) = (g,f)$ for all $g \in L^1 \cap L^2$; therefore, $\tilde{f} = f$. The general case now follows by the same considerations that were used to establish (2.2).

We see, therefore, that the problem of inverting the Fourier transform has a very simple and elegant solution in L^2. In view of the material developed in the last section it is natural to ask if there exists a solution involving summability concepts as well. For example, since $e^{-\delta|x|}$, as a function of x, is square integrable, the Abel means of the integral (1.10) are well defined. Is it true that they converge to f (in L^2 or a.e.)? The answer is "yes" and this follows immediately from Theorems 1.18 and 1.25 once the identity in Theorem 1.16 is established for $f \in L^2(E_n)$. But the latter can be proved in the same way as Theorem 1.16 by using the L^2 extension of the multiplication formula (Theorem 1.15).

Having defined the Fourier transform for functions in $L^1(E_n)$ and functions in $L^2(E_n)$ we can easily define it on the class $L^1(E_n) + L^2(E_n)$ consisting of all functions $f = f_1 + f_2$, with $f_1 \in L^1(E_n)$ and $f_2 \in L^2(E_n)$. We do this by putting $(f_1 + f_2)^\wedge = \hat{f}_1 + \hat{f}_2$. If $g_1 + g_2 = f_1 + f_2$, with $g_i \in L^i(E_n)$, $i = 1, 2$, then $g_1 - f_1 = f_2 - g_2 \in L^1 \cap L^2$. Since the two definitions of the Fourier transform coincide on $L^1 \cap L^2$ we have $\hat{g}_1 - \hat{f}_1 = \hat{f}_2 - \hat{g}_2$. Thus, $\hat{f}_1 + \hat{f}_2 = \hat{g}_1 + \hat{g}_2$. This shows that the Fourier transform is well-defined on $L^1(E_n) + L^2(E_n)$. Since the latter is easily seen to contain all the spaces $L^p(E_n)$, $1 \leqq p \leqq 2$, the Fourier transform is defined for all $f \in L^p(E_n)$. It is easily checked that the Fourier inversion problem can be solved in terms of the Abel or Gauss means in this case as well. Similarly, Theorem 1.4 has the following extension:

THEOREM 2.6. *If $f \in L^1(E_n)$ and $g \in L^p(E_n)$, $1 \leqq p \leqq 2$, then $h = f * g$ belongs to $L^p(E_n)$* (see Theorem 1.3) *and*

$$\hat{h}(x) = \hat{f}(x)\hat{g}(x)$$

for almost every x.

[11] We consider $L^2(E_n)$ as a Hilbert space with inner product $(f, g) = \int_{E_n} f\bar{g}\,dx$ and assume the reader is familiar with the fact that, if T is unitary, T^{-1} equals the *adjoint* operator T^* (i.e., the operator satisfying $(Tf, g) = (f, T^*g)$ for all f and g in $L^2(E_n)$). From this it follows that T preserves the inner product: $(Tf, Tg) = (f, g)$.

In the next section we extend the definition of the Fourier transform much further. We shall see that this extension is consistent with the one we have just given when it is applied to the spaces $L^p(E_n)$, $1 \leqq p \leqq 2$.

3. The Class of Tempered Distributions

The basic idea in the theory of distributions is to consider them as linear functionals on some space of "regular" functions—the so-called "testing functions." The space of testing functions is assumed to be well-behaved with respect to the operations (differentiation, Fourier transformation, convolution, translation, etc.) we have been studying, and this is then reflected in the properties of distributions. We are naturally led to the definition of such a space of testing functions by the following considerations. Suppose we want these operations to be defined on a function space, $\mathscr{S}$, and to preserve it. Then, it would certainly have to consist of functions that are indefinitely differentiable; this, in view of (1.9), part (i), indicates that each function of $\mathscr{S}$, after being multiplied by a polynomial, must still be in $\mathscr{S}$. We therefore make the following definition: The space $\mathscr{S}$ of *testing functions* is defined to be the class of all those C^∞ functions φ on E_n (i.e., all the partial derivatives of φ exist and are continuous) such that

$$\sup_{x \in E_n} |x^\alpha (D^\beta \varphi)(x)| < \infty \tag{3.1}$$

for all n-tuples $\alpha = (\alpha_1, \ldots, \alpha_n)$ and $\beta = (\beta_1, \ldots, \beta_n)$ of nonnegative integers (we are using the notation that was used in (1.9)). For example $\varphi(x) = e^{-\delta|x|^2}$, $\delta > 0$, belongs to $\mathscr{S}$; on the other hand, $\varphi(x) = e^{-\delta|x|}$ fails to be differentiable at the origin and, therefore, does not belong to $\mathscr{S}$. $\mathscr{S}$ contains the space $\mathscr{D}$ of all C^∞ functions with compact support.

It is not immediately clear that $\mathscr{D}$ is nonempty. To find a function in $\mathscr{D}$, consider the function f whose values are $f(t) = e^{-1/t}$ if $t > 0$ and $f(t) = 0$ when $t \leqq 0$. Then $f \in C^\infty$, is bounded and so are all its derivatives. Let $\varphi(t) = f(1 + t)f(1 - t)$; then $\varphi(t) = e^{-2/(1-t^2)}$ if $|t| < 1$, is zero otherwise and clearly belongs to $\mathscr{D} = \mathscr{D}(E_1)$. We can easily obtain n-dimensional variants from φ:

(a) For $x \in E_n$ define $\psi(x) = \varphi(x_1)\varphi(x_2) \cdots \varphi(x_n)$; then $\psi \in \mathscr{D} = \mathscr{D}(E_n)$

(b) For $x \in E_n$ define $\psi(x) = e^{-2/(1-|x|^2)}$, if $|x| < 1$, and 0 otherwise, then $\psi \in \mathscr{D} = \mathscr{D}(E_n)$.

(c) If $\eta \in C^\infty$ and ψ is the function in (b) then $\psi(\varepsilon x)\eta(x)$ defines a function in $\mathscr{D}(E_n)$; moreover $e^2\psi(\varepsilon x)\eta(x) \to \eta(x)$ as $\varepsilon \to 0$ (see (3.9) below).

We observe that the order of multiplication by powers of $x_1, \ldots, x_n$ and differentiation, in (3.1), could have been reversed. That is, $\varphi \in \mathscr{S}$ if and only if $\varphi \in C^\infty$ and $\sup_{x \in E_n} |D^\beta(x^\alpha \varphi(x))| < \infty$ for all n-tuples α, β of nonnegative integers. This shows that if P is a polynomial in n variables and $\varphi \in \mathscr{S}$ then $P(x)\varphi(x)$ and $P(D)\varphi(x)$ are again in $\mathscr{S}$.

The spaces C_0 and $L^p(E_n)$, $1 \leqq p < \infty$, contain $\mathscr{S}$. The latter is a dense subspace of each of the former; in fact, the same is true of $\mathscr{D}$. The L^p norm of $\varphi \in \mathscr{S}$ is bounded by a finite linear combination of L^∞ norms of terms of the form $x^\alpha\varphi(x)$. We see this in the following way: Let $A = \|\varphi\|_\infty$ and $B = \sup_{x\in E_n} |x|^{2n}|\varphi(x)|$ ($B < \infty$ by (3.1)) then, for $1 \leqq p < \infty$,

$$\left(\int_{E_n} |\varphi(x)|^p\,dx\right)^{1/p} \leqq \left(\int_{|x|\leqq 1} |\varphi(x)|^p\,dx\right)^{1/p} + \left(\int_{|x|>1} |\varphi(x)|^p\,dx\right)^{1/p}$$

$$\leqq A\left(\int_{|x|\leqq 1} 1\,dx\right)^{1/p} + B\left(\int_{|x|>1} |x|^{-2np}\,dx\right)^{1/p}$$

$$= A\left(\frac{\omega_{n-1}}{n}\right)^{1/p} + B\left(\frac{\omega_{n-1}}{(2p-1)n}\right)^{1/p}.$$

This shows that the coefficients and exponents α involved in the linear combination of L^∞ norms majorizing $\|\varphi\|_p$ are independent of φ.

From (1.9) part (ii) one easily deduces that $\hat{\varphi} \in C^\infty$ whenever $\varphi \in \mathscr{S}$. Combining this with part (i), it then follows that $\hat{\varphi} \in \mathscr{S}$. That is,

THEOREM 3.2. *If $\varphi \in \mathscr{S}$ then $\hat{\varphi} \in \mathscr{S}$.*

In fact, we see from the discussion on Fourier inversion in Section 1 that the Fourier transform must be a one to one mapping of $\mathscr{S}$ onto $\mathscr{S}$.

If φ and ψ belong to $\mathscr{S}$, then Theorem 3.2 implies that $\hat{\varphi}$ and $\hat{\psi}$ belong to $\mathscr{S}$ and, therefore, so does $\hat{\varphi}\hat{\psi}$. Since $(\varphi * \psi)\hat{} = \hat{\varphi}\hat{\psi}$, an application of the inverse Fourier transform shows that

THEOREM 3.3. *If φ and ψ are in $\mathscr{S}$ so is $\varphi * \psi$.*

We shall now introduce a metric on $\mathscr{S}$ that will make this space a topological vector space. Toward this end we introduce a countable family of norms, $\{\rho_{\alpha\beta}\}$, indexed by all the ordered pairs (α, β) of n-tuples of nonnegative integers. For such a pair we define, in view of (3.1),

$$\rho_{\alpha\beta}(\varphi) = \sup_{x\in E_n} |x^\alpha D^\beta\varphi(x)|$$

for $\varphi \in \mathscr{S}$. If we define $d'_{\alpha\beta}(\varphi, \psi) = \rho_{\alpha\beta}(\varphi - \psi)$ then $d'_{\alpha\beta}$ is a metric on $\mathscr{S}$. Let $d'_1, d'_2, \ldots$ be an ordering of these metrics and $d_n = d'_n/(1 + d'_n)$, $n = 1, 2 \ldots$. Then d_n is a metric that is equivalent to d'_n (that is, the two metrics define the same topology on $\mathscr{S}$). Furthermore, $d_n \leqq 1$. Therefore, $d = \sum_{n=1}^\infty 2^{-n}d_n$ is a metric on $\mathscr{S}$. This is the metric that will define the topology of $\mathscr{S}$. It is clear that $\varphi_k \to \varphi$, with respect to d, if and only if

$\varphi_k \to \varphi$, with respect to each d_n (as $k \to \infty$). From this it follows that the vector space operations $(\varphi, \psi) \to \varphi + \psi$ and $(a, \varphi) \to a\varphi$ (a being a complex number) are continuous; therefore, $(\mathcal{S}, d)$ is a topological vector space.

Some easily established properties of $\mathcal{S}$ and its topology, are the following:

(3.4) *The mapping* $\varphi(x) \to x^\alpha D^\beta \varphi(x)$ *is continuous.*

(3.5) *If* $\varphi \in \mathcal{S}$ *then* $\lim_{h\to 0} \tau_h\varphi = \varphi$.

(3.6) *Suppose* $\varphi \in \mathcal{S}$ *and* $h = (0, \ldots, h_i, \ldots, 0)$ *lies on the i-th coordinate axis of* E_n *then the difference quotient* $[\varphi - \tau_h\varphi]/h_i$ *tends to* $\partial\varphi/\partial x_i$ *as* $|h| \to 0$.

(3.7) $\mathcal{S}$ *is a complete metric space.*

(3.8) *The Fourier transform is a homeomorphism of* $\mathcal{S}$ *onto itself.*

(3.9) $\mathcal{D}$ *is a dense subset of* $\mathcal{S}$.

(3.10) $\mathcal{S}$ *is separable.*

The collection $\mathcal{S}'$ of all continuous linear functionals on $\mathcal{S}$ is called the *space of tempered distributions*. We describe some examples of tempered distributions:

(1) Let $f \in L^p(E_n)$, $1 \leqq p \leqq \infty$, and define $L = L_f$ by letting

$$L(\varphi) = L_f(\varphi) = \int_{E_n} f(x)\varphi(x)\, dx$$

for $\varphi \in \mathcal{S}$. It is clear that L_f is a linear functional on $\mathcal{S}$. To show that it is continuous, therefore, it suffices to show that it is continuous at the origin. Suppose then, $\varphi_k \to 0$ (in $\mathcal{S}$) as $k \to \infty$. We have seen that for any $q \geqq 1$, $\|\varphi_k\|_q$ is dominated by a finite linear combination of L^∞ norms of terms of the form $x^\alpha \varphi_k(x)$ (the coefficients of this linear combination and the exponents α depend only on n and q and not on φ_k). That is, $\|\varphi_k\|_q$ is dominated by a finite linear combination of norms $\rho_{\alpha 0}$ evaluated at φ_k. Thus $\|\varphi_k\|_q \to 0$ as $k \to \infty$. Choosing q so that $1/p + 1/q = 1$, Hölder's inequality shows $|L(\varphi_k)| \leqq \|f\|_p \|\varphi_k\|_q \to 0$ as $k \to \infty$. Thus, $L \in \mathcal{S}'$.

(2) If μ is a finite Borel measure, the linear functional $L = L_\mu$ defined by

$$L(\varphi) = L_\mu(\varphi) = \int_{E_n} \varphi(x)\, d\mu(x)$$

for $\varphi \in \mathcal{S}$ is a tempered distribution (choosing $q = \infty$, the proof is the same as the one just given in the first example).

(3) A measurable function f such that $f(x)/(1 + |x|^2)^k$ is in $L^p(E_n)$,

$1 \leqq p \leqq \infty$ (for some positive integer k), is called a *tempered L^p function* (when $p = \infty$ such a function is often also called a *slowly increasing function*). For each such function, $L(\varphi) = \int_{E_n} f(x)\varphi(x)\,dx$, $\varphi \in \mathscr{S}$, defines a member of $\mathscr{S}'$. We see this by writing

$$L(\varphi) = \int_{E_n} \{(1 + |x|^2)^k\varphi(x)\}\{f(x)/(1 + |x|^2)^k\}\,dx$$

and observing that the mapping $\varphi(x) \to (1 + |x|^2)^k\varphi(x)$ is continuous in $\mathscr{S}$; the result then follows from example (1).

(4) A *tempered measure* is a Borel measure μ such that

$$\int_{E_n} (1 + |x|^2)^{-k}\,d|\mu|(x) < \infty$$

for some integer k. Just as in example 3, we can show that $L(\varphi) = \int_{E_n} \varphi(x)\,d\mu(x)$ defines a tempered distribution.

(5) Fix $x_0 \in E_n$ and an n-tuple β of nonnegative integers. An immediate consequence of the continuity of the norm $\rho_{\alpha\beta}$ in $\mathscr{S}$ is the fact that $L(\varphi) = D^\beta\varphi(x_0)$, for $\varphi \in \mathscr{S}$, defines a tempered distribution. A special case is the *Dirac δ-function*: $L(\varphi) = \varphi(0)$. This distribution, however, is also obtainable from the tempered measure having mass 1 concentrated at the origin (example (4)). When $D^\beta = \partial/\partial x_i$ (that is, $L(\varphi) = (\partial\varphi/\partial x_i)(0)$) we obtain an example of a tempered distribution that is not a special case of one of the previous four types of distributions.

The tempered distributions of example (1) (or, more generally, example (3)) are called *functions*. Similarly, examples (2) and (4) define the distributions that are called *measures*. We shall write, in these cases, f and μ instead of L_f and L_μ. These functions and measures may be considered as embedded in $\mathscr{S}'$. If we put on $\mathscr{S}'$ the weakest topology such that the linear functionals $L \to L(\varphi)(\varphi \in \mathscr{S})$ are continuous, it is easy to see that the spaces $L^p(E_n)$, $1 \leqq p \leqq \infty$, are continuously embedded in $\mathscr{S}'$. The same is true for the space of all finite Borel measures on E_n (which is a Banach space with the norm $\|\mu\| = \int_{E_n} d|\mu|$).

There exists a simple and important characterization of tempered distributions:

THEOREM 3.11. *A linear functional L on $\mathscr{S}$ is a tempered distribution if and only if there exists a constant $C > 0$ and integers m and l such that*

$$|L(\varphi)| \leqq C \sum_{|\alpha| \leqq l,\, |\beta| \leqq m} \rho_{\alpha\beta}(\varphi)$$

for all $\varphi \in \mathscr{S}$.

PROOF. It is clear that the existence of C, l and m imply the continuity of L.

Suppose L is continuous. It follows from the definition of the metric that a basis for the neighborhoods of the origin in $\mathscr{S}$ is the collection of sets $N_{\varepsilon,l,m} = \{\varphi; \sum_{|\alpha| \leqq l, |\beta| \leqq m} \rho_{\alpha\beta}(\varphi) < \varepsilon\}$, where $\varepsilon > 0$ and l and m are integers (for in the topology induced by this system of neighborhoods—and their translates—$\varphi_k \to \varphi$ as $k \to \infty$ if and only if $\rho_{\alpha\beta}(\varphi_k - \varphi) \to 0$ for all (α, β)). Thus, there exists such a set $N_{\varepsilon,l,m}$ such that $|L(\varphi)| \leqq 1$ whenever $\varphi \in N_{\varepsilon,l,m}$. Let $\|\varphi\| = \sum_{|\alpha| \leqq l, |\beta| \leqq m} \rho_{\alpha\beta}(\varphi)$ for all $\varphi \in \mathscr{S}$. If $0 < \bar{\varepsilon} < \varepsilon$ then $\psi = (\bar{\varepsilon}/\|\varphi\|)\varphi \in N_{\varepsilon,l,m}$ (if $\varphi \neq 0$). From the linearity of L we obtain, therefore,

$$(\bar{\varepsilon}/\|\varphi\|)|L(\varphi)| = |L(\psi)| \leqq 1.$$

But this is the desired inequality with $C = 1/\bar{\varepsilon}$.

We shall now show that some important operations in analysis (differentiation, convolution, Fourier transformation, etc.) can be defined on the class $\mathscr{S}'$. We begin by defining the convolution of a distribution with a testing function.

Toward this end we make the following definition: If g is any function on E_n we define its *reflection*, $\tilde{g}$, by letting $\tilde{g}(x) = g(-x)$. A direct application of Fubini's theorem shows that if u, φ, and ψ are all in $\mathscr{S}$ then

$$\int_{E_n} (u * \varphi)(x)\psi(x)\, dx = \int_{E_n} u(x)(\tilde{\varphi} * \psi)(x)\, dx.$$

The mappings $\psi \to \int_{E_n} (u * \varphi)(x)\psi(x)\, dx$ and $\theta \to \int_{E_n} u(x)\theta(x)\, dx$ are linear functionals on $\mathscr{S}$. If we denote these functionals by $u * \varphi$ and u, the last equality can be written in the form:

$$(3.12) \qquad (u * \varphi)(\psi) = u(\tilde{\varphi} * \psi).$$

If $u \in \mathscr{S}'$ and φ, ψ belong to $\mathscr{S}$, the right side of (3.12) is well-defined since $\tilde{\varphi} * \psi \in \mathscr{S}$. Furthermore, the mapping $\psi \to u(\tilde{\varphi} * \psi)$, being the composition of two continuous functions, is continuous. Thus, we can define the *convolution of the distribution u with the testing function φ, $u * \varphi$*, by means of equality (3.12).

It is easy to show that this convolution is associative in the sense that $(u * \varphi) * \psi = u * (\varphi * \psi)$ whenever $u \in \mathscr{S}'$ and $\varphi, \psi \in \mathscr{S}$. The following result is a characterization of the convolution we have just described.

THEOREM 3.13. *If $u \in \mathscr{S}'$ and $\varphi \in \mathscr{S}$ then the convolution $u * \varphi$ is the function f, whose value at $x \in E_n$ is $f(x) = u(\tau_x\tilde{\varphi})$, where τ_x denotes the translation by x operator. Moreover, f belongs to the class C^∞ and it, as well as all its derivatives, are slowly increasing.*

PROOF. We first show that f is a C^∞ slowly increasing function. Let $h = (0, \ldots, h_j, \ldots, 0)$; then, by (3.6), $[\tau_{x+h}\tilde{\varphi} - \tau_x\tilde{\varphi}]/h_j \to -\tau_x(\partial\tilde{\varphi}/\partial x_j)$ as $|h| \to 0$, in the topology of $\mathcal{S}$. Thus, since u is continuous $[f(x+h) - f(x)]/h_j = u([\tau_{x+h}\tilde{\varphi} - \tau_x\tilde{\varphi}]/h_j) \to u[-\tau_x(\partial\tilde{\varphi}/\partial x_j)]$ as $h_j \to 0$. This, together with (3.5), shows that f has continuous first partial derivatives. Since $\partial\tilde{\varphi}/\partial x_j \in \mathcal{S}$ we can iterate this argument and show that $D^\beta f$ exists and is continuous for all n-tuples of nonnegative integers β. We observe that we obtain $(D^\beta f)(x) = (-1)^{|\beta|}u(\tau_x D^\beta\tilde{\varphi})$, where $|\beta| = \beta_1 + \cdots + \beta_n$. Consequently, since $D^\beta\tilde{\varphi} \in \mathcal{S}$ if f were slowly increasing, the same would hold for all the derivatives of f. The fact that f is slowly increasing, however, is an easy consequence of Theorem 3.11: There exist $C > 0$ and integers m and l such that

$$|f(x)| = |u(\tau_x\tilde{\varphi})| \leqq C \sum_{|\alpha| \leqq l, |\beta| \leqq m} \rho_{\alpha\beta}(\tau_x\tilde{\varphi}).$$

But $\rho_{\alpha\beta}(\tau_x\tilde{\varphi}) = \sup_{w \in E_n} |w^\alpha(D^\beta\tilde{\varphi})(w - x)| = \sup_{w \in E_n} |(w + x)^\alpha(D^\beta\tilde{\varphi})(w)|$ and the latter is clearly bounded by a polynomial in x.

In order to show that $u * \varphi$ is the function f we must show that $(u * \varphi)(\psi) = \int_{E_n} \psi(t)f(t)\,dt$. But,

$$(u * \varphi)(\psi) = u(\tilde{\varphi} * \psi) = u\left(\int_{E_n} \tilde{\varphi}(x - t)\psi(t)\,dt\right)$$
$$= u\left(\int_{E_n} (\tau_t\tilde{\varphi})(x)\psi(t)\,dt\right).$$

On the other hand, the Riemann sums of the last integral are easily shown to converge in the topology of $\mathcal{S}$. Thus,

$$u\left(\int_{E_n} (\tau_t\tilde{\varphi})(x)\psi(t)\,dt\right) = \int_{E_n} u(\tau_t\tilde{\varphi})\psi(t)\,dt = \int_{E_n} f(t)\psi(t)\,dt,$$

since u is continuous and linear, which is the desired equality.

We now examine differentiation in the class $\mathcal{S}'$. We first observe that integration by parts implies

$$\int_{E_n} (D^\beta u)(x)\varphi(x)\,dx = (-1)^{|\beta|} \int_{E_n} u(x)(D^\beta\varphi)(x)\,dx$$

for $u, \varphi \in \mathcal{S}$. The mappings $\varphi \to \int_{E_n} (D^\beta u)(x)\varphi(x)\,dx$ and

$$\psi \to \int_{E_n} u(x)\psi(x)\,dx$$

are continuous linear functionals on $\mathscr{S}$. Denoting them by $(D^\beta u)$ and u, the last equality can be written in the form

$$(D^\beta u)(\varphi) = (-1)^{|\beta|}u(D^\beta \varphi). \tag{3.14}$$

But the right side is well-defined whenever $u \in \mathscr{S}'$ and $\varphi \in \mathscr{S}$. Furthermore, $\varphi \to u(D^\beta \varphi)$ is continuous on $\mathscr{S}$ (being the composite of two continuous functions). Therefore, we can define the *partial derivative* $D^\beta u$ *of the distribution* u by means of equality (3.14). It is then clear that $D^\beta u$ is a member of $\mathscr{S}'$.

In a similar way, by first considering $u, \varphi \in \mathscr{S}$, we can motivate the following two definitions: The *translation operator* τ_h is defined on $\mathscr{S}'$ by letting $\tau_h u$, when u belongs to $\mathscr{S}'$, be the continuous linear functional on $\mathscr{S}$ whose value at φ is given by $(\tau_h u)(\varphi) = u(\tau_{-h}\varphi)$. Putting $\tilde{u}(\varphi) = u(\tilde{\varphi})$ for all $\varphi \in \mathscr{S}$ we obtain the *reflection*, $\tilde{u}$, *of the distribution* u.

By the multiplication formula (Theorem 1.15) we have $u(\hat{\varphi}) = \int_{E_n} u(x)\hat{\varphi}(x)\,dx = \int_{E_n} \hat{u}(x)\varphi(x)\,dx = \hat{u}(\varphi)$ for all $u, \varphi \in \mathscr{S}$. Therefore, we define the *Fourier transform* $\hat{u}$ *of the distribution* u to be the continuous linear functional whose value as $\varphi \in \mathscr{S}$ is

$$\hat{u}(\varphi) = u(\hat{\varphi}). \tag{3.15}$$

We observe that if f belongs to $L^p(E_n)$, $1 \leqq p \leqq 2$, its Fourier transform as a distribution coincides with the function $\hat{f}$ defined at the end of the second section.[12] It is easy to check that the Fourier transform we have just defined is an isomorphism of the topological vector space $\mathscr{S}'$ onto itself.

Having set down these facts of distribution theory we shall now apply them to the study of the basic class of linear operators that occur in Fourier analysis: the class of operators that commute with translations. While the problem of characterizing such operators in the general context of L^p spaces remains open, it can be solved in certain significant special cases, as we shall see below.

Suppose B is an operator mapping a linear space V of functions on E_n into another such space W. We say that B *commutes with translations* if $\tau_h B = B\tau_h$ for all $h \in E_n$. An example of such an operator is obtained as follows: Let us fix an $f \in L^p(E_n)$ and define $Bg = f * g$ for $g \in L^1(E_n)$. By Theorem 1.3 we then see that B is a bounded operator mapping $L^1(E_n)$ into $L^p(E_n)$ (we have $\|B\| \leqq \|f\|_p$) and a change of variables shows that $\tau_h(Bg) = B(\tau_h g)$ for all $g \in L^1(E_n)$. The following result shows that "all" bounded operators commuting with translations are of similar "convolution type":

[12] For the case $p > 2$, see (4.13) below.

THEOREM 3.16. *Suppose $B: L^p(E_n) \to L^q(E_n)$, $1 \leqq p, q \leqq \infty$, is linear, bounded and commutes with translations; then there exists a unique tempered distribution u such that $B\varphi = u * \varphi$ for all $\varphi \in \mathscr{S}$.*[13]

We shall show that the proof of this theorem is an easy consequence of the following lemma:

LEMMA 3.17. *If $f \in L^p(E_n)$ has derivatives in the L^p norm of all orders $\leqq n + 1$, then f equals almost everywhere a continuous function g satisfying*

$$|g(0)| \leqq C \sum_{|\alpha| \leqq n+1} \|D^\alpha f\|_p,$$

where C depends only on the dimension n and the exponent p.

PROOF. Let $x = (x_1, \ldots, x_n) \in E_n$. Then there exists $C' = C'_n$ such that

$$(1 + |x|^2)^{(n+1)/2} \leqq (1 + |x_1| + \cdots + |x_n|)^{n+1} \leqq C' \sum_{|\alpha| \leqq n+1} |x^\alpha|.$$

Let us first suppose $p = 1$. Then by (1.9), part (ii), and Theorem 1.1 (a)

$$\begin{aligned} |\hat{f}(x)| &\leqq C'(1 + |x|^2)^{-[(n+1)/2]} \sum_{|\alpha| \leqq n+1} |x^\alpha|\, |\hat{f}(x)| \\ &= C'(1 + |x|^2)^{-[(n+1)/2]} \sum_{|\alpha| \leqq n+1} |((2\pi)^{-|\alpha|} D^\alpha f)\hat{}\,(x)| \\ &\leqq C''(1 + |x|^2)^{-[(n+1)/2]} \sum_{|\alpha| \leqq n+1} \|D^\alpha f\|_1. \end{aligned}$$

Since $(1 + |x|^2)^{-[(n+1)/2]}$ defines an integrable function on E_n, it follows that $\hat{f} \in L^1(E_n)$ and, letting $C = C'' \int_{E_n} (1 + |x|^2)^{-[(n+1)/2]}\, dx$, we have

$$\|\hat{f}\|_1 \leqq C \sum_{|\alpha| \leqq n+1} \|D^\alpha f\|_1.$$

Thus, by Corollary 1.21 f equals almost everywhere a continuous function g and, by Theorem 1.1 (a),

$$|g(0)| \leqq \|f\|_\infty \leqq \|\hat{f}\|_1 \leqq C \sum_{|\alpha| \leqq n+1} \|D^\alpha f\|_1.$$

Suppose now that $p > 1$. Choose $\varphi \in C^\infty$ such that $\varphi(x) = 1$ if $|x| \leqq 1$ and $\varphi(x) = 0$ if $|x| > 2$. Then φf satisfies the conditions of the lemma for

[13] Consider, for simplicity, the case $n = 1$. By taking appropriate Riemann sums we see that if u is a well-behaved function, then $(B\varphi)(x)$ is approximated by a linear combination of the form $\sum_{k=1}^n u(t_k)(t_k - t_{k-1})\varphi(x - t_k) = \sum_{k=1}^n c_k \varphi(x - t_k)$. That is, B is approximated by a finite linear combination of translation operators. In a sense that is made precise by Theorem 3.16 all bounded linear operators commuting with translations can be so approximated.

$p = 1$. Thus, φf equals almost everywhere a continuous function h such that

$$|h(0)| \leqq C \sum_{|\alpha| \leqq n+1} \|D^\alpha(\varphi f)\|_1.$$

Since $D^\alpha(\varphi f) = \sum_{\mu+\nu=\alpha} C^{\mu\nu}(D^\mu f)(D^\nu \varphi)$, we have

$$\begin{aligned} \|D^\alpha(\varphi f)\|_1 &\leqq \int_{|x|\leqq 2} \sum_{\mu+\nu=\alpha} C^{\mu\nu}|D^\mu f|\,|D^\nu\varphi|\,dx \\ &\leqq \sum_{\mu+\nu=\alpha} C\Big\{\sup_{|x|\leqq 2}|(D^\nu\varphi)(x)|\Big\}\int_{|x|\leqq 2}|D^\mu f(x)|\,dx \\ &\leqq A \sum_{|\nu|\leqq|\alpha|}\int_{|x|\leqq 2}|D^\nu f(x)|\,dx \leqq AB\sum_{|\nu|\leqq|\alpha|}\|D^\nu f\|_p, \end{aligned}$$

where $A \geqq \|D^\nu\varphi\|_\infty$, $|\nu| \leqq |\alpha|$, and B depends only on p and n. Thus, we can find a constant K such that

$$|h(0)| \leqq K \sum_{|\alpha|\leqq n+1} \|D^\alpha f\|_p.$$

Since $\varphi(x) = 1$ if $|x| \leqq 1$ we see that f is equal almost everywhere to a continuous function g in the sphere of radius 1 about 0; moreover,

$$|g(0)| = |h(0)| \leqq K \sum_{|\alpha|\leqq n+1} \|D^\alpha f\|_p.$$

But, by choosing φ appropriately, the argument clearly shows that f equals almost everywhere a continuous function on any sphere about 0. This proves the lemma.

We now turn to the proof of Theorem 3.16. We first observe that if $\varphi \in \mathscr{S}$ then $B\varphi$ has derivatives in the L^q norm of all orders. For, if $h = (0, \ldots, h_j, \ldots, 0)$ lies on the j-th coordinate axis we have $(\tau_h(B\varphi) - B\varphi)/h_j = (B(\tau_h\varphi) - B\varphi)/h_j = B((\tau_h\varphi - \varphi)/h_j)$. Since $(\tau_h\varphi - \varphi)/h_j$ converges in $\mathscr{S}$ to $-\varphi_j = -\partial\varphi/\partial x_j$, it converges to this function in the L^p norm. Since B is a bounded operator from L^p to L^q it follows, therefore, that $(\tau_h(B\varphi) - B\varphi)/h_j$ converges to $-\partial(B\varphi)/\partial x_j = -B\varphi_j$ in the L^q norm. The fact that $B\varphi$ has derivatives in the L^q norm of all orders now follows by iteration. Clearly, $B(D^\alpha\varphi) = D^\alpha(B\varphi)$ for all n-tuples $\alpha = (\alpha_1 \ldots, \alpha_n)$ of nonnegative integers. By Lemma 3.17, therefore, $B\varphi$ equals almost everywhere a continuous function g_φ satisfying

$$\begin{aligned} |g_\varphi(0)| &\leqq C \sum_{|\alpha|\leqq n+1} \|D^\alpha(B\varphi)\|_q \\ &= C \sum_{|\alpha|\leqq n+1} \|B(D^\alpha\varphi)\|_q \leqq \|B\|C \sum_{|\alpha|\leqq n+1} \|D^\alpha\varphi\|_p. \end{aligned}$$

This last inequality shows that the mapping $\varphi \to g_\varphi(0)$ is a continuous linear functional, u_1, on $\mathscr{S}$ (this follows from Theorem 3.11 and the observation, made at the beginning of this section, that the L^p norm of a $\psi \in \mathscr{S}$ is bounded by a finite linear combination of L^∞ norms of terms of the form $x^\alpha\psi(x)$). We claim that $u = \tilde{u}_1$ is the linear functional we are seeking. Indeed, if $\varphi \in \mathscr{S}$, using Theorem 3.13 we obtain $(u * \varphi)(x) = u(\tau_x\tilde{\varphi}) = u([\tau_{-x}\varphi]^\sim) = \tilde{u}(\tau_{-x}\varphi) = u_1(\tau_{-x}\varphi) = (B(\tau_{-x}\varphi))(0) = (\tau_{-x}B\varphi)(0) = (B\varphi)(x)$. [14]

We note that it follows from this construction that u is unique. The theorem is therefore proved.

Combining this result with Theorem 3.13 we obtain the fact that $B\varphi$, for $\varphi \in \mathscr{S}$, is almost everywhere equal to a C^∞ function which, together with all its derivatives, is slowly increasing.

Let (L^p, L^q) be the collection of those tempered distributions u for which there exists an $A > 0$ such that $\|u * \varphi\|_q \leqq A\|\varphi\|_p$ for all $\varphi \in \mathscr{S}$. Theorem 3.16 when $p < \infty$ gives a one-to-one correspondence between this set and the bounded linear operators from L^p into L^q that commute with translations (that the operator $\varphi \to u * \varphi$ commutes with translations can be easily seen by using Theorem 3.13). When $p = 2 = q$ one can give a very simple characterization of this set of distributions:

THEOREM 3.18. *The distribution u belongs to (L^2, L^2) if and only if there exists $b \in L^\infty(E_n)$ such that $\hat{u} = b$. In this case, $\|b\|_\infty$ is the norm of the operator $B : L^2 \cap \mathscr{S} \to L^2$ defined by putting $B\varphi = u * \varphi$; moreover, $(u * \varphi)^\wedge = \hat{u}\hat{\varphi}$.*

PROOF. If $v \in \mathscr{S}'$ and $\psi \in \mathscr{S}$ we define their *product*, $v\psi$, to be the element of $\mathscr{S}'$ such that $(v\psi)(\varphi) = v(\psi\varphi)$ for all $\varphi \in \mathscr{S}$. With the product of a distribution with a testing function so defined we first observe that whenever $u \in \mathscr{S}'$ and $\varphi \in \mathscr{S}$ then

(i) $$(u * \varphi)^\wedge = \hat{u}\hat{\varphi}.$$

To see this we must show that $(u * \varphi)^\wedge(\psi) = (\hat{u}\hat{\varphi})(\psi)$ for all $\psi \in \mathscr{S}$. It follows immediately from Theorem 1.4 and the Fourier inversion formula that $\hat{\varphi}\psi$ is the inverse Fourier transform of $\tilde{\varphi} * \hat{\psi}$. Thus, by the above definitions of the Fourier transform and of convolution, $(u * \varphi)^\wedge(\psi) = (u * \varphi)(\hat{\psi}) = u(\tilde{\varphi} * \hat{\psi}) = \hat{u}(\hat{\varphi}\psi) = (\hat{u}\hat{\varphi})(\psi)$; hence, equality (i) is established.

Let $\varphi_0 = e^{-\pi|x|^2}$; then $\varphi_0 \in \mathscr{S}$ and $\hat{\varphi}_0 = \varphi_0$ (see Theorem 1.13 and footnote 4). If $u \in (L^2, L^2)$ then, by the Plancherel theorem, $\Phi_0 = \hat{u}\hat{\varphi}_0 = (u * \varphi_0)^\wedge \in L^2(E_n)$. Let $b(x) = e^{\pi|x|^2}\Phi_0(x) = \Phi_0(x)/\hat{\varphi}_0(x)$.

We claim that

(ii) $$(u * \varphi)^\wedge = b\hat{\varphi}$$

[14] We are identifying $B(\tau_{-x}\varphi)$ with the continuous function $g_{\tau_{-x}\varphi}$ that it equals a.e.

for all $\varphi \in \mathscr{S}$. By (i) it suffices to show that $(\hat{u}\hat{\varphi})(\psi) = (b\hat{\varphi})(\psi)$ for all $\psi \in \mathscr{D}$ (since $\mathscr{D}$ is dense in $\mathscr{S}$). But, if $\psi \in \mathscr{D}$, then $(\psi/\hat{\varphi}_0)(x) = \psi(x)e^{\pi|x|^2} \in \mathscr{D}$; thus,

$$\begin{aligned}(\hat{u}\hat{\varphi})(\psi) &= \hat{u}(\hat{\varphi}\psi) = \hat{u}(\hat{\varphi}\hat{\varphi}_0\psi/\hat{\varphi}_0) = (\hat{u}\hat{\varphi}_0)(\hat{\varphi}\psi/\hat{\varphi}_0)\\ &= \int_{E_n} \Phi_0(x)\hat{\varphi}(x)e^{\pi|x|^2}\psi(x)\,dx\\ &= \int_{E_n} b(x)\hat{\varphi}(x)\psi(x)\,dx = (b\hat{\varphi})(\psi).\end{aligned}$$

It follows immediately that $\hat{u} = b$: We have just shown that $\hat{u}(\hat{\varphi}\psi) = (b\hat{\varphi})(\psi) = b(\hat{\varphi}\psi)$ for all $\varphi \in \mathscr{S}$ and $\psi \in \mathscr{D}$. Selecting φ so that $\hat{\varphi}(x) = 1$ for x in the support of ψ, this shows that $\hat{u}(\psi) = b(\psi)$ for all $\psi \in \mathscr{D}$. Thus, $\hat{u} = b$.

Since $u \in (L^2, L^2)$, there exists $A > 0$ such that $\|b\hat{\varphi}\|_2 = \|(u * \varphi)\hat{\ }\|_2 = \|u * \varphi\|_2 \leqq A\|\varphi\|_2$ for all $\varphi \in \mathscr{S}$. But this clearly implies that $b \in L^\infty(E_n)$ with $\|b\|_\infty \leqq A$ (since $\mathscr{S}$ is dense in $L^2(E_n)$).

On the other hand, if $\hat{u} = b \in L^\infty(E_n)$ the Plancherel theorem and equality (i) immediately imply that $u \in (L^2, L^2)$ and $\|b\|_\infty$ is the operator norm of B.

Another simple characterization of the class (L^p, L^q) can be given when $p = 1 = q$:

THEOREM 3.19. *The distribution u belongs to (L^1, L^1) if and only if it is a finite Borel measure. In this case, the total variation of u equals the norm of the operator $B : L^1 \cap \mathscr{S} \to L^1$ defined by $B\varphi = u * \varphi$ for $\varphi \in \mathscr{S}$.*

PROOF. If u is a finite Borel measure it is clear that it belongs to (L^1, L^1) (see Remark following Theorem 1.3).

On the other hand, if $u \in (L^1, L^1)$ consider the family of L^1 functions $u_\varepsilon = (u * W(\cdot, \varepsilon))$, $\varepsilon > 0$ (the Gauss–Weierstrass kernel $W(\cdot, \varepsilon)$ belongs to $\mathscr{S}$ for each $\varepsilon > 0$). Then, by our assumption on u and Lemma 1.17, part (a), $\|u_\varepsilon\|_1 \leqq A\|W(\cdot, \varepsilon)\|_1 = A$. That is, the family $\{u_\varepsilon\}$ is uniformly bounded in the L^1 norm. Let us consider $L^1(E_n)$ as embedded in the Banach space $M = M(E_n)$ of finite Borel measures on E_n.[15] $M(E_n)$ can be identified with the dual of $C_0 = C_0(E_n)$ by making each $\mu \in M$ correspond to the linear functional assigning to $\varphi \in C_0$ the value $\int_{E_n} \varphi(x)\,d\mu(x)$. Thus, the unit sphere of M is compact in the weak* topology.[16] In particular, we

[15] If $f \in L^1(E_n)$ then $\mu(F) = \int_F f(x)\,dx$, F a measurable subset of E_n, defines a finite Borel measure on E_n. The total variation of μ equals $\|f\|_1$.

[16] That is, the weakest topology for which the members, φ, of C_0 can be considered as continuous linear functionals of M (if $\mu \in M$ is mapped onto $\int_{E_n} \varphi(x)\,d\mu(x)$ we obtain a linear functional on M). See bibliographical notes for details.

can find $\mu \in M$ and a null sequence $\{\varepsilon_k\}$ such that $u_{\varepsilon_k} \to \mu$ as $k \to \infty$ in this topology. That is, for each $\varphi \in C_0$,

$$\text{(i)} \qquad \lim_{k\to\infty} \int_{E_n} \varphi(x)u_{\varepsilon_k}(x)\,dx = \int_{E_n} \varphi(x)\,d\mu(x).$$

We now claim that μ, considered as a distribution, equals u.

We must show, therefore, that $u(\psi) = \int_{E_n} \psi(x)\,d\mu(x)$ for all $\psi \in \mathscr{S}$. Let $\psi_\varepsilon(x) = \int_{E_n} \psi(x - t)W(t, \varepsilon)\,dt$. Then, for all n-tuples $\alpha = (\alpha_1, \ldots, \alpha_n)$ of nonnegative integers, we have $(D^\alpha\psi_\varepsilon)(x) = \int_{E_n} (D^\alpha\psi)(x - t)W(t, \varepsilon)\,dt$. It follows from Theorem 1.18 that $(D^\alpha\psi_\varepsilon)(x)$ converges to $(D^\alpha\psi)(x)$ uniformly in x. Thus, $\psi_\varepsilon \to \psi$ as $\varepsilon \to 0$ (in $\mathscr{S}$) and this implies that $u(\psi_\varepsilon) \to u(\psi)$. But, since $W(\cdot, \varepsilon) = \tilde{W}(\cdot, \varepsilon)$,

$$u(\psi_\varepsilon) = u(W(\cdot, \varepsilon) * \psi) = (u * W(\cdot, \varepsilon))(\psi) = \int_{E_n} \psi(x)u_\varepsilon(x)\,dx.$$

Thus, putting $\varepsilon = \varepsilon_k$, letting $k \to \infty$ and applying (i) with $\varphi = \psi$ we obtain the desired equality $u(\psi) = \int_{E_n} \psi(x)\,d\mu(x)$. The remainder of the theorem now follows easily.

We mentioned before that a simple characterization of the general (L^p, L^q) spaces is not known (and probably does not exist). We do have, however, the following general result of duality:

THEOREM 3.20. *If* $1 \leqq p, q \leqq \infty$, $1/p + 1/p' = 1$ *and* $1/q + 1/q' = 1$ *then* $(L^p, L^q) = (L^{q'}, L^{p'})$.

PROOF. By the F. Riesz representation theorem we can identify the spaces $L^{p'}$ and $L^{q'}$ with the duals of L^p and L^q (this is true if $p, q < \infty$. If one of p or q is ∞, a slight modification of the argument we give will prove the theorem in this case as well). Let $u \in (L^p, L^q)$ and $B\colon L^p \to L^q$ denote the unique bounded linear extension to L^p of the mapping $\varphi \to u * \varphi$ $(\varphi \in \mathscr{S})$. If $B^*\colon L^{q'} \to L^{p'}$ denotes the adjoint to B we then have

$$\int_{E_n} (B\varphi)\psi\,dx = \int_{E_n} \varphi(B^*\psi)\,dx$$

for all $\varphi, \psi \in \mathscr{S}$. But, in terms of distributions, this is equivalent to the equality $(u * \varphi)(\psi) = (B^*\psi)(\varphi)$. That is, $(B^*\psi)(\varphi) = u(\tilde{\varphi} * \psi) = u([\tilde{\psi} * \varphi]^{\sim}) = \tilde{u}(\tilde{\psi} * \varphi) = (\tilde{u} * \psi)(\varphi)$ for all $\varphi, \psi \in \mathscr{S}$. Thus, $B^*\psi = \tilde{u} * \psi$ for all $\psi \in \mathscr{S}$. Since $\|B^*\psi\|_{p'} \leqq \|B^*\|\,\|\psi\|_{q'} = \|B\|\,\|\psi\|_{q'}$ we have $\tilde{u} \in (L^{q'}, L^{p'})$. Since $u \in (L^{q'}, L^{p'})$ if and only if $\tilde{u} \in (L^{q'}, L^{p'})$ this shows that $(L^p, L^q) \subset (L^{q'}, L^{p'})$. The opposite inclusion relation follows by interchanging the roles of (p, q) and (q', p').

4. Further Results

4.1. We remarked, after the proof of Theorem 1.2, that belonging to C_0 is not a sufficient condition for being the Fourier transform of an integrable function. Suppose, for simplicity, that $n = 1$; then one way of seeing this is the following: We first observe that if $\hat{f}$ is such a Fourier transform and, moreover, $\hat{f}$ is an odd function then, for $1 < b < \infty$, $|\int_1^b [\hat{f}(x)/x]\,dx| \leqq A$, where A does not depend on b. This is a consequence of the well-known corresponding property for the integrals of $\sin x/x$; that is $|\int_\alpha^\beta (\sin x/x)\,dx| \leqq B < \infty$ independently of α, β ($0 \leqq |\alpha| < |\beta| < \infty$). Indeed, because of the oddness of $\hat{f}$ we have $\hat{f}(x) = -i\int_{-\infty}^{\infty} f(t)\sin(2\pi xt)\,dt$. Thus, using Fubini's theorem one easily deduces that $|\int_1^b [\hat{f}(x)/x]\,dx| \leqq A$. To give an example, therefore, of a function in C_0 that is not a Fourier transform of an integrable function all we need to do is exhibit a continuous odd function g vanishing at ∞ and such that $\int_1^b (g(x)/x)\,dx$ is not bounded for $b \to \infty$. But this is clearly the case if, for example, $g(x) = 1/\log x$ for x large.

4.2. For $g \in L^1(E_n)$ fixed let us consider the operator B mapping $f \in L^p(E_n)$ to $f * g$. From Theorem 1.3 we see that B is a bounded linear transformation from $L^p(E_n)$ into $L^p(E_n)$ whose norm, $\|B\|^{(p)}$, is majorized by $\|g\|_1$. When $p = 1$ it is easy to show that $\|B\|^{(1)} = \|g\|_1$ (for, by Theorem 1.18 we see that $f_\varepsilon(t) = P(t, \varepsilon)$ is a family of functions of L^1 norm equal to 1 such that $f_\varepsilon * g \to g$ in the L^1 norm as $\varepsilon \to 0$). We have seen (Theorem 3.18) that $\|B\|^{(2)} = \|\hat{g}\|_\infty$. It is natural to ask, therefore, how the norm $\|B\|^{(p)}$ can be expressed in terms of g for $1 \leqq p \leqq \infty$. A satisfactory answer is not known in general. If $g \geqq 0$, however, then $\|g\|_1 = |\hat{g}(0)| \leqq \|\hat{g}\|_\infty \leqq \|g\|_1$; thus, $\|B\|^{(2)} = \|g\|_1$. From this, Theorem 3.20 and a result (the M. Riesz convexity theorem) in the theory of interpolation of operators (that we shall develop in Chapter V) it follows that $\|B\|^{(p)} = \|g\|_1$, for $1 \leqq p \leqq \infty$, provided $g \geqq 0$.

4.3. Theorem 1.3 has the following extension: *if* $f \in L^p(E_n)$ *and* $g \in L^r(E_n)$, $1 \leqq p, r$ *and* $(1/p) + (1/r) \geqq 1$ *then* $h = f * g \in L^q(E_n)$, *where* $(1/q) = (1/p) + (1/r) - 1$, *and*

$$\|h\|_q \leqq \|f\|_p \|g\|_r.$$

For a direct proof of this result (often called Young's inequality) see Zygmund [1], Chapter II, page 37. We shall show, in Chapter V, that this inequality also follows immediately from the M. Riesz convexity theorem.

4.4. We can obtain more insight into the meaning of the derivatives in the L^p norms by observing that (a) when the dimension n is 1, then $f \in L^p(E_1)$ has a derivative in the L^p norm if and only if f is equal almost everywhere to a locally absolutely continuous function whose derivative, f', belongs to $L^p(E_1)$ (see Bochner and Chandrasekharan [1], Section 6); (b) in general, $f \in L^p(E_n)$ has a k-th partial derivative in the L^p norm if and only if f, regarded as a tempered distribution, has a k-th partial derivative, $(\partial/\partial x_k)f$, in the distribution sense (see (3.14)) that is an L^p function.

4.5. If $f \in L^p(E_n)$, $1 < p \leqq \infty$, is regarded as a tempered distribution then all its first partial derivatives $(\partial/\partial x_k)f$, $k = 1, 2, \ldots, n$, which exist in the distribution sense, are L^p functions if and only if $\|\tau_h f - f\|_p = (\int_{E_n} |f(x - h) - f(x)|^p \, dx)^{1/p} = O(|h|)$. When $p = 1$, the first partial derivatives exist and are finite Borel measures if and only if $\|\tau_h f - f\|_1 = O(|h|)$.

4.6. The translation operators can be defined on the space of finite Borel measures by letting $\tau_h\mu$, where μ is in this space, be the measure whose value at a Borel set F is $\mu(F - h) = \mu(\{x \in E_n; x + h \in F\})$. The measure μ is absolutely continuous (with respect to Lebesgue measure) if and only if the total variation of $\tau_h\mu - \mu$ tends to 0 as $|h| \to 0$ (see Bochner [4]).

4.7. If μ is a discrete finite Borel measure then $\mu = \sum_{k=1}^{\infty} a_k \delta_{x_k}$, where δ_{x_k} denotes the Dirac δ measure concentrated at the point x_k. Thus, it follows from the definition of the Fourier transform of measures (or distributions) that $\hat{\mu}(t) = \sum_{k=1}^{\infty} a_k e^{-2\pi i t \cdot x_k}$. In this case it is true that

$$\text{(i)} \qquad \lim_{R \to \infty} \frac{1}{\Omega_n R^n} \int_{|t| \leqq R} |\hat{\mu}(t)|^2 \, dt = \sum_{k=1}^{\infty} |a_k|^2,$$

where Ω_n denotes the volume of the unit sphere of E_n. More generally, it can be shown that if μ is a general finite Borel measure and $x_1, x_2, \ldots, x_k, \ldots$ are the distinct points of E_n with nonzero measures $a_1, a_2, \ldots, a_k, \ldots$, then equality (i) still holds (see Wiener [1] for the one dimensional result).

4.8. Fourier transforms of nonnegative finite Borel measures can be characterized in terms of *positive definite* functions. These are continuous functions, f, defined on E_n having the property $\sum_{j=1}^{k} \sum_{i=1}^{k} f(x_i - x_j) \xi_i \bar{\xi}_j \geqq 0$ for any set of complex numbers $\{\xi_1, \xi_2, \ldots, \xi_k\}$ and points $\{x_1, x_2, \ldots, x_k\}$ in E_n. Bochner [3] has shown that f is the Fourier transform of a nonnegative finite Borel measure if and only if f is positive definite (see Rudin [1] for the extension of this result to locally compact Abelian groups).

4.9. If there existed constants $c_n, \lambda_n, n = 1, 2, \ldots, m$, such that $e^{-\delta} = \sum_{n=1}^{m} c_n e^{-(\lambda_n \delta)^2}$ this would obviously imply that if an integral is Gauss summable then it is Abel summable. Equality (i) in the proof of Theorem 1.14 almost asserts this. In view of this it is not surprising that it can be shown that if f is such that both the Abel and Gauss means, $A_\varepsilon(f)$ and $G_\varepsilon(f)$, exist and $\lim_{\varepsilon\to 0} G_\varepsilon(f) = l$ then $\lim_{\varepsilon\to 0} A_\varepsilon(f)$ exists and equals l (see Cartwright [1], Bochner and Chandrasekharan [1], Chapter I, Section 14).

4.10. The Poisson integral of a function $f \in L^p(E_n)$, $1 \leqq p \leqq \infty$, converges to $f(x)$ for every x satisfying (1.23) (for a proof of this fact see G. Weiss [2]). A more general statement that will be needed later is as follows. Suppose $\varphi(x)$ is radial (i.e., $\varphi(x) = \varphi_0(|x|)$. Assume $\varphi_0(r)$ is non-negative and decreasing in r and $\int_{E_n} \varphi(x)\,dx = 1$. Then $(f * \varphi_\varepsilon)(x) \to f(x)$ for every x satisfying (1.23), as $\varepsilon \to 0$. This applies in particular to the Gauss–Weierstrass kernels). The same argument can be used to show that if ν is a singular finite Borel measure then the Poisson–Stieltjes integral $v(x, \varepsilon) = \int_{E_n} \varphi_\varepsilon(x - t)\,d\nu(t)$ tends to 0 as $\varepsilon \to 0$ for almost every x (this argument makes use of the fact that, for almost every $x \in E_n$, $|S_x|^{-1} \times \int_{S_x} d\nu \to 0$ as the volume, $|S_x|$, of a sphere centered at x tends to zero—see (5.6) of Chapter II or Saks [1], Chapter IV). If μ is an arbitrary finite Borel measure we can decompose it into its singular part μ_1 and its absolutely continuous part μ_2. If f is the (Radon–Nikodym) derivative of μ_2 we then have $u(x, \varepsilon) = \int_{E_n} P(x - t, \varepsilon)\,d\mu(t) = \int_{E_n} P(x - t, \varepsilon)\,d\mu_1(t) + \int_{E_n} P(x - t, \varepsilon)f(t)\,dt = u_1(x, \varepsilon) + u_2(x, \varepsilon)$. Thus, $\lim_{\varepsilon \to 0} u(x, \varepsilon) = 0 + f(x) = f(x)$, for almost every $x \in E_n$.

4.11. The following additional facts concerning operators B which commute with translations are worth recording.

(a) The only interesting cases can occur when $p \leqq q$. Since if $p > q$, then B maps all elements of $\mathscr{S}$ to the zero element of $L^q(E_n)$.

(b) If $1 \leqq p \leqq \infty$, and $\|B(\varphi)\|_p \leqq M\|\varphi\|_p$, for all $\varphi \in \mathscr{S}$ (that is $B\varphi = u * \varphi$, with $B \in (L^p, L^p)$), then $B \in (L^q, L^q)$, with $\left|\frac{1}{q} - \frac{1}{2}\right| \leqq \left|\frac{1}{p} - \frac{1}{2}\right|$, and $\|B\varphi\|_q \leqq M\|\varphi\|_q$.

To prove (a) notice that if $\varphi \in \mathscr{D}$, and $f_h(x) = \varphi(x + h) + \varphi(x)$, then $\|f_h\|_p = 2^{1/p}\|\varphi\|_p$, for h sufficiently large. However, if $F_h = B(f_h)$, and $\Phi = B(\varphi)$, then $\|F_h\|_q \to 2^{1/q}\|\Phi\|_q$, as $|h| \to \infty$.

To prove (b) one applies the duality theorem (Theorem 3.20) and then the Riesz interpolation theorem (Theorem 1.3 of Chapter V).

4.12. Theorem 3.16 deals with bounded linear operators on L^p into L^q that commute with translations. There are many linear operators mapping $\mathscr{S}$ into $\mathscr{S}$ that commute with translations but do not have bounded linear extensions that map some L^p space into an L^q space. The operator mapping $\varphi \in \mathscr{S}$ onto its partial drivative $\partial\varphi/\partial x_k$ provides an example. It follows from Theorem 1.8 that this operator, B, applied to φ satisfies

$(B\varphi)\hat{}(x) = 2\pi i x_k \varphi(x)$. In view of Theorem 3.16 it is natural to inquire whether there exists a tempered distribution u such that $B\varphi = u * \varphi$. It is easy to check that the distribution mapping φ into $(-\partial\varphi/\partial x_k)(0)$ has this property: $(u * \varphi)(x) = u(\tau_x\varphi) = (-\partial\tau_x\varphi/\partial x_k)(0) = \partial\varphi/\partial x_k$. Moreover, we can easily verify that $\hat{u}$ is the function $2\pi i x_k$. More generally, by iteration we obtain the following result:

If $P(D)$ is a differential polynomial and $B\varphi = P(D)\varphi$ for $\varphi \in \mathcal{S}$, then there exists $u \in \mathcal{S}'$ such that $B\varphi = u * \varphi$, $\hat{u}(x) = P(2\pi i x)$ and $(B\varphi)\hat{}(x) = P(2\pi i x)\hat{\varphi}(x) = \hat{u}(x)\hat{\varphi}(x)$.

4.13. For any $p > 2$, there exists an $f \in L^p$, whose Fourier transform, as a tempered distribution, is not a function. If the opposite were true, the closed graph theorem would easily show that the following inequality would have to be valid for all $f \in L^p, p > 2: \int_{|x| \leqq 1} |\hat{f}(x)|\, dx \leqq A\|f\|_p$. To contradict this inequality we take, when $n = 1$, $\hat{f}(x) = e^{-\pi(1+i\delta)x^2}$, $f(x) = e^{-\pi x^2/(1+i\delta)}/(1 + i\delta)^{1/2}$, which we obtain from Theorem 1.13 by an easy argument of analytic continuation. But $\int_{|x| \leqq 1} |\hat{f}(x)|\, dx = A_1$, and $\|f\|_p \leqq A_2\delta^{1/p - 1/2}$, as $\delta \to \infty$, which is impossible if $p > 2$. A similar argument clearly holds for any n.

4.14. For further reference we state the integral identities involving the passage to polar coordinates and its relation with integration on the group of rotations $SO(n)$.

(a) Suppose f is an integrable function on E_n, then $\int_{E_n} f(x)\, dx = \int_{\Sigma_{n-1}} dx'(\int_0^\infty f(rx')r^{n-1}\, dr)$, where x' ranges over Σ_{n-1} and dx' is the induced Lebesgue measure on Σ_{n-1}.

(b) Suppose φ is integrable on Σ_{n-1}. Then $(1/\omega_{n-1}) \int_{\Sigma_{n-1}} \varphi(x')\, dx' = \int_{SO(n)} \varphi(\sigma\mathbf{1})\, d\sigma$, where $d\sigma$ is Haar measure on $SO(n)$ normalized by the condition $\int_{SO(n)} d\sigma = 1$ and $\mathbf{1} = (1, 0, \ldots, 0)$. To prove (b), note that since dx' is invariant under rotations $\sigma \in SO(n)$, $(1/\omega_{n-1}) \int_{\Sigma_{n-1}} \varphi(x')\, dx' = (1/\omega_{n-1}) \int_{SO(n)} \int_{\Sigma_{n-1}} \varphi(\sigma x')\, dx'\, d\sigma = (1/\omega_{n-1}) \int_{\Sigma_{n-1}} \int_{SO(n)} \varphi(\sigma\sigma_0\mathbf{1})\, d\sigma\, dx' = (1/\omega_{n-1}) \int_{\Sigma_{n-1}} \int_{SO(n)} \varphi(\sigma\mathbf{1})\, d\sigma\, dx' = \int_{SO(n)} \varphi(\sigma\mathbf{1})\, d\sigma$. Here σ_0 is any rotation mapping $\mathbf{1}$ to x'. The next to last equality follows by the right invariance of Haar measure $d\sigma$. For the basic properties of Haar measure see, e.g., Weil [1].

4.15. As a relatively straightforward consequence of Theorem 3.11, one can prove the following. Let u be a tempered distribution. Then for an appropriate integer m, and for each n-tuple $\alpha(|\alpha| \leqq m)$, there exist constants

a_α and slowly increasing continuous functions f_α, so that

$$u = \sum_{|\alpha| \leqq m} a_\alpha D^\alpha(L_{f_\alpha}).$$

Since the converse is a consequence of the material in Section 3, we see that this kind of representation characterizes the class of tempered distributions.

Bibliographical Notes

Some important treatises on classical Harmonic Analysis are Bochner [3], Wiener [1], Zygmund [1], Titchmarsh [2], Bochner and Chandrasekharan [1], Bari [1]. For an introductory treatment see R. Goldberg [1] and G. Weiss [3]. Much of the elementary analysis used in the first chapter, in particular the introduction of polar coordinates, can be found in Fleming [1]; in this connection also see Blumenson [1]. A general setting for most of the material in the second section and the first part of the first section is the theory of locally compact Abelian groups, for which see Weil [1], Rudin [1], Hewitt and Ross [1]. We have made no attempt to discuss the important topics originating with Wiener's Tauberian theorem which lead to the more extensive study of the L^1 algebra. For a treatment of these topics in the context of general locally compact Abelian groups the reader is referred to Naimark [1] or Rudin [1]. Our proof of Theorem 1.14 follows that in Bochner and Chandrasekharan [1]. The technique of the approximation to the identity (in particular, variants of Theorems 1.18 and 1.25) has a long history for which we can attempt no adequate description. The formulation of Theorem 1.25 presented here is based on a suggestion of R. Latzer; a similar result can be found in Calderón and Zygmund [3], Chapter II. For the results in Functional Analysis that were used in this chapter, in particular the weak* compactness of the unit sphere used in the proof of Theorem 3.19, we refer the reader to Royden [1].

For the general theory of distributions and their applications see L. Schwartz [1], Gelfand and Shilov [1], and Ehrenpreis [1]. There, the reader will also find the proofs that we omitted and the references to the classical literature of generalized functions. For rapid introduction to the theory, in connection with the Fourier transform, see Hörmander [2], Chapter I, and Yosida [1], Chapter VI. Theorems 3.16, 3.18, 3.19, 3.20, dealing with operators of type (L^p, L^q), have been known in various forms for some time. See, for example, Bochner [9], Hille and Phillips [1], Bochner and Chandrasekharan [1], Zygmund [1], Chapter IV, Hörmander [1], where, in addition, other references may be found.

CHAPTER II

Boundary Values of Harmonic Functions

A basic tool in the study of a function, f, on E_n is obtained by the passage to a harmonic function, defined on the $(n + 1)$-dimensional upper half-space, which has f as its boundary value. In the first section we set down some of the useful basic facts concerning harmonic functions of several variables. The connection between an f, defined on E_n, and a harmonic function of $(n + 1)$ variables (its Poisson integral) is studied in Section 2. Sections 3 and 4 deal with nontangential convergence at the boundary (for harmonic functions) and with harmonic majorization. While these notions are interesting in themselves, they will give us some important tools used in Chapters III and VI.

1. Basic Properties of Harmonic Functions

We assume the reader is familiar with the properties of harmonic functions that are usually presented in a course in complex variables. Many of these results extend to higher dimensions; however, we can no longer rely on the theory of analytic functions (as is often done in two dimensions) to establish these properties. In this section, therefore, we develop some of the basic theory of harmonic functions of n variables.

A function, u, defined in a *domain* (an open connected subset of E_n) is said to be *harmonic* if it is twice differentiable (i.e., its second partial derivatives exist and are continuous) and satisfies the *Laplace equation* $\Delta u = \sum_{k=1}^{n} \partial^2 u/\partial x_k^2 = 0$ at every point of the domain. From the first two of the following examples of harmonic functions we can easily deduce that the Abel means and Poisson integrals introduced in the first section of Chapter I are harmonic functions:

(1) For $t \in E_n$ fixed $e^{-2\pi|t|y}e^{2\pi i x\cdot t} = u(x, y) = u(x_1, x_2, \ldots, x_n, y)$ is harmonic in E_{n+1};

(2) $P(x, y) = c_n[y/(|x|^2 + y^2)^{(n+1)/2}]$ is also harmonic in $E_{n+1}^{+} = \{(x, y) = (x_1, \ldots, x_n, y) \in E_{n+1}\,; y > 0\}$ (this domain will often be called the *upper half-space*)[1];

[1] We denote the points of E_{n+1}^{+} by the ordered pairs (x, y), where $x \in E_n$ and y is a positive real, in order to emphasize the fact that this domain is a generalization of the upper half-plane (the case $n = 1$).

(3) If $n \geqq 3$ then $u(x) = |x|^{2-n}$ is harmonic in any domain not containing the origin. The corresponding example in two dimensions is $u(x) = \log |x|$. In a sense that will become clearer later, these functions "generate" all harmonic functions;

(4) We shall show later (Theorem 1.7) that a harmonic function is infinitely differentiable, and, in fact, is real analytic (see (5.5) Chapter IV). It is clear that any partial derivative of a harmonic function is also harmonic. As an illustration we observe that $P(x, y)$ is the partial derivative with respect to y of $c_n/(1 - n)\cdot 1/(|x|^2 + y^2)^{(n-1)/2}$, when $n > 1$, and of $c_n \log (|x|^2 + y^2)^{1/2}$, when n is 1.

(5) If u is harmonic in a domain $\mathscr{D}$ then $\tau_h u$ is harmonic in $\mathscr{D} + h = \{x + h; x \in \mathscr{D}\}$. That is, harmonicity is preserved under translations. Similarly, harmonicity is preserved under rotations and dilations. (This can be computed directly, but will be more evident after we establish the characterization of harmonic functions in terms of the mean value property—see Theorem 1.1 below);

(6) $\sin \sqrt{n}\, y \prod_{k=1}^{n} \cosh x_k = u(x, y)$ is harmonic in $E_{n+1} = \{(x, y); x \in E_n, y \in E_1\}$.

We shall denote the *mean value of a function u taken over the surface of a sphere of radius* $r > 0$ *about the point* x by the symbols $\mathscr{M}(r) = \mathscr{M}_{x,u}(r)$ (depending on the necessity of showing its dependence on u, x and r). That is, letting Σ denote the surface of the unit sphere in E_n and dt' the element of surface area on Σ (this was the notation used in the proof of Lemma 1.17 in Chapter I)

$$\mathscr{M}_{x,u}(r) = \frac{1}{\omega_{n-1}} \int_{\Sigma} u(x + rt')\, dt'.$$

THEOREM 1.1 (Mean-Value Theorem for Harmonic Functions). *If u is harmonic in a domain $\mathscr{D}$ and if the sphere of radius r_0 about $x \in \mathscr{D}$ is contained in $\mathscr{D}$ then, for* $0 < r \leqq r_0$,

$$u(x) = \mathscr{M}_{x,u}(r).$$

PROOF. Because of (5) we see that we can assume $x = 0$. Let $\mathscr{E}$ be a subdomain with a boundary $\partial\mathscr{E} \subset \mathscr{D}$ that is sufficiently smooth. Then, if we apply Green's theorem to the functions u and 1 in $\mathscr{E}$, we obtain

$$\int_{\partial\mathscr{E}} \frac{\partial u}{\partial n}\, ds = 0, \tag{1.2}$$

where $\partial/\partial n$ is differentiation in the direction of the outward directed normal to $\partial\mathscr{E}$ and ds is the element of surface area on $\partial\mathscr{E}$.

Let Σ_ε and Σ_r, $0 < \varepsilon < r \leqq r_0$, be the surfaces of the spheres of radii ε and r about 0. Applying Green's theorem to the functions u and

$$v(x) = \begin{cases} |x|^{-(n-2)} & \text{if } n > 2 \\ \log |x| & \text{if } n = 2 \end{cases}$$

in the shell domain between these two spheres we obtain[2]

$$0 = \left(\int_{\Sigma_r} - \int_{\Sigma_\varepsilon}\right)(-(n-2)|x|^{-(n-1)}u)\, ds - \left\{\int_{\Sigma_r} v\frac{\partial u}{\partial n}\, ds + \int_{\Sigma_\varepsilon} v\frac{\partial u}{\partial n}\, ds\right\}.$$

But v is constant on Σ_r (where it equals $r^{-(n-2)}$) and on Σ_ε (where it equals $\varepsilon^{-(n-2)}$). Since u is harmonic in $\mathscr{D} \supset \{x \in E_n;\ |x| \leqq r_0\}$, applying (1.2) to $\partial\mathscr{E} = \Sigma_r$ and to $\partial\mathscr{E} = \Sigma_\varepsilon$, we see that the term in curly brackets must be 0. Consequently

$$\varepsilon^{1-n}\int_{\Sigma_\varepsilon} u\, ds = r^{1-n}\int_{\Sigma_r} u\, ds.$$

Hence,

$$\begin{aligned}\mathscr{M}_{0,u}(r) &= \frac{1}{\omega_{n-1}}\int_{\Sigma} u(rt')\, dt' = \frac{1}{\omega_{n-1}r^{n-1}}\int_{\Sigma_r} u\, ds \\ &= \frac{1}{\omega_{n-1}\varepsilon^{n-1}}\int_{\Sigma_\varepsilon} u\, ds.\end{aligned}$$

But the last expression obviously tends to $u(0)$ as $\varepsilon \to 0$; this proves the theorem.

Corollary 1.3 (Maximum Principle for Harmonic Functions). *Suppose a real-valued harmonic function u, defined in a domain $\mathscr{D}$, satisfies $A = \sup_{x\in\mathscr{D}} u(x) < \infty$; then $u(x) < A$ for all $x \in \mathscr{D}$, provided u is not a constant function.*

Proof. Suppose $u(x) = A$ for some $x \in \mathscr{D}$. Then, by Theorem 1.1, there exists $r_0 > 0$ such that $A = u(x) = \mathscr{M}_{x,u}(r)$ for $0 < r \leqq r_0$. But since u is continuous and satisfies $u \leqq A$, we must have $u \equiv A$ on $\Sigma_r(x) = \{y \in E_n;\ |y - x| = r\}$. That is, $u(t) = A$ for all $t = x + h$, with $|h| < r_0$. This shows that the set $\{t \in E_n;\ u(t) = A\}$ is open. On the other hand, the continuity of u implies that this set is also closed (relative to $\mathscr{D}$). Since $\mathscr{D}$ is connected it follows that this set and $\mathscr{D}$ coincide. This shows that u is the constant function $u(x) \equiv A$.

[2] The proof will be written out for the case $n > 2$. Obvious modifications give us the result when $n = 2$. When $n = 1$ the theorem is obvious since the harmonic functions are the linear ones in this case.

By applying this result to $-u$ we obtain the *minimum principle for harmonic functions*: Under the hypotheses of Corollary 1.3, if $B = \inf_{x \in \mathscr{D}} u(x) > -\infty$ then $u(x) > B$ for all $x \in \mathscr{D}$, provided u is not a constant function.

An equivalent form of the maximum (minimum) principle is the following:

COROLLARY 1.3′. *If u is continuous on the closure $\bar{\mathscr{D}}$ of the bounded domain $\mathscr{D}$ and is harmonic in $\mathscr{D}$ then the maximum (minimum) of u is attained only on the boundary $\partial\mathscr{D} = \bar{\mathscr{D}} - \mathscr{D}$ of $\mathscr{D}$, provided u is not a constant function.*

Applying this to the difference $u_1 - u_2$ we obtain:

COROLLARY 1.4. *If u_1 and u_2 are continuous on the closure $\bar{\mathscr{D}}$ of the bounded domain $\mathscr{D}$, are harmonic in $\mathscr{D}$ and agree on the boundary $\partial\mathscr{D} = \bar{\mathscr{D}} - \mathscr{D}$. then $u_1(x) = u_2(x)$ for all $x \in \bar{\mathscr{D}}$.*

Another consequence of the mean valuc property is the following result, known as *Liouville's theorem*:

THEOREM 1.5. *If v is harmonic on E_n and bounded then v is a constant function.*

PROOF. Let $\Omega_n = \omega_{n-1}/n$ be the volume of the solid unit sphere of E_n. From the mean value property Theorem 1.1 we obtain, for all $x \in E_n$ and $t > 0$:

$$\begin{aligned} v(x) &= v(x)\frac{n}{t^n}\int_0^t \rho^{n-1}\,d\rho = \frac{n}{t^n}\int_0^t \mathscr{M}_{x,v}(\rho)\rho^{n-1}\,d\rho \\ &= \frac{n}{\omega_{n-1}}\frac{1}{t^n}\int_0^t \left\{\int_\Sigma v(x+\rho y')\,dy'\right\}\rho^{n-1}\,d\rho \\ &= \frac{1}{t^n\Omega_n}\int_{|y|\leqq t} v(x+y)\,dy. \end{aligned}$$

Thus, if $S_t(x)$ denotes the solid sphere about x of radius t,

$$\begin{aligned} |v(x_1) - v(x_2)| &= \frac{1}{t^n\Omega_n}\left|\int_{S_t(x_1)} v(x)\,dx - \int_{S_t(x_2)} v(x)\,dx\right| \\ &\leqq \frac{1}{t^n\Omega_n}\left|\left(\int_{S_t(x_1)-S_t(x_2)} dx + \int_{S_t(x_2)-S_t(x_1)} dx\right)\right|\|v\|_\infty. \end{aligned}$$

But the measure of the symmetric difference $(S_t(x_1) - S_t(x_2)) \cup$

$(S_t(x_2) - S_t(x_1))$ divided by t^n clearly tends to 0 as $t \to \infty$, and it follows that $v(x_1) - v(x_2) = 0$. Thus, v must be a constant function.

The mean value property completely characterizes harmonic functions. To see this we first establish the following lemma:

LEMMA 1.6. *Suppose u is twice differentiable in the domain $\mathscr{D}$ and $\mathscr{M}_{x,u}(r) = u(x)$ whenever the sphere $\{t \in E_n;\ |t - x| \leqq r\}$ belongs to $\mathscr{D}$, then u is harmonic in $\mathscr{D}$.*

PROOF. Fix $x \in \mathscr{D}$ and let $\mathscr{M}''(0)$ $(= \mathscr{M}''_{x,u}(0))$ be the limit, as $r \to 0$, of the second derivative $(d^2/dr^2)\mathscr{M}(r)$ (the continuity of the second derivatives of u implies the existence of this limit). The lemma will then be an immediate consequence of the identity.

$$\mathscr{M}''(0) = \frac{1}{n}(\Delta u)(x).$$

Letting $t' = (t_1', t_2', \ldots, t_n') \in \Sigma$ we have

$$\frac{d^2}{dr^2}\mathscr{M}(r) = \frac{1}{\omega_{n-1}} \int_{\Sigma} \left(\sum_{j,k=1}^{n} u_{jk}(x + rt')t_j't_k' \right) dt'.$$

Hence,

$$\begin{aligned}\mathscr{M}''(0) &= \frac{1}{\omega_{n-1}} \sum_{j,k=1}^{n} \int_{\Sigma} u_{jk}(x)t_j't_k'\, dt' \\ &= \frac{1}{\omega_{n-1}} \sum_{j,k=1}^{n} \left(\int_{\Sigma} t_j't_k'\, dt' \right) u_{jk}(x).\end{aligned}$$

But, by an easy symmetry argument, $\int_{\Sigma} t_j't_k'\, dt' = 0$ if $j \neq k$ while $\int_{\Sigma} (t_j')^2\, dt' = \omega_{n-1}/n$ is independent of j, $1 \leqq j \leqq n$. Thus, we obtain $\mathscr{M}''(0) = (1/n) \sum_{j=1}^{n} u_{jj}(x)$ and the lemma is proved.

The following result shows, in particular, that Lemma 1.6. is valid even if we weaken the hypothesis of second differentiability by assuming only local integrability of u. Moreover, together with Theorem 1.1, the following result implies that a harmonic function on a domain $\mathscr{D}$ belongs to the class $C^\infty(\mathscr{D})$. We consider first the case when u is continuous.

THEOREM 1.7. *Suppose a continuous function, u, on a domain $\mathscr{D} \subset E_n$ satisfies the mean value property (as in Lemma 1.6); then u is harmonic and has continuous partial derivatives of all orders in $\mathscr{D}$.*

PROOF. Since the problem is local we can assume, by restricting u to a sphere whose closure belongs to $\mathscr{D}$, that $\mathscr{D}$ is a sphere and that u is integrable on $\bar{\mathscr{D}}$. We can extend u to all of E_n by defining it to be 0 outside

$\bar{\mathscr{D}}$. Choose a radial $\varphi \in C^\infty(E_n)$ such that $\int_{E_n} \varphi(x)\,dx = 1$ and having support contained in the solid sphere of radius 1. We then let $u_\varepsilon = u * \varphi_\varepsilon$, where $\varphi_\varepsilon(x) = \varepsilon^{-n}\varphi(x/\varepsilon)$ for $\varepsilon > 0$. It is clear that $u_\varepsilon \in C^\infty(E_n)$. We have, using polar coordinates,

$$u_\varepsilon(x) = \int_{E_n} \varphi_\varepsilon(t)u(x - t)\,dt = \int_0^\varepsilon \varphi_\varepsilon(r)\left\{\int_\Sigma u(x - rt')\,dt'\right\}r^{n-1}\,dr$$

$$= \int_0^\varepsilon \varphi_\varepsilon(r)\omega_{n-1}\mathscr{M}_{x,u}(r)r^{n-1}\,dr.$$

If ε is less than the distance from x to the boundary of $\mathscr{D}$ then $\mathscr{M}_{x,u}(r) = u(x)$ for $0 \leqq r \leqq \varepsilon$. Thus, $u_\varepsilon(x) = u(x)\omega_{n-1}\int_0^\varepsilon \varphi_\varepsilon(r)r^{n-1}\,dr = u(x)$. [3] In particular, if $2d$ is the radius and x_0 the center of $\mathscr{D}$, then $u(x) = u_d(x)$ for all x whose distance from x_0 is less than d. This shows that in a neighborhood of each $x_0 \in \mathscr{D}$, u coincides with a C^∞ function, and the theorem is proved since the harmonicity of u follows from Lemma 1.6.

If u were merely locally integrable we would need to assume the mean-value property in the form

$$u(x) = \frac{1}{t^n\Omega_n}\int_{|y|<t} u(x + y)\,dy,$$

for t sufficiently small so that the sphere of center x and radius t lies in $\mathscr{D}$. By an argument analogous to the one given above this would show that $u(x) = u_d(x)$, and therefore u would still satisfy the conclusions of Theorem 1.7.

COROLLARY 1.8. *Suppose $\{u_n\}$ is a sequence of harmonic functions defined in the region $\mathscr{D}$. If this sequence converges uniformly to a function u in each compact subset of $\mathscr{D}$ then u is also harmonic.*

PROOF. Since u is, then, continuous it follows from Theorem 1.7 that it suffices to show that u satisfies the mean value property. By Theorem 1.1 $\mathscr{M}_{x,u_n}(r) = u_n(x)$ for all $x \in \mathscr{D}$ and r such that $\{t\ ;\ |x - t| \leqq r\} \subset \mathscr{D}$. Since u_n converges uniformly to u on the sphere $\{t\ ;\ |x - t| = r\}$, the

[3] We already have seen in the first chapter (see Theorems 1.18 and 1.25) how such convolutions $u_\varepsilon = u * \varphi_\varepsilon$ approximate u. With the added hypothesis that u satisfies the mean value property we obtain the fact that $u_\varepsilon(x) = u(x)$ if ε is small enough. In general, when φ is smooth it follows that so is $u_\varepsilon = u * \varphi_\varepsilon$; in this case, such a family $\{u_\varepsilon\}$ is often called a *regularization* of u.

integrals defining the means $\mathscr{M}_{x,u_n}(r)$ must converge to the integral defining $\mathscr{M}_{x,u}(r)$. It follows that $\mathscr{M}_{x,u}(r) = \lim_{n\to\infty} \mathscr{M}_{x,u_n}(r) = \lim_{n\to\infty} u_n(x) = u(x)$, which is the desired result.

A classical problem involving harmonic functions is the well-known Dirichlet problem. Its most common formulation is the following one:

Suppose $\mathscr{D}$ is a region with compact closure $\bar{\mathscr{D}}$ and f a continuous function defined on the boundary $\partial\mathscr{D} = \bar{\mathscr{D}} - \mathscr{D}$. Does there exist a continuous function u on $\bar{\mathscr{D}}$ such that

(i) *u is harmonic when restricted to $\mathscr{D}$.*
(ii) *$u(x) = f(x)$ when $x \in \partial\mathscr{D}$?*

Corollary 1.4 tells us that if such a function u exists then u is unique.

Our main interest lies in a variant of the Dirichlet problem, where $\mathscr{D}$ is the upper half-space E^+_{n+1} (see Section 2). Nevertheless, the solution of the problem as stated here, when $\mathscr{D}$ is the interior of a sphere, is of basic importance and has many applications. In this special case, we obtain the solution by making use of the *Poisson kernel for the unit sphere*:

$$p(s, x) = \frac{1}{\omega_{n-1}} \frac{1 - |x|^2}{|x - s|^n} = \frac{1}{\omega_{n-1}} \frac{1 - r^2}{(1 - 2r\cos\gamma + r^2)^{n/2}},$$

where $r = |x| < 1 = |s|$ and γ is the angle between the vectors x and s; that is, $r\cos\gamma = x\cdot s$.

Three basic properties of this kernel are the following:

THEOREM 1.9. *If $\Sigma = \Sigma_{n-1} = \{s \in E_n;\ |s| = 1\}$ denotes the surface of the unit sphere of E_n and ds the element of surface area, then*

(a) *$p(s, x) \geqq 0$ for all $s \in \Sigma$ and $|x| < 1$;*
(b) *$\int_\Sigma p(s, x)\, ds = 1$ whenever $|x| < 1$;*
(c) *for $x' \in \Sigma$ let $x = rx'$ $(0 \leqq r < 1)$ then, for each $\delta > 0$,*

$$\int_{s\in\Sigma, |s - x'| > \delta} p(s, rx')\, ds \to 0$$

as $r \to 1$, uniformly in x'.

PROOF. Property (a) is obvious and so is (c) once we observe that $|s - x| = |s - rx'|$ is bounded away from 0 (uniformly in x') when $s \in \Sigma$ satisfies $|s - x'| > \delta$.

In order to show (b) we first note that $p(s, x)$ is a harmonic function of x, $|x| < 1$, for each $s \in \Sigma$. Thus, by the mean value property (Theorem 1.1),

$$1 = \omega_{n-1} p(s, 0) = \frac{1}{\omega_{n-1}} \int_\Sigma \omega_{n-1} p(s, rx')\, dx' = \int_\Sigma p(s, rx')\, dx'$$

for $0 \leqq r < 1$ and $s, x' \in \Sigma$. But $|rx' - s| = |rs - x'|$ and, consequently, $p(s, rx') = p(x', rs)$. We obtain (b), therefore, by interchanging the roles of s and x'.

Using these properties we can now solve the Dirichlet problem for the interior of the unit sphere:

THEOREM 1.10. *Suppose f is a continuous function on Σ, then the function u defined by*

$$u(x) = \int_{\Sigma} f(s)p(s, x)\, ds$$

when $|x| < 1$ and $u(x) = f(x)$ when $|x| = 1$ is harmonic for $|x| < 1$ and continuous for $|x| \leqq 1$.

PROOF. The fact that u is harmonic for $|x| < 1$ follows from the fact that the Laplacian of u is equal to the integral of the product of f with the Laplacian (with respect to x) of $p(s, x)$. This requires justifying the change in order of differentiation and integration. A simpler proof is obtained by observing that it follows immediately from Fubini's theorem that the mean value of u taken over a sphere of radius $r < 1 - |x|$ about the point x equals the integral of $f(s)$ times the mean value of $p(s, \cdot)$ over the same sphere. Since $p(s, \cdot)$ is harmonic the latter equals $p(s, x)$. Thus, u satisfies the mean value property and, by Theorem 1.7, must, therefore, be harmonic in the interior of the unit sphere.

To show that u is continuous on the closed unit sphere it suffices to show that this is true on the boundary Σ. Let $x' \in \Sigma$, $0 \leqq r < 1$, $x = rx'$ and $A_\delta = \{s \in \Sigma;\ |s - x'| > \delta\}$. Then, by (b) and (a) of Theorem 1.9:

$$\begin{aligned} |u(x) - u(x')| &= \left| \int_{\Sigma} [f(s) - f(x')]p(s, rx')\, ds \right| \\ &\leqq \int_{A_\delta} |f(s) - f(x')|p(s, rx')\, ds \\ &\quad + \int_{\Sigma - A_\delta} |f(s) - f(x')|p(s, rx')\, ds. \end{aligned}$$

Again making use of (b), Theorem 1.9, we see that the second summand is majorized by

$$\left\{ \sup_{|s - x'| \leqq \delta} |f(s) - f(x')| \right\} \int_{\Sigma - A_\delta} p(s, x)\, ds \leqq \left\{ \sup_{|s - x'| \leqq \delta} |f(s) - f(x')| \right\} \cdot 1,$$

which is small, say less than $\varepsilon > 0$, if δ is chosen close to 0. On the other hand, for δ fixed, the first summand is majorized by

$$2\left\{\sup_{t\in\Sigma} |f(t)|\right\} \int_{A_\delta} p(s, x)\, ds;$$

but, by Theorem 1.9 part (c), the latter tends to 0 as $r \to 1$, uniformly in x'. Thus, $|u(x) - u(x')| < \varepsilon$, uniformly in x', if r is close enough to 1.

Consequently, if $x_0 \in \Sigma$ and $x = rx'$ we then have $|u(x_0) - u(x)| \leqq |u(x_0) - u(x')| + |u(x') - u(x)| = |f(x_0) - f(x')| + |u(x') - u(x)| < \varepsilon + \varepsilon = 2\varepsilon$ if x is sufficiently close to x_0 (since this implies that both $1 - r$ and $|x' - x|$ are small). This proves the theorem.

By applying an appropriate translation and dilation, this result gives us the solution of the Dirichlet problem for the general sphere in E_n:

COROLLARY 1.11. *Let $\mathscr{D}$ be the interior of the sphere of radius a about x_0 and f a continuous function on the boundary, $\partial\mathscr{D}$, of $\mathscr{D}$. Then the function u defined by*

$$u(x) = \frac{1}{\omega_{n-1}a^{2-n}} \int_{\Sigma} f(x_0 + as) \frac{a^2 - |x - x_0|^2}{|(x - x_0) - as|^n}\, ds$$

if $|x - x_0| < a$ and $u(x) = f(x)$ if $|x - x_0| = a$, is harmonic in $\mathscr{D}$ and continuous on $\bar{\mathscr{D}}$.

We shall give two applications of this result. The first is the following basic theorem concerning uniformly bounded sequences of harmonic functions:

THEOREM 1.12. *Suppose $\{u_m\}$ is a sequence of harmonic functions defined in a domain $\mathscr{D} \subset E_n$ such that the functions u_m are uniformly bounded in the closure of a bounded subdomain $\mathscr{S}$, such that $\bar{\mathscr{S}} \subset \mathscr{D}$. Then there exists a subsequence $\{u_{m_k}\}$ that converges uniformly to a harmonic function defined on $\mathscr{S}$.*

PROOF. If we can show that the functions u_m are equicontinuous on $\bar{\mathscr{S}}$ then this theorem is a consequence of Ascoli's theorem and Corollary 1.8. In order to show this equicontinuity it suffices to show that the first derivatives of the u_m are uniformly bounded in each closed subset of $\mathscr{S}$. But, if $x_0 \in \mathscr{S}$ then $u_m(x)$ has the integral expression of Corollary 1.11, for $|x - x_0| < a$, as long as a is small enough so that the sphere of radius a about x_0 lies in $\mathscr{S}$. Differentiating with respect to x_i $(i = 1, 2, .,, n)$ and evaluating this partial derivative at x_0, we obtain

$$\frac{\partial u_m}{\partial x_i}(x_0) = \frac{n}{\omega_{n-1}a} \int_{\Sigma} u_m(x_0 + as)s_i\, ds.$$

Thus, $|(\partial u_m/\partial x_i)(x_0)| \leqq (n/a) \sup_{y \in \mathscr{S}} |u_m(y)|$ from which we obtain the desired uniform boundedness of the first partial derivatives of the u_m's in each closed subset of $\mathscr{S}$.

The second application of Corollary 1.11 is the following *reflection principle* for harmonic functions:

THEOREM 1.13. *Suppose $\mathscr{D} \subset E_{n+1}$ is a domain that is symmetric with respect to E_n (that is, if $(x, y) = (x_1, x_2, \ldots, x_n, y) \in \mathscr{D}$ then $(x, -y) \in \mathscr{D}$). If a continuous function, u, defined on $\mathscr{D}$ satisfies $u(x, y) = -u(x, -y)$ and is harmonic in the "upper half," $\mathscr{D}^+ = \{(x, y) \in \mathscr{D}; y > 0\}$ of $\mathscr{D}$ then it is harmonic in all of $\mathscr{D}$.*

PROOF. The function u is harmonic in $\mathscr{D}^+$ by assumption. Since $-(\Delta u)(x, -y) = (\Delta u)(x, y)$, it is harmonic in $\mathscr{D}^- = \{(x, y) \in \mathscr{D}; y < 0\}$ as well. Consequently, it suffices to show that u is harmonic in a neighborhood of each point of $\mathscr{D} \cap E_n = \{(x, y) \in \mathscr{D}; y = 0\}$. If $x_0 = (x_0, 0)$ is such a point then we can find a closed (solid) sphere, $S_a(x_0)$, of radius $a > 0$ centered at this point that is contained in $\mathscr{D}$. For (x, y) in this sphere let

$$(1.14) \quad w(x, y) = \frac{a^{n-1}}{\omega_n} \int_{\Sigma_n} u(x_0 + as, at) \frac{a^2 - |(x - x_0, y)|^2}{|(x - x_0 - as, y - at)|^{n+1}} \, d\sigma,$$

where $\sigma = (s, t) = (s_1, s_2, \ldots, s_n, t)$ is a point on the surface, Σ_n, of the unit sphere of E_{n+1}. By Corollary 1.11, w is the harmonic function agreeing with u on the boundary of $S_a(x_0)$. On the other hand, the integral (1.14) vanishes when $y = 0$, since $u(x_0 + as, at) = -u(x_0 + as, -at)$,

$$\frac{a^2 - |(x - x_0, 0)|^2}{|(x - x_0 - as, -at)|^{n+1}} = \frac{a^2 - |(x - x_0, 0)|^2}{|(x - x_0 - as, at)|^{n+1}},$$

while by continuity, $u(x_0, 0) = 0$ whenever $(x_0, 0) \in \mathscr{D}$. Thus, w agrees with the continuous function u on the upper hemisphere $\{(x, y) \in \bar{E}^+_{n+1}; |x - x_0|^2 + y^2 = a^2\}$, consisting of the upper half of the boundary of $S_a(x_0)$, and on $S_a(x_0) \cap E_n$. Since both w and u are harmonic in the interior of this hemisphere, by Corollary 1.4 they must agree there. Applying the same argument to the lower hemisphere we obtain $u \equiv w$ on $S_a(x_0)$. Since w is harmonic in the interior of $S_a(x_0)$, so is u and the theorem is proved.

The following immediate consequence of this reflection principle and Liouville's theorem (see Theorem 1.5) will be useful in the next section:

COROLLARY 1.15. *Suppose u is a function that is continuous on $\bar{E}^+_{n+1} = E^+_{n+1} \cup E_n = \{(x, y) \in E_{n+1}; y \geqq 0\}$ that is harmonic in E^+_{n+1} and vanishes on E_n. Then, if u is bounded in E^+_{n+1} it follows that u is identically* 0.

Let us note that unless a restriction, such as boundedness, is imposed on

u this result does not hold. For example, the function $u(x, y) = y$ satisfies all the conditions of the corollary except that it is not bounded. This example also shows that the Dirichlet problem, posed in the case when $\mathscr{D}$ is the unbounded domain E_{n+1}^+, does not have a unique solution (for $u(x, y) = y$ and $v(x, y) = 0$ are harmonic in $\mathscr{D}$ and both have boundary value $f \equiv 0$ on E_n). Corollary 1.15, on the other hand, tells us that the solution is unique if we impose on it the added condition that it be bounded. In the next section we make a detailed study of some variants of the Dirichlet problem associated with the region E_{n+1}^+.

2. The Characterization of Poisson Integrals

Now that we have disposed of the required preliminaries we come to that part of the theory of harmonic functions connected with Fourier analysis on E_n.

We pose the following problem: Given a function $f \in L^p(E_n)$, $1 \leqq p \leqq \infty$, does there exist a harmonic function, u, defined on the upper half space E_{n+1}^+ such that

$$\|u(\cdot, y) - f\|_p = \left(\int_{E_n} |u(x, y) - f(x)|^p \, dx\right)^{1/p} \to 0$$

as $y \to 0$? If f were continuous and $p = \infty$ then this would be a direct generalization of the Dirichlet problem posed for the unbounded domain $\mathscr{D} = E_{n+1}^+$. We have just seen that the solution need not be unique in this case. It is also clear that if f is not continuous then there could not exist a harmonic (and, in particular, continuous) function u such that $\|u(\cdot, y) - f\|_\infty \to 0$ as $y \to 0$. Nevertheless, a solution of this problem does exist when $p < \infty$; if the problem is properly posed, a solution also exists if $p = \infty$.

The following result, in large part a collection of results that have been already established, tells us that the Poisson integral of f is the harmonic function we are looking for.

THEOREM 2.1. (a) *If $f \in L^p(E_n)$, $1 \leqq p \leqq \infty$, and*

$$u(x, y) = \int_{E_n} f(t) P(x - t, y) \, dt$$

is the Poisson integral of f then u is harmonic in E_{n+1}^+, $\lim_{y \to 0} u(x, y) = f(x)$ for almost every $x \in E_n$ and

$$\|u(\cdot, y)\|_p = \left(\int_{E_n} |u(x, y)|^p \, dx\right)^{1/p} \leqq \|f\|_p \tag{2.2}$$

for all $y > 0$. If $1 \leqq p < \infty$ then $u(\cdot, y)$ converges to f in the L^p norm as

$y \to 0$; that is,

$$\|u(\cdot, y) - f\|_p = \left(\int_{E_n} |u(x, y) - f(x)|^p \, dx\right)^{1/p} \to 0$$

as $y \to 0$.

(b) *If $f \in C_0 \subset L^\infty(E_n)$ then the Poisson integral of f, $u(x, y)$, converges to f uniformly: $\|u(\cdot, y) - f\|_\infty = \sup_{x \in E_n} |u(x, y) - f(x)| \to 0$ as $y \to 0$. If we only assume f to be continuous and bounded the convergence is uniform on compact subsets of E_n. In either case, we can extend $u(x, y)$ to $\overline{E_{n+1}^+} = E_{n+1}^+ \cup E_n$ by letting $u(x, 0) = f(x)$ and we obtain a continuous function.*

PROOF. The almost everywhere convergence and the convergence in the norm in part (a) are Theorems 1.25 and 1.18 in Chapter I. Inequality (2.2) is an immediate consequence of Theorem 1.3 in that chapter, with $g(x) = P(x, y)$ and $h(x) = u(x, y)$ (recall that Lemma 1.17 asserts that $g = P(\cdot, y) \in L^1(E_n)$ and $\|g\|_1 = 1$). The fact that u is harmonic can be shown by using the harmonicity of the Poisson kernel and an argument completely analogous to that used in the beginning of the proof of Theorem 1.10.

If f is continuous and bounded $u(x, y)$ converges to $f(x)$, as $y \to 0$, for all $x \in E_n$, since condition (1.24) of Chapter I obviously holds everywhere (see Theorem 1.25 in that chapter). The uniform convergence on a compact subset, F, of E_n needs only the following simple argument: given $\varepsilon > 0$ there exists $\delta > 0$ such that $|f(x - t) - f(x)| < \varepsilon$ whenever $x \in F$ and $|t| < \delta$. Thus,

$$\begin{aligned}
|u(x, y) - f(x)| &= \left|\int_{E_n} f(x - t)P(t, y)\,dt - f(x)\right| \\
&= \left|\int_{E_n} f(x - t)P(t, y)\,dt - \int_{E_n} f(x)P(t, y)\,dt\right| \\
&\leq \int_{|t| \leq \delta} |f(x - t) - f(x)|P(t, y)\,dt \\
&\quad + \int_{|t| > \delta} |f(x - t) - f(x)|P(t, y)\,dt \\
&\leq \varepsilon \int_{E_n} P(t, y)\,dt + 2\|f\|_\infty \int_{|t| > \delta} P(t, y)\,dt \\
&= \varepsilon \cdot 1 + 2\|f\|_\infty \int_{|t| > \delta} P(t, y)\,dt.
\end{aligned}$$

But, for $\delta > 0$ fixed, $\int_{|t|>\delta} P(t, y)\, dt \leqq c_n y \int_{|t|>\delta} |t|^{-(n+1)}\, dt$ tends to 0 as $y \to 0$. Consequently, $|u(x, y) - f(x)| \leqq \varepsilon$, if y is sufficiently small, for all $x \in F$. If $f \in C_0$ then it is uniformly continuous and, consequently, we can find $\delta > 0$ such that $|f(x - t) - f(x)| < \varepsilon$ for all $x \in E_n$ when $|t| < \delta$; we therefore obtain the uniform convergence of $u(x, y)$ to $f(x)$ as $y \to 0$. This proves the theorem.[4]

The space $L^1(E_n)$ is naturally embedded in the space $M(E_n)$ of finite Borel measures on E_n (see the proof of Theorem 3.19 in the last chapter). Many results and concepts associated with the former space can be extended to the latter. In the first chapter for example, we have done so with the Fourier transform, the operation of convolution and Theorem 1.3. Similarly, we can do this for Theorem 2.1:

THEOREM 2.3. *If $\mu \in M(E_n)$ and $u(x, y) = \int_{E_n} P(x - t, y)\, d\mu(t)$ is the Poisson–Stieltjes integral of μ then u is harmonic in E^+_{n+1} and*

$$\|u(\cdot, y)\|_1 = \int_{E_n} |u(x, y)|\, dx \leqq \|\mu\|, \tag{2.4}$$

where $\|\mu\|$ is the total variation, $|\mu|(E_n)$, of μ. Furthermore,

$$\lim_{y \to 0} \int_{E_n} u(x, y)\varphi(x)\, dx = \int_{E_n} \varphi(x)\, d\mu(x)$$

for all $\varphi \in C_0$; that is, $u(\cdot, y)$ converges to μ in the weak topology.*

PROOF. The fact that u is harmonic can be shown by the same argument that was used in the proof of Theorem 1.10 (and of Theorem 2.1). Inequality (2.4) is an immediate consequence of the extension to Borel measures of Theorem 1.3 (in Chapter I) that was described after the

[4] The convergence $\|u(\cdot, y) - f\|_\infty \to 0$ when $f \in C_0$ is a special case of Theorem 1.18 of Chapter I. The proof we have given of part (b) was based on the following three properties of the Poisson kernel: (i) $P(x, y) \geqq 0$; (ii) $\int_{E_n} P(x, y)\, dx = 1$ for all $y > 0$; (iii) if $\delta > 0$ then $\int_{|x|>\delta} P(x, y)\, dx \to 0$ as $y \to 0$ ((i) is obvious, (iii) was just proved and (ii) is part (b) of Lemma 1.17 of Chapter I). A kernel satisfying these three properties is often called an *approximation to the identity*. It is clear that the Gauss–Weierstrass kernel also satisfies these properties. The reader will note that a slight modification of the last part of the proof of part (b) (replacing the L^∞ norm by the L^p norm, $1 \leqq p < \infty$) gives us another proof of Theorem 1.18 when $\varphi(x) = P(x, 1)$. An advantage of this treatment is that it does not make use of the operation of convolution (see, for example, Theorem 1.9, where analogous properties (a), (b), and (c) enable us to obtain the solution of the Dirichlet problem for the sphere: Theorem 1.10). Later, in Theorem 5.6 of Chapter III this type of L^p convergence is established by these methods for another kind of Poisson integral.

statement of that theorem. The weak* convergence can be shown in the following way: If $\varphi \in C_0$ and $v(x, y)$ is its Poisson integral then

$$\int_{E_n} u(x, y)\varphi(x)\, dx = \int_{E_n} \left(\int_{E_n} P(x - t, y)\, d\mu(t) \right) \varphi(x)\, dx$$
$$= \int_{E_n} \left(\int_{E_n} P(x - t, y)\varphi(x)\, dx \right) d\mu(t)$$
$$= \int_{E_n} v(t, y)\, d\mu(t).$$

Consequently, since, by part (b) of Theorem 2.1, $\|v(\cdot, y) - \varphi\|_\infty \to 0$ as $y \to 0$,

$$\left| \int_{E_n} u(x, y)\varphi(x)\, dx - \int_{E_n} \varphi(t)\, d\mu(t) \right| = \left| \int_{E_n} \{v(t, y) - \varphi(t)\}\, d\mu(t) \right|$$
$$\leqq \int_{E_n} |v(t, y) - \varphi(t)|\, |d\mu|(t)$$
$$\leqq \|v(\cdot, y) - \varphi\|_\infty |\mu|(E_n) \to 0$$

as $y \to 0$ and the theorem is established.

The principal result of this section is the proof of a converse of Theorem 2.1. Inequality (2.2) asserts that the L^p norms $\|u(\cdot, y)\|_p$ are bounded for $y > 0$. We show that this is sufficient for u to be a Poisson integral.

THEOREM 2.5. *If $u(x, y)$ is harmonic in E^+_{n+1} and there exists a constant $c > 0$ and a p, $1 \leqq p \leqq \infty$, such that*

$$\|u(\cdot, y)\|_p = \left(\int_{E_n} |u(x, y)|^p\, dx \right)^{1/p} \leqq c < \infty$$

for all $y > 0$ then

(a) *when $1 < p \leqq \infty$, $u(x, y)$ is the Poisson integral of a function f in $L^p(E_n)$;*

(b) *if $p = 1$, $u(x, y)$ is the Poisson–Stieltjes integral of a finite Borel measure; if, in addition, $u(\cdot, y)$ is Cauchy in the L^1 norm as $y \to 0$ then $u(x, y)$ is the Poisson integral of a function f in $L^1(E_n)$.*

PROOF. If $1 < p \leqq \infty$ and $\|u(\cdot, y)\|_p \leqq c < \infty$ for all $y > 0$ then there exists a sequence $\{y_k\}$, with $\lim_{k\to\infty} y_k = 0$, and a function f in $L^p(E_n)$ such that $u(\cdot, y_k)$ converges weakly to f as $k \to \infty$. That is, for each $g \in L^{p'}(E_n)$,

$1/p + 1/p' = 1$,

$$\lim_{k\to\infty} \int_{E_n} u(x, y_k)g(x)\,dx = \int_{E_n} f(x)g(x)\,dx.$$

If $p = 1$ there exists a finite Borel measure μ that is the weak* limit of a sequence $\{u(\cdot, y_k)\}$. That is, for each g in C_0

$$\lim_{k\to\infty} \int_{E_n} u(x, y_k)g(x)\,dx = \int_{E_n} g(x)\,d\mu(x).$$ [5]

Since $P(\cdot, y)$, for each $y > 0$, belongs to $L^{p'}(E_n)$ for $1 \leqq p' \leqq \infty$ and also belongs to C_0 we have, in particular,

$$\lim_{k\to\infty} \int_{E_n} P(x - t, y)u(t, y_k)\,dt = \int_{E_n} P(x - t, y)f(t)\,dt = v(x, y)$$

when $1 < p \leqq \infty$, and

$$\lim_{k\to\infty} \int_{E_n} P(x - t, y)u(t, y_k)\,dt = \int_{E_n} P(x - t, y)\,d\mu(t) = v(x, y)$$

when $p = 1$.

If we show that, in either case, $v(x, y) = u(x, y)$ then we will have established (a), and the first part of (b). But this is an immediate consequence of the following two lemmas (since they imply that $u(x, y + y_k) = \int_{E_n} P(x - t, y)u(t, y_k)\,dt$):

LEMMA 2.6. *If $u(x, y)$ satisfies the hypotheses of Theorem 2.5 then there exists a constant $A = A_{n,p} > 0$ such that*

$$\|u(\cdot, y)\|_\infty = \sup_{x\in E_n} |u(x, y)| \leqq Acy^{-n/p}.$$

In particular, u is bounded in each proper sub-half-space

$$E^+_{n+1,y_0} = \{(x, y) \in E_{n+1}; y \geqq y_0 > 0\} \subset E^+_{n+1}.$$

LEMMA 2.7. *If $u(x, y)$ is harmonic in E^+_{n+1} and bounded in each proper sub-half-space then*

$$u(x, y_1 + y_2) = \int_{E_n} u(t, y_1)P(x - t, y_2)\,dt$$

for all $y_1, y_2 > 0$.

[5] These results are special cases of the fact that the unit sphere of the dual of a Banach space is compact in the weak* topology. The last assertion, concerning $p = 1$, is often known as Helly's theorem (see bibliographical notes to Chapter I).

The first of these lemmas is an easy consequence of the mean value property for harmonic functions: Let Ω_{n+1} be the volume of the solid unit sphere of E_{n+1}, then $\Omega_{n+1} = \omega_n/(n+1)$ and

$$\begin{aligned}
u(x, y) &= u(x, y)\frac{n+1}{(y/2)^{n+1}} \int_0^{y/2} r^n \, dr \\
&= [(n+1)2^{n+1}/y^{n+1}] \int_0^{y/2} \mathscr{M}_{(x,y),u}(r) r^n \, dr \\
&= [(n+1)2^{n+1}/y^{n+1}\omega_n] \\
&\quad \times \int_0^{y/2} \left\{ \int_{\Sigma_n} u(x_1 + rt_1', \ldots, x_n + rt_n', y + rt_{n+1}') \, dt' \right\} r^n \, dr \\
&= (2^{n+1}/\Omega_{n+1}) y^{-(n+1)} \int_{|t| \leqq y/2} u(x_1 + t_1, \ldots, x_n + t_n, y + t_{n+1}) \, dt.
\end{aligned}$$

Consequently, letting $(\xi_1, \ldots, \xi_n, \eta) = (\xi, \eta)$ and $A = (2^{n+1}/\Omega_{n+1})^{1/p}$,

$$\begin{aligned}
|u(x, y)| &\leqq (2^{n+1}/\Omega_{n+1}) y^{-(n+1)} \int_{|(x,y)-(\xi,\eta)|<y/2} |u(\xi, \eta)| \, d\xi \, d\eta \\
&\leqq (2^{n+1}/\Omega_{n+1}) y^{-(n+1)} \left(\int_{|(x,y)-(\xi,\eta)|<y/2} |u(\xi, \eta)|^p \, d\xi \, d\eta \right)^{1/p} \\
&\qquad \times ((\Omega_{n+1} y^{n+1}/2^{n+1})^{1-(1/p)} \\
&\leqq A y^{-(n+1)/p} \left[\int_{y/2}^{3y/2} \left\{ \int_{\xi \in E_n} |u(\xi, \eta)|^p \, d\xi \right\} d\eta \right]^{1/p} \\
&\leqq A y^{-(n+1)/p} \left[\int_{y/2}^{3y/2} c^p \, d\eta \right]^{1/p} = Acy^{-(n+1)/p} y^{1/p} = Acy^{-n/p}.
\end{aligned}$$

In order to prove (2.7) we fix $y_0 > 0$ and put $w(x, y) = u(x, y + y_0)$ for $y \geqq 0$. Let $w_1(x, y) = \int_{E_n} u(t, y_0) P(x - t, y) \, dt$ for $(x, y) \in E_{n+1}^+$. The lemma will be established if we show that $w \equiv w_1$.

It follows from Lemma 2.6 that the function w is harmonic in E_{n+1}^+ and, on $\overline{E_{n+1}^+}$, it is continuous and bounded. From Theorem 2.1 (see parts (a) and (b)) w_1 can be extended to $\overline{E_{n+1}^+}$ by letting $w_1(x, 0) = u(x, y_0)$ and we obtain a continuous and bounded function that is harmonic in E_{n+1}^+. Thus, the function $h = w_1 - w$ is also continuous and bounded on $\overline{E_{n+1}^+}$ and harmonic in E_{n+1}^+. Since $h(x, 0) = 0$ for all $x \in E_n$ it follows from Corollary 1.15 that h is identically zero; thus, we obtain the desired result $w \equiv w_1$.

We still need to show the last part of Theorem 2.5. If $u(\cdot, y)$ is Cauchy in the L^1 norm it follows from the completeness of $L^1(E_n)$ that there exists $f \in L^1(E_n)$ such that $\|u(\cdot, y) - f\|_1 \to 0$ as $y \to 0$. Therefore, $\int_{E_n} u(x, y)g(x)\,dx \to \int_{E_n} f(x)g(x)\,dx$ whenever $g \in L^\infty(E_n)$. Now, the same argument that was used to establish part (a) of Theorem 2.5 applies and we can conclude that u is the Poisson integral of the L^1 function f.

3. The Hardy–Littlewood Maximal Function and Nontangential Convergence of Harmonic Functions

Up to this point we have considered only one type of convergence at the boundary for harmonic functions defined in E_{n+1}^+. We have shown, for example, that when u is a Poisson integral then, for almost every point $x_0 \in E_n$, there exists a boundary value

$$\lim_{(x,y)\to(x_0,0)} u(x, y) \tag{3.1}$$

as the point (x, y) approaches $(x_0, 0)$ "downward" along the line $x = x_0$ (see Theorem 2.1). An important more general result holds, however; the limits (3.1) exist almost everywhere even when (x, y) is allowed to approach $(x_0, 0)$ along more general paths. These are the *nontangential* paths. Before making this notion more precise we shall study an important operator, the Hardy–Littlewood maximal operator, whose basic properties can be used to show the existence of these more general limits.

This operator assigns to each function $f \in L^p(E_n)$, $1 \leqq p \leqq \infty$, its *Hardy–Littlewood maximal function* m_f defined at $x \in E_n$ by

$$\begin{aligned} m_f(x) &= \sup_{r>0} \frac{1}{\Omega_n r^n} \int_{|t|\leqq r} |f(x - t)|\,dt \\ &= \sup_{S_x} \frac{1}{|S_x|} \int_{S_x} |f(t)|\,dt, \end{aligned}$$

where Ω_n is the "volume" (Lebesgue measure) of the solid unit sphere $\{t \in E_n : |t| \leqq 1\}$; the last supremum is taken over all spheres S_x having center x and positive radius, and $|S_x|$ is the Lebesgue measure of S_x (thus, if the radius of S_x is r we have $|S_x| = \Omega_n r^n$). We sometimes refer to m_f as the *spherical* maximal function in order to distinguish it from the *cubic maximal function* $\bar{m}_f$ defined by

$$\bar{m}_f(x) = \sup_{Q_x} \frac{1}{|Q_x|} \int_{Q_x} |f(t)|\,dt,$$

where the supremum is taken over all nondegenerate cubes Q_x centered at x having sides parallel to the axes and $|Q_x|$ denotes the Lebesgue measure of Q_x. It is clear that, if f is essentially bounded, then $m_f(x)$ and $\overline{m}_f(x)$ do not exceed $\|f\|_\infty$. We shall show that $m_f(x)$ and $\overline{m}_f(x)$ are finite almost everywhere whenever $f \in L^p(E_n)$, $1 \leqq p \leqq \infty$.

Given a sphere S_x of radius $r > 0$ and center x, let q_x be the inscribed cube (it has, therefore, diagonals of length $2r$) and Q_x the circumscribed cube (it has, therefore, sides of length $2r$), both cubes having sides parallel to the axes. Then, clearly, there exist constants a_n and A_n, depending only on the dimension, such that

$$|Q_x| \leqq A_n|S_x| \quad \text{and} \quad |S_x| \leqq a_n|q_x|.$$

Thus,

$$\frac{1}{|S_x|}\int_{S_x} |f(t)|\, dt \leqq \frac{A_n}{|Q_x|}\int_{Q_x} |f(t)|\, dt.$$

But this implies

$$m_f(x) \leqq A_n\overline{m}_f(x). \tag{3.2}$$

Similarly, we have

$$\overline{m}_f(x) \leqq a_n m_f(x). \tag{3.2$'$}$$

We shall be more interested in the spherical maximal function m_f. In some cases, however, it is easier to establish needed estimates for $\overline{m}_f$; inequalities (3.2) and (3.2$'$) can then be used to obtain corresponding results for m_f. The following lemma will enable us to show that $\overline{m}_f(x) < \infty$ almost everywhere (and, hence, $m_f(x) < \infty$ a.e.).

LEMMA 3.3. *Let $R_1 \subset R_2 \subset \cdots \subset R_k$ be nonempty open rectangles in E_n with sides parallel to the axes and centered at 0. Suppose $S \subset E_n$ is a bounded set and, for each x in S, let $i(x)$ be an integer satisfying $1 \leqq i(x) \leqq k$. If*

$$R^x = \{u \in E_n : u - x \in R_{i(x)}\}$$

then there exist a finite number of points, $x_1, x_2, \ldots, x_l$, of S such that $S \subset R^{x_1} \cup \cdots \cup R^{x_l}$, and each v in E_n belongs to at most 2^n of the sets $R^{x_1}, \ldots, R^{x_l}$.

PROOF. Choose x_1 so that $i(x_1)$ is the largest possible (we can do this since $i(x) \leqq k$). Having done so, we then choose $x_2 \in S - R^{x_1}$ so that $i(x_2)$ is largest possible and continue in this way by choosing $x_3 \in S - \{R^{x_1} \cup R^{x_2}\}$ so that $i(x_3)$ is maximal, etc. We thus obtain a sequence of open rectangles $R^{x_1}, R^{x_2}, \ldots$ centered at $x_1, x_2, \ldots$ with the property that each center x_j belongs to no other rectangle R^{x_i}, $i \neq j$. But this means that the distance between x_i and x_j, $i \neq j$, must exceed half the length of the shortest side of R_1 (the smallest rectangle). Since, by assumption, this half-length is positive and the set S is bounded it follows that the sequence $x_1, x_2, \ldots$ is finite and, after reaching the, say, l-th stage of our process we have obtained a covering $\{R^{x_1}, \ldots, R^{x_l}\}$ of S.

It now remains to be shown that a point $v \in E_n$ can belong to no more than 2^n of the rectangles $R^{x_1}, \ldots, R^{x_l}$. In order to see this, consider the hyperplanes through v that are parallel to the coordinate hyperplanes. These form 2^n "octants" with vertex v that cover all of E_n. Suppose two of the points, x_i and x_j $(1 \leqq i, j \leqq l)$, belong to the same octant in such a way that R^{x_i} and R^{x_j} both contain v. Since each of these two rectangles is a translate of one of the rectangles $R_1, \ldots, R_k$, one of these two, say R^{x_i}, must have sides that are at least as long as those of the other, R^{x_j}. But this, together with the fact that $v \in R^{x_i} \cap R^{x_j}$, implies that $x_j \in R^{x_i}$, contradicting the fact that each center x_j belongs to no other rectangle R^{x_i}. Thus, at most one rectangle with center in any given "octant" can contain v and the lemma is proved.

THEOREM 3.4. *There exists a constant $c = c(n)$, depending only on the dimension, such that whenever $f \in L^1(E_n)$ and*

$$F_s = \{x \in E_n : m_f(x) > s > 0\}$$

then

$$|F_s| \leqq \frac{c\|f\|_1}{s},$$

where $|F_s|$ denotes the Lebesgue measure of F_s. In particular, $m_f(x) < \infty$ for almost every $x \in E_n$.

PROOF. We shall show that

$$|\{x \in E_n : \overline{m}_f(x) > s\}| \leqq 2^n \frac{\|f\|_1}{s} \tag{3.5}$$

for each $s > 0$. Since, by (3.2), $F_s \subset \{x \in E_n : \overline{m}_f(x) > s/A_n\}$, we would then have

$$|F_s| \leqq \left|\left\{x \in E_n : \overline{m}_f(x) > \frac{s}{A_n}\right\}\right| \leqq \frac{2^n A_n \|f\|_1}{s}$$

and the theorem follows with $c = 2^n A_n$.

Let S be any compact subset of $\{x \in E_n : \overline{m}_f(x) > s\}$. For each $x \in S$, therefore, there exists a cube Q_x such that

$$\frac{1}{|Q_x|} \int_{Q_x} |f(t)|\, dt > s.$$

By continuity of the integral, there exists a neighborhood U of x such that for each $u \in U$

$$\frac{1}{|Q_u|} \int_{Q_u} |f(t)|\, dt > s,$$

where $Q_u = \{(w - x) + u : w \in Q_x\}$ is the cube of the same size as Q_x that is centered at u. Since S is compact there are a finite number of such neighborhoods covering S. Hence, there are a finite number of cubes $Q_1 \subset \cdots \subset Q_k$, centered at 0, such that for each $x \in S$ there exists an integer $i(x)$, $1 \leqq i(x) \leqq k$, for which

$$\frac{1}{|Q^x|} \int_{Q^x} |f(t)|\, dt > s,$$

where $Q^x = \{u \in E_n : u - x \in Q_{i(x)}\}$. By Lemma 3.3, we can find a finite number of points $x_1, \ldots, x_l$ in S such that

$$S \subset Q^{x_1} \cup \cdots \cup Q^{x_l}$$

and each $v \in E_n$ belongs to at most 2^n of these cubes. If, for a set A, χ_A denotes the characteristic function of A, this last property is equivalent to the inequality

$$\sum_{j=1}^{l} \chi_{Q^{x_j}} \leqq 2^n \chi_{\bigcup_{j=1}^{l} Q^{x_j}}.$$

Thus,

$$|S| \leqq \left| \bigcup_{j=1}^{l} Q^{x_j} \right| \leqq \sum_{j=1}^{l} |Q^{x_j}| \leqq \frac{1}{s} \sum_{j=1}^{l} \int_{Q^{x_j}} |f(t)|\, dt$$

$$= \frac{1}{s} \int_{E_n} \left\{ \sum_{j=1}^{l} \chi_{Q^{x_j}}(t) \right\} |f(t)|\, dt \leqq \frac{2^n}{s} \int_{E_n} \chi_{\bigcup_{j=1}^{l} Q^{x_j}}(t) |f(t)|\, dt \leqq \frac{2^n}{s} \|f\|_1.$$

Since S is an arbitrary compact subset of $\{x \in E_n : \overline{m}_f(x) > s\}$ inequality (3.5) and, hence, the theorem now follow immediately.

We have already observed that $m_g(x) \leqq \|g\|_\infty < \infty$ for any $g \in L^\infty(E_n)$. From Theorem 3.4 we have $m_f(x) < \infty$ for almost every x when $f \in L^1(E_n)$. It follows from these two facts that $m_h(x) < \infty$ a.e. whenever $h \in L^p(E_n)$, $1 < p < \infty$. In order to see this we first observe that the maximal operator is an example of a *sublinear operator*. By this we mean an operator T mapping a linear space of measurable functions defined on a measure space into measurable functions defined on (possibly) another measure space satisfying

(i) $$|[T(f + g)](x)| \leqq |(Tf)(x)| + |(Tg)(x)|$$

almost everywhere (this property is called *subadditivity*), and

(ii) $$|T(af)| = |a|\,|Tf|$$

whenever a is a scalar and f is in the domain of T. Thus, if, for $h \in L^p(E_n)$, we put

$$g(x) = \begin{cases} h(x) & \text{when } |h(x)| \leqq 1 \\ 0 & \text{when } |h(x)| > 1 \end{cases}$$

and $f = h - g$, then $g \in L^\infty(E_n)$ and $f \in L^1(E_n)$. By property (i) (subadditivity), therefore,

$$m_h(x) = m_{f+g}(x) \leqq m_f(x) + m_g(x).$$

Consequently, $m_h(x) < \infty$ for almost every x. All these observations are, of course, equally valid for the cubic maximal function.

The maximal operator, as a mapping defined on $L^1(E_n)$ is not a bounded transformation into $L^1(E_n)$ (e.g., if f is the characteristic function of the interval $[0, 1]$ then m_f is not even integrable). Theorem 3.4, however, tells us that it is "almost" bounded. In order to make this statement more precise we introduce the notion of the *distribution function* $\lambda = \lambda_g$ of a function $g \in L^p(E_n)$: if $s > 0$ then $\lambda(s)$ is the Lebesgue measure of the set $F_s = \{x \in E_n : |g(x)| > s\}$.[6] We thus obtain a nonincreasing function λ on the positive real axis. It is clear that $s^p\lambda(s) = \int_{F_s} s^p\,dx \leqq \int_{F_s} |g|^p\,dx \leqq \|g\|_p^p$. That is, $\lambda(s)$ is majorized by a constant times s^{-p}. This condition is clearly not sufficient to insure the integrability of $|g|^p$. On the other hand, the equality

(3.6) $$\|g\|_p^p = -\int_0^\infty s^p\,d\lambda(s) = p\int_0^\infty s^{p-1}\lambda(s)\,ds,$$

which is immediate for g simple (the general case then follows by approximating $|g|$ from below by simple functions), indicates that the condition $\lambda(s) \leqq (\text{constant})\,s^{-p}$ is close to being equivalent with the finiteness of $\int |g|^p$. Theorem 3.4 tells us that m_f has a distribution function λ satisfying $\lambda(s) \leqq c\|f\|_1 s^{-1}$. It is in this sense that we say that the maximal operator is almost bounded as an operator on $L^1(E_n)$.

The maximal operator, however, is bounded when restricted to $L^p(E_n)$, $1 < p \leqq \infty$:

[6] The distribution function of g can obviously be defined more generally for any measurable function on an arbitrary measure space. In Chapter V we study many of the properties of these more general distribution functions.

THEOREM 3.7. *There exists a constant $b = b(p, n)$, depending on the dimension n and the exponent $p > 1$, such that*

$$\|m_f\|_p \leqq b\|f\|_p$$

for all $f \in L^p(E_n)$.

PROOF. Suppose $f \in L^p(E_n)$, $1 < p \leqq \infty$, and $s > 0$. For each $x \in E_n$, let

$$f^s(x) = \begin{cases} f(x) & \text{if } |f(x)| > s \\ 0 & \text{if } |f(x)| \leqq s \end{cases}$$

and

$$f_s(x) = \begin{cases} 0 & \text{if } |f(x)| > s \\ f(x) & \text{if } |f(x)| \leqq s. \end{cases}$$

Then $f = f^s + f_s$, with $f^s \in L^1(E_n)$ and $f_s \in L^\infty(E_n)$.

We let λ, λ^s, λ_s be the distribution functions of m_f, m_{f^s}, and m_{f_s}. Since, by subadditivity, $m_f \leqq m_{f^s} + m_{f_s}$ it follows that $\lambda(2s) \leqq \lambda^s(s) + \lambda_s(s)$. We already observed that the maximal function of an essentially bounded function g is bounded by $\|g\|_\infty$. Thus, $m_{f_s}(x) \leqq \|f_s\|_\infty \leqq s$ and it follows that $\lambda_s(s) = 0$. Therefore $\lambda(2s) \leqq \lambda^s(s)$.

From (3.6), this last inequality and Theorem 3.4 we therefore obtain

$$\begin{aligned} \|m_f\|_p^p &= p2^p \int_0^\infty s^{p-1}\lambda(2s)\,ds \leqq p2^p \int_0^\infty s^{p-1}\lambda^s(s)\,ds \\ &\leqq p2^p \int_0^\infty s^{p-1}cs^{-1}\left\{\int_{E_n} |f^s(x)|\,dx\right\} ds \\ &= p2^p c \int_0^\infty s^{p-2}\left\{\int_{|f(x)|>s} |f(x)|\,dx\right\} ds \\ &= p2^p c \int_{E_n} |f(x)|\left\{\int_0^{|f(x)|} s^{p-2}\,ds\right\} dx \\ &= \frac{p2^p c}{p-1} \int_{E_n} |f(x)|\,|f(x)|^{p-1}\,dx, \end{aligned}$$

which proves the theorem with $b = b(p, n) = 2[pc/(p-1)]^{1/p}$.

Another way of expressing m_f is the following one:

Let φ be the characteristic function of the solid unit sphere $\{x \in E_n: |x| \leqq 1\}$ divided by Ω_n (so that $\int_{E_n} \varphi(x)\,dx = 1$). Let $\varepsilon > 0$ and put $\varphi_\varepsilon(t) =$

$\varepsilon^{-n}\varphi(t/\varepsilon)$. Then, if $f \in L^p(E_n)$, $1 \leqq p \leqq \infty$, and if, for the sake of notational simplicity, we assume $f \geqq 0$, we have

$$(f * \varphi_\varepsilon)(x) = \frac{1}{\varepsilon^n \Omega_n} \int_{|t| \leqq \varepsilon} f(x - t)\, dt;$$

therefore,

$$m_f(x) = \sup_{\varepsilon > 0} (f * \varphi_\varepsilon)(x). \tag{3.8}$$

Similarly, if φ is the characteristic function of the cube $\{x = (x_1, \ldots, x_n) \in E_n\colon -\frac{1}{2} \leqq x_j \leqq \frac{1}{2}, j = 1, 2, \ldots, n\}$, then the right side of (3.8) gives us $\overline{m}_f(x)$. We have seen that $\overline{m}_f$ and m_f are "equivalent" in the sense that inequalities (3.2) and (3.2′) hold. It is natural, therefore, to ask about the relation of m_f with the supremum in (3.8) when $\varphi \geqq 0$ is a more general function.

For example, suppose $\varphi = \sum_{k=1}^m c_k \chi_k$, $c_k > 0$, where χ_k is the characteristic function of the sphere $\sigma_k = \{x \in E_n\colon |x| \leqq r_k\}$. Then, for $f \geqq 0$,

$$\begin{aligned}(f * \varphi_\varepsilon)(x) &= \sum_{k=1}^m c_k \varepsilon^{-n} \int_{|t| \leqq \varepsilon r_k} f(x - t)\, dt \\ &= \sum_{k=1}^m c_k r_k^n \Omega_n \frac{1}{\Omega_n (\varepsilon r_k)^n} \int_{|t| \leqq \varepsilon r_k} f(x - t)\, dt \\ &\leqq m_f(x) \sum_{k=1}^m c_k |\sigma_k| = m_f(x) \|\varphi\|_1.\end{aligned}$$

That is,

$$\sup_{\varepsilon > 0} (f * \varphi_\varepsilon)(x) \leqq m_f(x) \|\varphi\|_1. \tag{3.9}$$

This result clearly extends to any $\varphi \in L^1(E_n)$ of the form $\varphi(t) = \psi(|t|)$ (i.e., φ is radial) with ψ a nonnegative decreasing function on $[0, \infty)$. We need only approximate φ from below by an ascending sequence of simple functions of the type just considered and apply the Lebesgue monotone convergence theorem.

In particular, if we choose $\varphi(t) = c_n/(|t|^2 + 1)^{(n+1)/2}$ then $u(x, y) = (f * \varphi_y)(x)$, $y > 0$, is the Poisson integral of f and we have the theorem:

THEOREM 3.10. *If $u(x, y)$, $y > 0$, is the Poisson integral of $f \in L^p(E_n)$, $1 \leqq p \leqq \infty$, then $|u(x, y)| \leqq m_f(x)$.*

It is not hard to establish the converse to this last result:

THEOREM 3.11. *If $f \geqq 0$ and its Poisson integral*

$$u(x, y) = \int_{E_n} f(x - t) P(t, y)\, dt, \qquad y > 0,$$

is defined (for example, if f belongs to $L^p(E_n)$, $1 \leqq p \leqq \infty$), then there exists a constant $A = A(n)$, depending only on the dimension, such that

$$m_f(x) \leqq A\left\{\sup_{y>0} u(x, y)\right\}.$$

PROOF. Fix $r > 0$, then

$$\begin{aligned}\sup_{y>0} u(x, y) \geqq u(x, r) &= c_n \int_{E_n} f(x - t) \frac{r}{(|t|^2 + r^2)^{(n+1)/2}}\, dt \\ &\geqq c_n' \int_{|t|\leqq r} f(x - t) r^{-n}\, dt \\ &= c_n'\Omega_n\left\{\frac{1}{\Omega_n r^n} \int_{|t|\leqq r} f(x - t)\, dt\right\}.\end{aligned}$$

Thus, taking the supremum over all $r > 0$ of the expressions on the right we obtain the theorem with $A(n) = 1/c_n'\Omega_n$.

The importance of the maximal function operator lies in the fact that it majorizes many important operators in analysis. We have just seen an example of this situation in the case of the Poisson integral. The following theorem gives us a method for using this majorization property in order to obtain pointwise convergence results for many of the families of operators that we shall encounter.

THEOREM 3.12. *Let $\{T_\varepsilon\}$, $0 < \varepsilon$, be a family of linear operators mapping $L^p(E_n)$, $1 \leqq p \leqq \infty$, into the space of measurable functions on E_n. For each $h \in L^p(E_n)$ define Mh by letting $(Mh)(x) = \sup_{\varepsilon>0} |(T_\varepsilon h)(x)|$ for each $x \in E_n$. Suppose that there exists a constant $a > 0$ and a real number $q \geqq 1$ such that*

$$|\{x: (Mh)(x) > \lambda\}| \leqq (a\|h\|_p\lambda^{-1})^q \tag{3.13}$$

for all $\lambda > 0$ and $h \in L^p(E_n)$. If there exists a dense subset $\mathscr{D}$ of $L^p(E_n)$ such that $\lim_{\varepsilon\to 0} (T_\varepsilon g)(x)$ exists and is finite almost everywhere whenever $g \in \mathscr{D}$, then for each $f \in L^p(E_n)$, $\lim_{\varepsilon\to 0} (T_\varepsilon f)(x)$ exists and is finite almost everywhere.

PROOF. Let f be a function in $L^p(E_n)$. For each positive k, let F_k be the set of all $x \in E_n$ such that $|(T_{\varepsilon'}f)(x) - (T_{\varepsilon''}f)(x)| > 2k^{-1}$ for infinitely many pairs $(\varepsilon', \varepsilon'')$ tending to $(0, 0)$. For each $\eta > 0$ we can find g and h in $L^p(E_n)$ such that $f = g + h$ with $g \in \mathscr{D}$ and $\|h\|_p < \eta$. We shall show that $|F_k| \leqq (2a(k + 1)\eta)^q$; since η is arbitrary this means that F_k has measure 0. Thus, $F = \bigcup_{k=1}^{\infty} F_k$ has measure 0 and, for $x \in E_n - F$, $\lim_{\varepsilon\to 0} (T_\varepsilon f)(x)$ exists and is finite. This, then, would prove the theorem.

Let G be the set of all $x \in E_n$ for which $\lim_{\varepsilon \to 0} (T_\varepsilon g)(x)$ exists and is finite. Since $E_n - G$ has measure zero it suffices to show that $|F_k \cap G| \leqq (2a(k+1)\eta)^q$. Since $(T_{\varepsilon'} f)(x) - (T_{\varepsilon''} f)(x) = \{(T_{\varepsilon'} h)(x) - (T_{\varepsilon''} h)(x)\} + \{(T_{\varepsilon'} g)(x) - (T_{\varepsilon''} g)(x)\}$ and, for $x \in G$, $\lim_{\varepsilon', \varepsilon'' \to 0} \{(T_{\varepsilon'} g)(x) - (T_{\varepsilon''} g)(x)\} = 0$, it follows that $F_k \cap G$ is contained in the set of all $x \in E_n$ such that $|(T_\varepsilon h)(x) - (T_{\varepsilon''} h)(x)| \geqq k^{-1}$ for infinitely many pairs $(\varepsilon', \varepsilon'')$ tending to $(0, 0)$. For each such x we must certainly have $(Mh)(x) \geqq (2k)^{-1}$; therefore, $F_k \cap G \subset \{x \in E_n; (Mh)(x) > 2^{-1}(k+1)^{-1}\}$. The desired inequality $|F_k \cap G| \leqq (a\eta 2(k+1))^q$ now follows from our hypothesis.

An example of such a family is given by defining $T_\varepsilon f$, for $\varepsilon > 0$, by letting

$$(T_\varepsilon f)(x) = (\varepsilon^n \Omega_n)^{-1} \int_{|t| < \varepsilon} f(x - t)\, dt$$

for each $x \in E_n$. In this case we have $Mf = m_f$ and, by Theorem 3.4, we have

$$|\{x: (Mf)(x) > \lambda\}| \leqq c \|f\|_1 \lambda^{-1}$$

for $f \in L^1(E_n)$. Furthermore, if g belongs to the class of continuous functions with compact support (a dense subset of $L^1(E_n)$) then it is clear that $\lim_{\varepsilon \to 0} (T_\varepsilon g)(x) = g(x)$. Thus, by Theorem 3.12, $\lim_{\varepsilon \to 0} (T_\varepsilon f)(x)$ exists and is finite for almost all x. From this we easily deduce that almost every x is a point of differentiability of the integral of f (see Chapter I, (1.23) and (1.24)); more precisely, we have:

COROLLARY 3.14. *If f is locally integrable on E_n (i.e., f is integrable over any bounded subset of E_n) then*

$$\lim_{\varepsilon \to 0} \varepsilon^{-n} \int_{|t| < \varepsilon} [f(x - t) - f(x)]\, dt = 0$$

for almost every $x \in E_n$.

PROOF. Since the problem is local we can assume $f \in L^1(E_n)$ (otherwise we can multiply f by the characteristic function of a sphere of radius r about 0 and obtain the desired convergence at almost every point within this sphere; letting r tend to ∞ we then obtain it for almost all points of E_n). We already know that

$$\lim_{\varepsilon \to 0} (T_\varepsilon f)(x) = \lim_{\varepsilon \to 0} (\varepsilon^n \Omega_n)^{-1} \int_{|t| < \varepsilon} f(x - t)\, dt$$

exists and is finite for almost all x. We only need to show that this limit is $f(x)$ a.e. But

$$\|T_\varepsilon f - f\|_1 = \int_{E_n} \left| \frac{1}{\varepsilon^n \Omega_n} \int_{|t|<\varepsilon} [f(x-t) - f(x)]\, dt \right| dx$$

$$\leq \frac{1}{\varepsilon^n \Omega_n} \int_{|t|<\varepsilon} \left\{ \int_{E_n} |f(x-t) - f(x)|\, dx \right\} dt \to 0$$

as $\varepsilon \to 0$ since the L^1 modulus of continuity $\omega_{1,f}(t) \to 0$ as $|t| \to 0$ (see the proof of Theorem 1.18 in Chapter I). Consequently, there exists a sequence, $\{\varepsilon_k\}$, tending to 0 such that $(T_{\varepsilon_k} f)(x) \to f(x)$ a.e. as $k \to \infty$. It follows, therefore, that $\lim_{\varepsilon \to 0} (T_\varepsilon f)(x) = f(x)$ a.e.

This method extends immediately to Poisson integrals and we can obtain from it a direct proof of Theorem 2.1. Our intention, however, is to extend this theorem to include the more general nontangential approach to the boundary that was mentioned briefly at the beginning of this section. We first make this notion more precise and then will show that Poisson integrals have these more general boundary values.

If $\alpha > 0$ and $x_0 \in E_n$ the *cone in* E_{n+1}^+ *with vertex* $(x_0, 0)$ *and aperture* α is the region

$$\Gamma_\alpha(x_0) = \{(x, y) \in E_{n+1}^+ : |x - x_0| < \alpha y\}.$$

If u is defined in E_{n+1}^+ we shall say that it has a *nontangential limit* l at $x_0 \in E_n$ if, for each $\alpha > 0$,

$$\lim_{(x,y)\to(x_0,0)} u(x, y) = l, \tag{3.15}$$

where (x, y) tends to $(x_0, 0)$ within the cone $\Gamma_\alpha(x_0)$.

The importance of the nontangential approach to the boundary will become clearer later. In several basic results that we shall develop, the vertical approach to the boundary is not sufficient (see, e.g. (5.4)).

THEOREM 3.16. *If* $f \in L^p(E_n)$, $1 \leq p \leq \infty$, *then the Poisson integral*

$$\int_{E_n} f(x - t) P(t, y)\, dt = u(x, y)$$

has the nontangential limit $f(x_0)$ *at each* x_0 *belonging to the Lebesgue set of* f*; in particular, it has these limits at almost all points of* E_n.

PROOF. We first observe that the proof of Theorem 1.25 in the first chapter consisted in showing that

$$0 = \lim_{y\to 0} \int_{E_n} |f(x_0 - t) - f(x_0)|\, |\varphi_y(t)|\, dt$$
$$= \lim_{y\to 0} \int_{E_n} |f(t) - f(x_0)|\, |\varphi_y(x_0 - t)|\, dt$$

whenever x_0 belongs to the Lebesgue set of f. In case $\varphi_y(t)$ is the Poisson kernel $P(t, y) = c_n[y/(y^2 + |t|^2)^{(n+1)/2}]$ we have the easily established inequality

(3.17) $P(x - t, y) = \varphi_y(x - t) \leqq d_\alpha \varphi_y(x_0 - t) = d_\alpha P(x_0 - t, y)$

for $(x, y) \in \Gamma_\alpha(x_0)$, where $d_\alpha^{2/(n+1)} = \max\{1 + 2\alpha^2, 2\}$. Consequently, using Lemma 1.17, part (b) (in Chapter I),

$$|u(x, y) - f(x_0)| = \left| \int_{E_n} f(t)P(x - t, y)\, dt - f(x_0) \int_{E_n} P(x - t, y)\, dt \right|$$
$$\leqq \int_{E_n} |f(t) - f(x_0)| P(x - t, y)\, dt$$
$$\leqq d_\alpha \int_{E_n} |f(t) - f(x_0)| P(x_0 - t, y)\, dt$$
$$= d_\alpha \int_{E_n} |f(t) - f(x_0)| \varphi_y(x_0 - t)\, dt.$$

But, since the last term tends to 0 as y tends to 0, the theorem is proved.

An immediate consequence of the result we have just proved is that the Poisson integral u is *nontangentially bounded* at almost every point of E_n. That is, it is a function u defined in E_{n+1}^+ that is bounded in the truncated cones $\Gamma_\alpha(x_0) \cap \{(x, y) \in E_{n+1}; 0 < y \leqq 1\}$ for almost all $x_0 \in E_n$ (we truncate these cones at height 1; since the functions we shall consider are continuous in E_{n+1}^+, the height at which we truncate these cones is of no importance). In fact, the maximal function m_f can be used to give the following quantitative expression of this:

(3.18) $$\sup_{(x,y)\in\Gamma_\alpha(x_0)} |u(x, y)| \leqq d_\alpha m_f(x_0).$$

This is a simple consequence of Theorem 3.10 and inequality 3.17 involving the Poisson kernel.

The following theorem (a basic tool used in the theory of the H^p spaces developed in Chapter VI) shows that, for harmonic functions, nontangential limits and nontangential boundedness are essentially equivalent.

THEOREM 3.19. *Suppose u is harmonic in E^+_{n+1} and is nontangentially bounded at all the points of a subset $S \subset E_n$ of positive measure, then u has nontangential limits at almost every point of S.*

PROOF. First, we observe that it suffices to show that whenever $S \subset E_n$ is a subset of positive measure at which u is nontangentially bounded then there exists a subset of S, also of positive measure, at whose points u has nontangential limits. Considering only cones $\Gamma_\alpha(q)$ with rational apertures α we see that S is a denumerable union of subsets $F = F_{\alpha,m}$ such that $|u(x, y)| \leqq m$ for all

$$(x, y) \in \left\{\bigcup_{q \in F} \Gamma_\alpha(q)\right\} \cap \{p = (x, y); 0 < y < 2\} = \mathscr{A},$$

$m = 1, 2, \ldots$.[7] That is, u is uniformly bounded in the region $\mathscr{A}$ formed by the union of congruent vertical cones whose vertices are the points of F and are bounded above by the hyperplane $y = 2$.

Now, it is easy to see that it is sufficient for us to show that $u(p)$ tends to a limit when p tends to q within $\Gamma_\alpha(q)$, for almost all $q \in F$. For, if we discard the sets where this does not occur, for each $F = F_{\alpha,m}$ (recall that α is rational and m is a positive integer), we will have discarded a set of measure zero; if q is a point of S not in this set of measure 0 and $\beta > 0$ any rational aperture, then there exists a positive integer k such that $|u(p)| = |u(x, y)| \leqq k$ for $p \in \Gamma_\beta(q)$, $0 < y < 2$; that is, $q \in F_{\beta,k}$. Consequently, $u(p)$ tends to a limit as p tends to q within $\Gamma_\beta(q)$. Thus, a nontangential limit exists for u at all such q's.

A further reduction is obtained by observing that, dividing by m, we can assume that $|u| \leqq 1$ in $\mathscr{A}$. Moreover, we can also assume that S is contained in a cube of side 1; since it suffices to prove the theorem for each intersection of S with a member of a mesh of cubes of side 1.

Having made all these simplifications, we let $\mathscr{D} = \{\bigcup_{q \in F} \Gamma_\alpha(q)\} \cap \{p = (x, y); 0 < y < 1\}$, $\mathscr{B} = \partial\mathscr{D}$ (= boundary of $\mathscr{D}$), $\mathscr{D}_j$ denote the set $\mathscr{A}$ translated by $-1/j$ in the direction of the y axis and $\mathscr{G}_j = \mathscr{D}_j \cap E_n$. Moreover, we put $u_j(x, y) = u(x, y + 1/j)$, the characteristic function of $\mathscr{G}_j$ will be denoted by χ_j, the Poisson integral of $\chi_j(x)u_j(x, 0)$ by $\varphi_j(x, y)$ and $\psi_j(x, y) = u_j(x, y) - \varphi_j(x, y)$. We observe that $|\varphi_j| \leqq 1$ (this follows from Lemma 1.17, part (b), of Chapter I and the fact that the Poisson kernel is nonnegative).

The sequence of functions $f_j(s) = \chi_j(s)u_j(s, 0)$ is uniformly bounded in the $L^2(E_n)$ norm (for $\|f_j\|_2 \leqq \|\chi_j\|_2 = (\int_{\mathscr{G}_j} ds)^{1/2} \leqq$ (volume of cube of side $1 + 2\alpha/j)^{1/2} \leqq (1 + 2\alpha)^{n/2}$). Thus, there exists a subsequence $\{f_{j_k}\}$ that

[7] Throughout this proof we shall consistently use this notation: the points of E_n (i.e., those points of E_{n+1} with last coordinate 0) will be denoted by the letter q. The letter p will be used to denote points of E^+_{n+1}.

converges weakly to a function $f \in L^2(E_n)$. In particular,

$$\varphi_{j_k}(x, y) = \int_{E_n} f_{j_k}(s)P(x - s, y)\, ds \to \int_{E_n} f(s)P(x - s, y)\, ds = \varphi(x, y)$$

as $k \to \infty$, for each $(x, y) \in E^+_{n+1}$.

Since it is obviously true that $\lim_{j \to \infty} u_j(x, y) = u(x, y)$, we have the existence of the limit

$$\psi(x, y) = \lim_{k \to \infty} \psi_{j_k}(x, y) = u(x, y) - \varphi(x, y).$$

Since φ is the Poisson integral of a function in $L^2(E_n)$ it has, by Theorem 3.16, nontangential limits at almost all points of E_n. Since $u = \varphi + \psi$ we shall, therefore, want to show that $\lim \psi(p) = 0$ when p tends to q within $\Gamma_\alpha(q)$, for almost all $q \in F$.

Toward this end we first observe that the functions ψ satisfy

(1) $|\psi_j(p)| \leqq 2$ for $p \in \mathscr{D}$ (since $|\psi_j| \leqq |u_j| + |\varphi_j| \leqq 1 + 1 = 2$ in $\mathscr{D}$);

(2) $\psi_j(p) \to 0$ as $p \to q \in F$, $p \in \mathscr{D}$ (in fact, it follows from Theorem 3.16 that this occurs at all interior points of $\mathscr{G}_j$).

Now we claim that our theorem would be established if we could find a function w, harmonic in E^+_{n+1}, satisfying

(a) $w(x, y) \geqq 0$ in E^+_{n+1};
(b) $w(x, y) \geqq 2$ in $\mathscr{B} - F$;
(c) w has nontangential limits 0 at almost every point of F.

If this were the case, $w(p) \pm \psi_j(p) \geqq 0$ in $\mathscr{B} - F$ because of (1) and (b). Furthermore, (2) and (a) imply

$$\liminf_{p \to q \in F,\, p \in \mathscr{D}} \{w(p) \pm \psi_j(p)\} \geqq \liminf_{p \to q \in F,\, p \in \mathscr{D}} w(p) \geqq 0.$$

It follows that $w(p) \pm \psi_j(p) \geqq 0$ for all $p \in \mathscr{D}$. For, if not, the minimum principle implies the existence of a sequence $\{p_k\} \subset \mathscr{D}$, tending to a point of F as $k \to \infty$, such that $w(p_k) \pm \psi_j(p_k) < -\varepsilon$ for each k, where ε is some positive number. But, by (2), $\psi_j(p_k) \to 0$ as $k \to \infty$ and, thus, for k large $w(p_k)$ would be negative, contradicting property (a). Consequently, letting j tend to ∞ (through a subsequence), we obtain $w(p) \pm \psi(p) \geqq 0$ for all $p \in \mathscr{D}$. That is, $|\psi(p)| \leqq w(p)$ for all $p \in \mathscr{D}$. Hence, the desired result, $\lim \psi(p) = 0$ when p tends to q within $\Gamma_\alpha(q)$, for almost all $q \in F$, is a consequence of property (c).

We shall now continue the proof by constructing a function w satisfying the three properties (a), (b), and (c). Let ξ be the characteristic function of $E_n - F$ and let, for $y > 0$,

$$w(x, y) = 2y + cy \int_{E_n} \frac{\xi(s)}{(y^2 + |x - s|^2)^{(n+1)/2}}\, ds,$$

where c will be a positive constant (to be chosen later). It is clear that $w \geqq 0$ in E_{n+1}^+. Property (c) is an immediate consequence of Theorem 3.16. Thus, we need only show that w satisfies property (b). We first observe that if $(x, y) \in \mathscr{B} - F$ with $y = 1$ then $w(x, y) \geqq 2y = 2$. Now suppose that $p = (x, y) \in \mathscr{B}$ with $0 < y < 1$. Consider the inverted cone with vertex p and aperture α (see figure). If C is the interior of the sphere

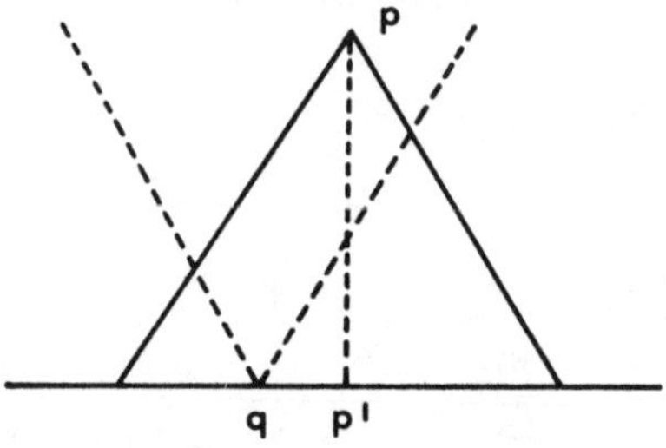

formed by the intersection of this cone with the hyperplane E_n then C and F are disjoint. For, if this were not the case, there would exist a point $q \in C \cap F$ and, consequently, p would be in the interior of $\Gamma_\alpha(q)$. But this is impossible since $p \in \mathscr{B}$. Thus,

$$w(p) \geqq cy \int_C \frac{ds}{(y^2 + |x - s|^2)^{(n+1)/2}} = c\omega_{n-1} y \int_0^{\alpha y} \frac{r^{n-1}\, dr}{(y^2 + r^2)^{(n+1)/2}}$$

$$= c\omega_{n-1} \int_0^{\alpha} \frac{r^{n-1}\, dr}{(1 + r^2)^{(n+1)/2}}.$$

This last expression can obviously be made to exceed 2 by choosing c appropriately.

Having constructed the function w we have, therefore, shown that $\psi(p)$ has the limit 0 as p tends to q within $\Gamma_\alpha(q)$, for almost every q in F. We next claim that the majorization $|\psi| \leqq w$ in $\bigcup_{q \in F} \Gamma_\alpha(q)$ implies that for almost every q in F, $|\psi(p)| \leqq w(p)$, where p may lie in a cone with vertex q of *arbitrary* aperture, as long as p is sufficiently close to q. A simple geometric argument shows that this is, indeed, the case for each

q which is a *point of density of* F.[8] Consequently, $\lim \psi(p) = 0$ as p tends to $q \in F$ nontangentially for almost every q in F. This means that u and φ have the same nontangential limits at almost every q in F and, therefore, the theorem is proved.

We shall be interested in extending this result to functions that are harmonic in sets of variables (these will be called *multiply harmonic* functions). By this we mean twice differentiable functions, u, defined on a domain, $\mathscr{D}$, of points $(x^{(1)}, x^{(2)}, \ldots, x^{(k)}) = (x_1^{(1)}, \ldots, x_{n_1}^{(1)}, x_1^{(2)}, \ldots, x_{n_2}^{(2)}, \ldots, x_1^{(k)}, \ldots, x_{n_k}^{(k)})$ belonging to the Cartesian product $E_{n_1} \times E_{n_2} \times \cdots \times E_{n_k}$ of k Euclidean spaces, that are harmonic in each of the variables $x^{(j)}$, $1 \leqq j \leqq k$; that is,

$$\sum_{i=1}^{n_j} \frac{\partial^2 u}{(\partial x_i^{(j)})^2} = 0$$

for $1 \leqq j \leqq k$. We shall encounter such functions in the next chapter, where we deal with analytic functions of several variables. In this case, each of the spaces E_{n_j} will be two-dimensional and can be identified with a copy of the complex plane.

It is not hard to state this extension and to indicate the changes in the proof of Theorem 3.19 that are needed in order to establish it. One of these changes involves results analogous to Theorems 2.5 and 3.16 for "iterated Poisson integrals" (i.e., solutions of a variant of the Dirichlet problem involving multiply harmonic functions).

These iterated Poisson integrals are examples of a family of operators somewhat more general than that occurring in Theorem 3.12, in that they involve several independent parameters tending to 0. That is, the family $\{T_\varepsilon\}$ is parametrized by the set $\{\varepsilon = (\varepsilon_1, \varepsilon_2, \ldots, \varepsilon_m) \in E_m : \varepsilon_j > 0, j = 1, 2, \ldots, m\}$ (the *first octant* in E_m); their behavior as ε tends to $0 \in E_m$ will be the object of our study. In the case of iterated Poisson integrals let us, for simplicity consider the case when $m = 2$. That is, suppose we identify E_n with the Cartesian product of E_{n_1} and E_{n_2}, where $n_1 + n_2 = n$, and let

$$P_j(t, \varepsilon_j) = c_{n_j} \frac{\varepsilon_j}{(|t|^2 + \varepsilon_j^2)^{(n_j+1)/2}}$$

for $t = (t_1, \ldots, t_{n_j}) \in E_{n_j}, j = 1, 2$. Then, for $f \in L^p(E_n)$ we define $T_\varepsilon f = T_{(\varepsilon_1, \varepsilon_2)} f$ by letting

$$(T_\varepsilon f)(x) = \int_{E_{n_1}} P_1(x^{(1)} - t, \varepsilon_1) \left\{ \int_{E_{n_2}} P_2(x^{(2)} - s, \varepsilon_2) f(t, s)\, ds \right\} dt,$$

[8] Let $\xi_F = \xi$ be the characteristic function of a measurable set $F \subset E_n$, then x is a *point of density of* F if $\Omega_n^{-1} r^{-n} \int_{|x-t|<r} \xi(t)\, dt \to 1$ as $r \to 0$. By Corollary 3.14 almost every point of F is a point of density.

where

$$x = (x_1^{(1)}, \ldots, x_{n_1}^{(1)}, x_1^{(2)}, \ldots, x_{n_2}^{(2)}) = (x^{(1)}, x^{(2)}) \in E_{n_1} \times E_{n_2} = E_n.$$

If $f(x) = f_1(x^{(1)})f_2(x^{(2)})$ with $f_j \in L^p(E_{n_j}) \cap C_0(E_{n_j}), j = 1, 2,$ it follows from Theorem 2.1 that $(T_\varepsilon f)(x)$ converges to $f(x)$ uniformly as $\varepsilon = (\varepsilon_1, \varepsilon_2) \to (0, 0)$. The class of finite linear combinations of such functions is dense in $L^p(E_n)$, $1 \leqq p < \infty$; thus, if we can show that inequality (3.13) holds we can then conclude that $\lim_{\varepsilon \to 0} (T_\varepsilon f)(x)$ exists and is finite almost everywhere. It is not hard to establish the following stronger result for $1 < p \leqq \infty$ for $(Mf)(x) = \sup_{\varepsilon_1, \varepsilon_2 > 0} |(T_{(\varepsilon_1, \varepsilon_2)} f)(x)|$

$$\|Mf\|_p \leqq A\|f\|_p, \tag{3.20}$$

where A depends on p, n_1 and n_2, but is independent of $f \in L^p(E_n)$ (see the discussion preceding (3.6)). In order to do this we define $M^{(1)}$ to be the operator of $L^p(E_n)$ assigning to f the maximal function of $f(\cdot, x^{(2)})$; that is,

$$(M^{(1)}f)(x) = (M^{(1)}f)(x^{(1)}, x^{(2)}) = \sup_{r>0} \Omega_{n_1}^{-1} r^{-n_1} \int_{|t| \leqq r} |f(x^{(1)} - t, x^{(2)})| \, dt.$$

Similarly, we let

$$(M^{(2)}f)(x) = \sup_{r>0} \Omega_{n_2}^{-1} r^{-n_2} \int_{|s| \leqq r} |f(x^{(1)}, x^{(2)} - s)| \, ds.$$

Thus, by Theorem 3.7,

$$\begin{aligned} \|M^{(2)}M^{(1)}f\|_p &\leqq b(p, n_2)\|M^{(1)}f\|_p \leqq b(p, n_2)b(p, n_1)\|f\|_p \\ &= B\|f\|_p. \end{aligned} \tag{3.21}$$

On the other hand, it follows from (3.10) that

$$\begin{aligned} |(T_\varepsilon f)(x)| &= |(T_{(\varepsilon_1, \varepsilon_2)} f)(x)| \\ &\leqq \int_{E_{n_2}} P_2(x^{(2)} - s, \varepsilon_2)(M^{(1)}f)(x^{(1)}, s) \, ds \\ &\leqq (M^{(2)}M^{(1)}f)(x^{(1)}, x^{(2)}). \end{aligned}$$

Consequently, $(Mf)(x) = \sup_{\varepsilon_1, \varepsilon_2 > 0} |(T_{(\varepsilon_1, \varepsilon_2)} f)(x)| \leqq (M^{(2)}M^{(1)}f)(x)$ and inequality (3.20) follows from (3.21).

We can, therefore, conclude that $\lim (T_\varepsilon f)(x)$ as $\varepsilon = (\varepsilon_1, \varepsilon_2) \to (0, 0)$ exists and is finite for almost all $x \in E_n$. It is clear that we can modify the above argument to apply to m-fold iterated Poisson integrals. More precisely, we have the following theorem:

THEOREM 3.22. *If $f \in L^p(E_n)$, $1 < p < \infty$, and*

$$u(x, \varepsilon) = u(x^{(1)}, x^{(2)}, \ldots, x^{(m)}; \varepsilon_1, \varepsilon_2, \ldots, \varepsilon_m) =$$

$$\int_{E_{n_1}} \cdots \int_{E_{n_m}} P_1(x^{(1)} - t^{(1)}, \varepsilon_1) \cdots P_m(x^{(m)} - t^{(m)}, \varepsilon_m) f(t^{(1)}, \ldots, t^{(m)})\, dt^{(1)} \cdots dt^{(m)},$$

where $t^{(j)}$ and $x^{(j)}$ belong to E_{n_j} and $\varepsilon = (\varepsilon_1, \varepsilon_2, \ldots, \varepsilon_m)$ belong to the first octant of E_m (that is, $\varepsilon_j > 0$ for $j = 1,2, \ldots, m$), then

$$\lim_{\varepsilon \to 0} u(x, \varepsilon) = \lim_{\varepsilon_1, \ldots, \varepsilon_m \to 0} u(x^{(1)}, x^{(2)}, \ldots, x^{(m)}; \varepsilon_1, \ldots, \varepsilon_m) = f(x)$$

for almost every $x \in E_n$.

PROOF. We have already seen that $\lim_{\varepsilon \to 0} u(x, \varepsilon)$ exists and is finite almost everywhere. We need only show that this limit is $f(x)$ a.e. As in the proof of Corollary 3.14 it suffices to show that $\|u(\cdot, \varepsilon) - f\|_p \to 0$ as $\varepsilon \to 0$. This can be done in much the same way as was done in the proof of Theorem 1.18 in Chapter I. In order to simplify the notation we shall, again, present the proof when $m = 2$; the method extends in an obvious way to the general case. We have, by Lemma 1.17, part (b), of the first chapter and by a change of variables:

$$\begin{aligned} u(x, \varepsilon) - f(x) &= \int_{E_{n_1}} \int_{E_{n_2}} P_1(t, \varepsilon_1) P_2(s, \varepsilon_2) \{f(x^{(1)} - t, x^{(2)} - s) - f(x^{(1)}, x^{(2)})\}\, dt\, ds \\ &= \int_{E_{n_1}} \int_{E_{n_2}} P_1(t, 1) P_2(s, 1) \{f(x^{(1)} - \varepsilon_1 t, x^{(2)} - \varepsilon_2 s) - f(x^{(1)}, x^{(2)})\}\, dt\, ds. \end{aligned}$$

For $(h, k) \in E_{n_1} \times E_{n_2}$ let

$$\begin{aligned} \omega_{p,f}(h, k) &= \omega(h, k) \\ &= \left(\int_{E_{n_1}} \int_{E_{n_2}} |f(x^{(1)} + h, x^{(2)} + k) - f(x^{(1)}, x^{(2)})|^p\, dx^{(1)}\, dx^{(2)} \right)^{1/p} \end{aligned}$$

be the L^p modulus of continuity of f. We have seen that $\omega(h, k) \to 0$ as $|h|, |k| \to 0$ and that $\omega(h, k) \leqq 2\|f\|_p$. Thus, by Minkowski's inequality for integrals and the Lebesgue dominated convergence theorem,

$$\left(\int_{E_n} |u(x, \varepsilon) - f(x)|^p\, dx \right)^{1/p} = \int_{E_{n_1}} \int_{E_{n_2}} \omega(-\varepsilon_1 t, -\varepsilon_2 s) P_1(t, 1) P_2(s, 1)\, dt\, ds \to 0$$

as $\varepsilon_1, \varepsilon_2 \to 0$. This completes the proof.

A modification of the argument we have just given shows that the theorem (as well as its nontangential version that follows below) is valid also for $p = \infty$. Since, in any case, this is a consequence of Theorem 3.24 proved below, we need not present this modification.

Theorem 3.22 asserts that the iterated Poisson integral of a function in $L^p(E_n)$ converges to the function almost everywhere if we approach the points $x_0 = (x_0^{(1)}, x_0^{(2)}, \ldots, x_0^{(m)})$ in E_n "downward" along the products of the lines $x^{(j)} = x_0^{(j)}$ in $E_{n_j+1}^+, j = 1, 2, \ldots, m$ (compare with the discussion at the beginning of this section). As was the case before, more general limits exist. These are generalizations of nontangential limits and are defined in the following way: Suppose $n_1 + n_2 + \cdots + n_k = n$ and

$$\Gamma_{\alpha_j}^{(j)}(x_0^{(j)}) = \{(x^{(j)}, y_j) \in E_{n_j+1}^+; |x^{(j)} - x_0^{(j)}| < \alpha_j y_j\}, \qquad j = 1, 2, \ldots, k.$$

A function, u, defined on $E_{n_1+1}^+ \times \cdots \times E_{n_k+1}^+$ is said to have the *nontangential limit l in each set of variables*[9] at the point $x_0 = (x_0^{(1)}, x_0^{(2)}, \ldots, x_0^{(k)})$ if $u(x^{(1)}, y_1; x^{(2)}, y_2; \ldots; x^{(k)}, y_k) = u(x^{(1)}, x^{(2)}, \ldots, x^{(k)}; y_1, y_2, \ldots, y_k)$ tends to l as the point $(x^{(1)}, y_1; x^{(2)}, y_2; \ldots; x^{(k)}, y_k)$ tends to x_0 within the Cartesian product $\Gamma_{\alpha_1}^{(1)} \times \Gamma_{\alpha_2}^{(2)} \times \cdots \times \Gamma_{\alpha_k}^{(k)}$, for all k-tuples $(\alpha_1, \alpha_2, \ldots, \alpha_k)$ of positive real numbers (i.e., each of the points $(x^{(j)}, y_j)$ tends nontangentially to $x_0^{(j)}$, $1 \leqq j \leqq k$). By the argument used to prove Theorem 3.22 and the inequality (3.17) applied to each of the kernels P_j, $1 \leqq j \leqq m$, we then obtain

COROLLARY 3.23. *If $f \in L^p(E_n)$, $1 < p < \infty$, and $u(x, \varepsilon)$ is the function defined in Theorem* (3.22), *then $u(x, \varepsilon)$ has the nontangential limit $f(x_0)$ in each set of variables at almost all $x_0 \in E_n$.*

We say that u, defined on $E_{n_1+1}^+ \times E_{n_2+1}^+ \times \cdots \times E_{n_k+1}^+$, is *nontangentially bounded in each set of variables* at $x_0 \in E_n$ if u is bounded in the Cartesian product of the truncated cones $\Gamma_{\alpha_j}(x_0^{(j)}) \cap \{(x^{(j)}, y_j) \in E_{n_j+1}^+; 0 < y_j \leqq 1\}$, $1 \leqq j \leqq k$, for all k-tuples $(\alpha_1, \alpha_2, \ldots, \alpha_k)$ of positive real numbers. Suppose u is the iterated Poisson integral of a function $f \in L^p(E_n)$, $1 < p < \infty$. The fact that, at almost all points of E_n, u is nontangentially bounded in each set of variables is a somewhat more elementary result than Corollary 3.23 (it is, of course, an immediate consequence of that corollary).

In complete analogy to the case of harmonic functions on E_{n+1}^+, this nontangential boundedness is almost everywhere equivalent to the existence of nontangential limits in each set of variables, even for harmonic

[9] We are grouping the n variables $x_1, x_2, \ldots, x_n$ in k mutually disjoint sets. The j-th set consists of the variables $x_{N_{j-1}+1}, \ldots, x_{N_j}$, where $N_l = n_1 + n_2 + \cdots + n_l$ for $1 \leqq l \leqq k$. The concept we have just defined involves this grouping and we should, properly, refer to it in the definition. On the other hand, we shall see that in all our applications it will be clear which grouping is involved; therefore, we avoid referring to it for the sake of simplicity.

functions that are not Poisson integrals. More precisely, we have the following extension of Theorem 3.19:

THEOREM 3.24. *Suppose* $n_1 + n_2 + \cdots + n_k = n$ *and* u *is a function defined in the Cartesian product* $E^+_{n_1+1} \times E^+_{n_2+1} \times \cdots \times E^+_{n_k+1}$ *that is harmonic in each set of variables* $(x^{(j)}, y_j) = (x^{(j)}_1, \ldots, x^{(j)}_{n_j}, y_j)$ *in* $E^+_{n_j+1}$, $1 \leqq j \leqq k$. *(That is, u is twice differentiable and satisfies*

$$\Delta_j u = \sum_{i=1}^{n_j} \frac{\partial^2 u}{(\partial x^{(j)}_i)^2} + \frac{\partial^2 u}{\partial y_j^2} = 0).$$

If u is nontangentially bounded in each set of variables at all the points of a subset $S \subset E_n$ *of positive measure, then u has nontangential limits in each set of variables at almost every point of S.*

PROOF. The main ideas in establishing this theorem were developed in the proof of Theorem 3.19. Therefore, we shall only indicate the changes in that proof that are needed to obtain the present theorem. We confine ourselves to the case $k = 2$ and $n_1 = n_2 = n$; (the argument extends easily to the general case).

In analogy with the notation used in the proof of Theorem 3.19 we let $p = (p_1, p_2) = (x^{(1)}, y_1; x^{(2)}, y_2)$ denote the general point of $E^+_{n+1} \times E^+_{n+1}$; and let $q = (q_1, q_2) = (x^{(1)}, 0; x^{(2)}, 0)$ denote the general point of $E_n \times E_n = E_{2n}$, (the latter is the *distinguished boundary* of $E^+_{n+1} \times E^+_{n+1}$).[10] By reasoning similar to that used before, we essentially reduce the problem to showing that if $|u| \leqq 1$ in the region

$$\mathscr{A} = \left\{ \bigcup_{(q_1,q_2)\in F} \gamma_\alpha(q_1, q_2) \right\} \cap \{p = (x^{(1)}, y_1; x^{(2)}, y_2); 0 < y_1, y_2 < 2\},$$

then $u(p)$ tends to a limit when p tends to $q = (q_1, q_2)$ within $\gamma_\alpha(q) = \gamma_\alpha(q_1, q_2)$, for almost all of $q \in F \subset E_{2n}$. Here we have written $\gamma_\alpha(q_1, q_2)$ for the Cartesian product cone $\Gamma_{\alpha_1}(q_1) \times \Gamma_{\alpha_2}(q_2)$. By taking appropriate subsets we can assume that F is closed and, in addition, that F is contained in the unit sphere of E_{2n}.

We let

$$\mathscr{D} = \left\{ \bigcup_{q\in F} \gamma_\alpha(q) \right\} \cap \{p = (x^{(1)}, y_1; x^{(2)}, y_2); 0 < y_1, y_2 < 1\},$$

$\mathscr{A}_j$ denote the set translated by $-1/j$ in the direction of the y_1 and y_2 axes, and $\mathscr{G}_j = \mathscr{A}_j \cap E_{2n}$. Moreover, we put $u_j(p) = u(x^{(1)}, y_1 + 1/j;$

[10] The terminology "distinguished boundary" is often used in the literature; it indicates that we are considering only a part of the topological boundary of $E^+_{n+1} \times E^+_{n+1}$.

$x^{(2)}, y_2 + 1/j)$; the characteristic function of $\mathscr{G}_j$ will be denoted by χ_j, the iterated Poisson integral of $\chi_j(x^{(1)}; x^{(2)})u_j(x^{(1)}, 0; x^{(2)}, 0)$ by φ_j and $\psi_j = u_j - \varphi_j$.

As was the case in the proof of Theorem 3.19 we can find a sequence $\{\varphi_{j_k}\}$ converging to an iterated Poisson integral (of a function in $L^2(E_{2n})$), at each point of $E_{n+1}^+ \times E_{n+1}^+$. We denote this limit by φ. Since $u_j \to u$ as $j \to \infty$, it follows that $\psi_{j_k}(p) = u_{j_k}(p) - \varphi_{j_k}(p)$ tends to the limit $\psi(p) = u(p) - \varphi(p)$, for each $p \in E_{n+1}^+ \times E_{n+1}^+$. Since φ is the Poisson integral of a function in $L^2(E_{2n})$, by Corollary 3.23 it has nontangential limits in each set of variables at almost all points of E_{2n}. We shall show that $\lim \psi(p) = 0$, when p tends to q within $\gamma_\alpha(q)$, for almost all $q \in F$.

In keeping with the argument used in the proof of Theorem 3.19 the crucial point consists in constructing a multiply harmonic function w in $E_{n+1}^+ \times E_{n+1}^+$, satisfying

(a) $w \geqq 0$ in $E_{n+1}^+ \times E_{n+1}^+$,

(b) $\liminf_{p' \to p,\, p' \in \mathscr{D}} w(p') \geqq |\psi_j(p)|, j = 1, 2, \ldots$ wherever p is a point of the boundary of $\mathscr{D}$,

(c) w has zero nontangential limits in each set of variables at almost every point of F.

Let ξ be the characteristic function of that part of $E_{2n} - F$ contained in a fixed sphere of sufficiently large radius centered at the origin (the lower bounds required for the size of this sphere are more easily described in the context of the proof and will be obvious). We then define $w = 2(y_1 + y_2) + cW(x^{(1)}, y_1; x^{(2)}, y_2)$, where W is the repeated Poisson integral of ξ, and c is a positive constant to be determined later.

The condition (a) is obvious, and condition (c) follows from Corollary 3.23 (since ξ belongs to all the spaces $L^p(E_{2n})$, $1 < p < \infty$). Thus, we need only prove (b). But $\psi_j(p) = 0$, for $p \in F$, so inequality (b) is true for $p \in F$. We consider separately the following remaining parts of the boundary of $\mathscr{D}$:

(i) points where either y_1 or y_2 is 1;

(ii) points where $0 < y_1 < 1$, and $0 < y_2 < 1$;

(iii) points such that $y_1 > 0$ and $y_2 = 0$, or such that $y_2 > 0$ and $y_1 = 0$.

Since $|\psi_j| \leqq 2$, property (b) is obvious in case (i) holds.

Suppose $p = (p_1, p_2) = (x^{(1)}, y_1; x^{(2)}, y_2)$ satisfies (ii). Then consider the Cartesian product of the inverted cones with vertices p_1 and p_2 and apertures α_1 and α_2. These cones intersect the hyperplanes $y_1 = 0$, and $y_2 = 0$ (in each E_{n+1}^+) in two open spheres C_1 and C_2. We claim that

$C_1 \times C_2$ does not contain a point of F. (If this were not the case p would belong to $\mathcal{D}$.) Thus ξ equals 1 at all points $C_1 \times C_2$, provided we chose the sphere containing F which was used to define ξ to be large enough to include these Cartesian products. The spheres, C_1 and C_2 are centered at $x^{(1)}$ and $x^{(2)}$ and have radii $\alpha_1 y_1$ and $\alpha_2 y_2$, respectively. Therefore,

$$w(p) \geqq cW(p) \geqq c \prod_{j=1}^{2} \left(c_n y_j \int_{C_j} \frac{ds^{(j)}}{(y_j^2 + |x^{(j)} - s^{(j)}|^2)^{(n+1)/2}} \right)$$

$$= c \prod_{j=1}^{2} \left(y_j c_n \omega_{n-1} \int_0^{\alpha_j y_j} \frac{r^{n-1}\, dr}{(y_j^2 + r^2)^{(n+1)/2}} \right) \geqq 2,$$

if c is large enough.

Now suppose $p = (p_1, p_2) = (x^{(1)}, y_1; x^{(2)}, 0)$ satisfies the condition (iii), (the case $y_1 = 0$, $y_2 > 0$ is similar) in this case $w(p)$ is not defined and so we must estimate lim inf $w(p')$, as $p' \in \mathcal{D}$ approaches p. Letting $p' = (p_1', p_2')$, and writing y_1' and y_2' for the y coordinates of p_1', p_2', respectively we have

$$w(p') \geqq 2y_1' + cc_n^2 \int_{E_n} \frac{y_1'}{|p_1' - q_1|^{(n+1)}} \left\{ \int_{E_n} \frac{y_2'}{|p_2' - q_2|^{(n+1)/2}}\, \xi(q_1, q_2)\, dq_2 \right\} dq_1.$$

It follows from Fatou's lemma that

$$\liminf_{p' \to p,\, p' \in \mathcal{D}} w(p') \geqq$$

$$2y_1 + cc_n^2 \int_{E_n} \frac{y_1}{|p_1 - q_1|^{(n+1)}} \left\{ \liminf_{p_2' \to p_2} c_n \int_{E_n} \frac{y_2'}{|p_2' - q_2|^{(n+1)/2}}\, \xi(q_1 q_2)\, dq_2 \right\} dq_1.$$

We now claim that

$$\text{(3.25)} \qquad \liminf_{p_2' \to p_2} c_n \int_{E_n} \frac{y_2'}{|p_2' - q_2|^{(n+1)}}\, \xi(q_1, q_2)\, dq_2 \geqq \xi(q_1, p_2).$$

To see this we first note that the inequality certainly holds when $\xi(q_1, p_2) = 0$, since the integrand on the left is nonnegative. The only other value assumed by ξ is 1 on the open set which is the intersection of $E_{2n} - F$ with an open sphere in E_{2n}. Thus ξ is continuous at these points. It follows that the integral (3.25) actually converges to $\xi(q, p_2)$ (this can be shown by the same argument that was used to establish the last part of (b) in Theorem 2.1). Thus (3.25) is proved and, consequently,

$$\liminf_{p' \to p,\, p' \in \mathcal{D}} w(p') \geqq 2y_1 + cc_n \int_{E_n} \frac{y_1}{|p_1 - q_1|^{(n+1)/2}}\, \xi(q_1, p_2)\, dq_1.$$

The rest of the proof is then essentially a repetition of the last part of the argument we used in order to establish Theorem 3.19 applied to the functions $\psi_j(\bar{p}_1, p_2)$ and

$$v(\bar{p}_1) = 2y_1 + cc_n \int_{E_n} \frac{y_1}{|\bar{p}_1 - q_1|^{(n+1)}} \xi(q_1, p_2)\, dq_1$$

in the region $\mathscr{D}_{p_2}$ consisting of those $\bar{p}_1 \in E_{n+1}^+$ that are contained in some truncated cone $\Gamma_{\alpha_1}(q_1)$ with $(q_1, p_2) \in F$. (At this stage p_2 is a fixed point which lies on the boundary E_n of E_{n+1}^+.) This completes the proof of (b), and with it the majorization

$$|\psi(p)| \leqq w(p),\ p \in \mathscr{D}. \tag{3.26}$$

To conclude the proof of the theorem, we shall need to extend (3.26) so that for almost all $q \in F$, $|\psi(p)| \leqq w(p)$ where p may lie in product cones $\Gamma_{\beta_1}(q_1) \times \Gamma_{\beta_2}(q_2)$, with arbitrary apertures β_1 and β_2, as long as p is sufficiently close to $q = (q_1, q_2)$. This will indeed be the case at all points of *strong density* of F. These are the points which satisfy the conclusion of the following lemma:

LEMMA 3.27. *Let ξ be the characteristic function of an arbitrary measurable set F of $E_{2n} = E_n \times E_n$. Then for almost every $x = (x^{(1)}, x^{(2)}) \in F$*

$$\lim_{\substack{h_1 \to 0\\ h_2 \to 0}} \Omega_n^{-2} h_1^{-n} h_2^{-n} \iint_{|s^{(1)}| < h_1,\, |s^{(2)}| < h_2} \xi(x^{(1)} - s^{(1)}, x^{(2)} - s^{(2)})\, ds^{(1)}\, ds^{(2)} = 1.$$

PROOF. For any $f \in L^p(E_n \times E_n)$, we define the *strong maximal function* μ_f, as

$$\mu_f(x^{(1)}, x^{(2)}) =$$

$$\sup_{h > 0,\, h_2 > 0} \Omega_n^{-2} h_1^{-n} h_2^{-n} \iint_{|s^{(1)}| < h_1,\, |s^{(2)}| < h_2} |f(x^{(1)} - s^{(1)}, x^{(2)} - s^{(2)})|\, ds^{(1)}\, ds^{(2)}.$$

Then, as in the proof of (3.20) $\mu_f(x) \leqq (M^{(2)} M^{(1)} f)(x)$, and so

$$\|\mu_f(x)\|_p \leqq A_p \|f\|_p, \qquad 1 < p < \infty. \tag{3.28}$$

By the same argument as in the proof of Theorem 3.22 it then follows that

$$\lim_{\substack{h_1 \to 0\\ h_2 \to 0}} \Omega_n^{-2} h_1^{-n} h_2^{-n} \iint_{|s^{(1)}| < h_1,\, |s^{(2)}| < h_2} f(x^{(1)} - s^{(1)}, x^{(2)} - s^{(2)})\, ds^{(1)}\, ds^{(2)} = f(x^{(1)}, x^{(2)}) \tag{3.29}$$

for almost every $(x^{(1)}, x^{(2)}) \in E_{2n}$, for any fixed $f \in L^p(E_{2n})$, $1 < p < \infty$. The validity of the assertion (3.29), however, depends only on the values of f in a neighborhood of the point $(x^{(1)}, x^{(2)})$, and so (3.29) holds almost everywhere when f is merely locally in $L^p(E_{2n})$, $1 < p < \infty$. Thus it holds for bounded functions and, in particular, for the function ξ. This proves Lemma 3.27.

We can now conclude the proof of Theorem 3.24. Suppose that (q_1, q_2) is a point of strong density of F. For fixed $y_1, y_2 > 0$, consider the set

$$A = \{(x^{(1)}, x^{(2)}): |x^{(1)} - q_1| < \beta_1 y_1, |x^{(2)} - q_2| < \beta_2 y_2\}.$$

For any $(x^{(1)}, x^{(2)}) \in A$ we also consider the set

$$B = \{(\bar{x}^{(1)}, \bar{x}^{(2)}); |x^{(1)} - \bar{x}^{(1)}| < \alpha_1 y_1, |x^{(2)} - \bar{x}^{(2)}| < \alpha_2 y_2\}.$$

If y_1 and y_2 are sufficiently small, then B must intersect F, otherwise we would obtain a contradiction with the fact that $q = (q_1, q_2)$ is a point of strong density of F. Thus, if $p = (p_1, p_2)$, $p_1 = (x^{(1)}, y_1)$, $p_2 = (x^{(2)}, y_2)$, and p is sufficiently close to q, then $p \in \Gamma_{\beta 1}(q_1) \times \Gamma_{\beta_2}(q_2)$ implies that $p \in \Gamma_{\alpha_1}(\bar{x}^{(1)}) \times \Gamma_{\alpha_2}(\bar{x}^{(2)})$, with $(\bar{x}^{(1)}, \bar{x}^{(2)}) \in F$.

The estimate (3.26) then also holds for all $p \in \Gamma_{\beta_1}(q_1) \times \Gamma_{\beta_2}(q_2)$, if p is sufficiently close to $q = (q_1, q_2)$ and the desired nontangential limit of ψ is therefore established.

4. Subharmonic Functions and Majorization by Harmonic Functions

In later chapters we will need to bound from above, with appropriate functions that are harmonic, functions having the form $|F|^p$, where F is either a holomorphic function of several complex variables (see Chapter III) or a vector-valued function whose components are harmonic (see Chapter VI). These functions $|F|^p$ will turn out to be subharmonic. We can give an indication of the nature of subharmonic functions and of this type of majorization by considering the banal case of functions of a single real variable. In this case the harmonic functions are the linear ones and, as will be apparent, the subharmonic functions are the convex functions. The latter are generally defined to be continuous[11] functions φ, defined on an interval of the real line, satisfying

$$\varphi\left(\frac{x+y}{2}\right) \leqq \frac{\varphi(x) + \varphi(y)}{2} \tag{4.1}$$

whenever the interval $[x, y]$ is contained in the domain of φ. This condition can easily be shown to be equivalent to the following one: If the interval $[x, y]$ is contained in the domain of φ then the linear (harmonic)

[11] It can be shown that it suffices, in this one dimensional case, to suppose the function to be measurable. The general theory of subharmonic functions can be carried out for upper semicontinuous functions; however, though this is important in some applications we will have occasion to apply it only to continuous functions and, therefore, we shall limit ourselves to this somewhat (technically) simpler case.

function having the same values as φ at the end points x and y majorizes[12] φ in this interval. In case the convex function φ has a continuous second derivative at all points of its domain D, then (4.1) is equivalent to the condition $\varphi''(x) \geqq 0$ for all $x \in D$.

In the first part of this section we generalize these facts and deal with functions defined on bounded domains; in the second part (as well as in later chapters) we consider these problems on certain unbounded domains that arise naturally in our study.

We say that a function, s, defined and continuous in a domain $\mathscr{D} \subset E_n$ is *subharmonic* if

$$s(x_0) \leqq \mathscr{M}_{x_0,s}(r) = \frac{1}{\omega_{n-1}} \int_{\Sigma_{n-1}} s(x_0 + rt')\, dt' \tag{4.2}$$

whenever the sphere $\{x \in E_n; |x - x_0| \leqq r\}$ is contained in $\mathscr{D}$. This inequality clearly generalizes (4.1).

The fact that a nonnegative second derivative implies convexity has the following extension to n dimensions:

THEOREM 4.3. *Suppose s has continuous second partial derivatives in a domain $\mathscr{D}$ and $(\Delta s)(x) \geqq 0$ for all $x \in \mathscr{D}$, then s satisfies the mean value inequality* (4.2) *at all points of $\mathscr{D}$.*

PROOF. Suppose the closed solid sphere of radius r_0 about x_0 belongs to $\mathscr{D}$ and $\Sigma_\varepsilon, \Sigma_r, 0 < \varepsilon < r \leqq r_0$, denote the surfaces of the spheres of radii ε and r about x_0. Applying Green's theorem to the functions s and

$$v(x) = \begin{cases} |x - x_0|^{-(n-2)} - r^{-(n-2)}, & \text{if } n > 2 \\ -\log |x - x_0| + \log r, & \text{if } n = 2, \end{cases}$$

in the shell domain, S, between Σ_ε and Σ_r we obtain (since $\Delta v = 0$ and $\Delta s \geqq 0$ in S)

$$0 \geqq \int\int_S (s\Delta v - v\Delta s)\, dx = \int_{\partial S} \left(s \frac{\partial v}{\partial n} - v \frac{\partial s}{\partial n}\right) d\sigma,$$

where ∂S is the boundary of S, $\partial/\partial n$ denotes differentiation in the direction of the outward directed normal to ∂S, and $d\sigma$ is the element of surface area on ∂S. Thus, as in the proof of Theorem 1.1,

$$0 \geqq \left(\int_{\Sigma_r} + \int_{\Sigma_\varepsilon}\right) \frac{\partial v}{\partial n} s\, d\sigma - \left(\int_{\Sigma_r} + \int_{\Sigma_\varepsilon}\right) v \frac{\partial s}{\partial n}\, d\sigma.$$

[12] We say that a function f *majorizes* a function g if the two functions have the same domain D, are real valued, and $f(x) \geqq g(x)$ for all $x \in D$.

But v is 0 on Σ_r and $\int_{\Sigma_\varepsilon} \partial s/\partial n \, v \, d\sigma \leqq 0$. (Here $\partial/\partial n$ denotes differentiation in the direction of the normal to Σ_ε directed toward the center of this sphere. Thus, applying Green's theorem to the functions s and 1 in the closed region $\{x; |x - x_0| \leqq \varepsilon\}$ we see that $\int_{\Sigma_\varepsilon} \partial s/\partial n \, d\sigma \leqq 0$; moreover, $v \geqq 0$ on Σ_ε whenever $\varepsilon \leqq r$.) Hence, when $n > 2$,

$$0 \geqq \left(\int_{\Sigma_r} - \int_{\Sigma_\varepsilon} \right) (-(n-2)|x - x_0|^{-(n-1)} s(x)) \, d\sigma(x)$$

and, when $n = 2$,

$$0 \geqq \left(\int_{\Sigma_r} - \int_{\Sigma_\varepsilon} \right) (-|x - x_0|^{-1} s(x)) \, d\sigma(x).$$

In either case,

$$\frac{1}{\omega_{n-1}\varepsilon^{n-1}} \int_{\Sigma_\varepsilon} s(x) \, d\sigma(x) \leqq \frac{1}{\omega_{n-1} r^{n-1}} \int_{\Sigma_r} s(x) \, d\sigma(x) = \mathcal{M}_{x_0,s}(r).$$

Letting ε tend to 0 the left side tends to $s(x_0)$ and we obtain inequality (4.2).

In later chapters we will want to show that certain functions are subharmonic. In general, they will not be twice differentiable; thus, Theorem 4.3 cannot be applied in order to establish the desired subharmonicity. With the appropriate interpretation of the statement $\Delta s \geqq 0$ (in the sense of distributions), Theorem 4.3 remains valid for all continuous s. Further details can be found in (5.8) below. For our purposes, however, the following more specific variant of the theorem we just proved will be more useful.

THEOREM 4.4. *Suppose $s \geqq 0$ is continuous in a domain $\mathcal{D}$, has continuous second partial derivatives in the subset $\mathcal{R} = \{x \in \mathcal{D}; s(x) > 0\}$ and satisfies $\Delta s \geqq 0$ in $\mathcal{R}$. Then s is a subharmonic function in $\mathcal{D}$.*

PROOF. We have to show that s satisfies the mean value inequality (4.2). Thus, let $x_0 \in \mathcal{D}$ and suppose the closed sphere $S_r = \{x; |x - x_0| \leqq r\}$ is contained in $\mathcal{D}$. Let u be the solution of the Dirichlet problem for the closed domain S_r having boundary values $s(x)$ for $x \in \partial S_r = \{t; |t - x_0| = r\}$ (see Corollary 1.11). If we can show that $s(x) \leqq u(x)$ for $x \in S_r$, inequality (4.2) is a consequence of the mean-value property (Theorem 1.1):

$$s(x_0) \leqq u(x_0) = \mathcal{M}_{x_0,u}(r) = \frac{1}{\omega_{n-1}} \int_{\Sigma} u(x_0 + rt') \, dt'$$

$$= \frac{1}{\omega_{n-1}} \int_{\Sigma} s(x_0 + rt') \, dt'.$$

Suppose, on the contrary, that $\sup_{x \in S_r} \{s(x) - u(x)\} = c > 0$, and let $F = \{x \in S_r; s(x) - u(x) = c\}$. Then the set F is obviously closed and is contained in the interior of S_r. On the other hand, $s(x) > 0$ at each point of F (since $u \geqq 0$). We can, therefore, apply Theorem 4.3 to the function $s - u$ in the set $\mathscr{R}$ and we obtain, for each $x \in F(\subset \mathscr{R})$

$$c = s(x) - u(x) \leqq \frac{1}{\omega_{n-1}} \int_{\Sigma} \{s(x + \rho t') - u(x + \rho t')\} \, dt'$$

for all $\rho > 0$ sufficiently small. Since $s - u \leqq c$ in S_r and x is interior to S_r, the continuity of $s - u$ implies that $s(x + \rho t') - u(x + \rho t') = c$ for all $t' \in \Sigma$ and ρ sufficiently small. Thus, F is open. Since S_r is connected F must be either empty or all of S_r; but, as we have already observed, the boundary of S_r is disjoint from F and, consequently, F is empty. This shows that $s(x) \leqq u(x)$ for all x in S_r and the theorem is proved.

It is useful to note that we have actually proved the following more general result:

THEOREM 4.4′. *Suppose $s \geqq 0$ is continuous in a domain $\mathscr{D}$ and is subharmonic in the set $\mathscr{R} = \{x \in \mathscr{D}; s(x) > 0\}$, then s is subharmonic in $\mathscr{D}$.*

The argument we gave at the end of the last proof applies generally to subharmonic functions defined on an arbitrary bounded domain. More precisely, we have the following theorem (that justifies the use of the word "subharmonic"):

THEOREM 4.5. *Suppose s is a continuous function on the closure of a bounded domain $\mathscr{D}$. If s is subharmonic in $\mathscr{D}$ and u is a continuous function on $\overline{\mathscr{D}}$ that is harmonic in $\mathscr{D}$ and satisfies $s(x) \leqq u(x)$ for all $x \in \partial\mathscr{D} = \overline{\mathscr{D}} - \mathscr{D}$ then $s(x) \leqq u(x)$ for all $x \in \overline{\mathscr{D}}$.*

PROOF. Consider the function $s - u$. If $s(x) - u(x) > 0$ for some $x \in \mathscr{D}$ let $c = \sup_{x \in \overline{\mathscr{D}}} \{s(x) - u(x)\} > 0$ and $F = \{x \in \overline{\mathscr{D}}; s(x) - u(x) = c\}$ ($c < \infty$ since $\overline{\mathscr{D}}$ is compact). Then F is obviously closed. On the other hand, $-u$, being a harmonic function, satisfies the mean-value property; therefore,

$$c = s(x) - u(x) \leqq \frac{1}{\omega_{n-1}} \int_{\Sigma} \{s(x + \rho t') - u(x + \rho t')\} \, dt'$$

whenever $x \in F$ and the sphere $\{t; |t - x| \leqq \rho\}$ is contained in $\mathscr{D}$. As in the previous proof we can, therefore, conclude that F is open. It follows that F must be empty (since $\overline{\mathscr{D}}$ is connected and $s(x) - u(x) \leqq 0 < c$ when $x \in \overline{\mathscr{D}} - \mathscr{D}$) and it follows that $s(x) - u(x) \leqq 0$ for all $x \in \overline{\mathscr{D}}$.

Before proceeding further in our brief study of subharmonic functions it will be useful to present some examples:

(1) It is clear that a harmonic function is subharmonic. Moreover, it follows immediately from the definition that $|u|$ is subharmonic whenever u is harmonic.

(2) If s is subharmonic and φ is a nondecreasing convex function defined on an interval containing the range of s then the composition $\varphi \circ s$ is subharmonic. This follows from Jensen's inequality[13] applied to the right member of inequality (4.2):

$$\varphi(s(x_0)) \leqq \varphi\left(\int_\Sigma s(x_0 + rt')\, dt'/\omega_{n-1}\right) \leqq \int_\Sigma \varphi(s(x_0 + rt'))\, dt'/\omega_{n-1}.$$

Since $\varphi(x) = x^p, p \geqq 1$, defines a convex function for $x \geqq 0$, this observation, combined with the first example, gives us the fact that $|u|^p$ is subharmonic whenever u is subharmonic.

(3) It is not true, in general, that $|u|^p$ is subharmonic for $p < 1$ when u is a harmonic function. For example, no power less than 1 of $|u|$ is subharmonic if u is the harmonic function with values $u(x, y) = x$, for $(x, y) \in E_2$. In two dimensions, however, a considerably stronger result than that stated at the end of the previous example is true if we consider a harmonic function together with a *harmonic conjugate*.[14] More precisely, if $F = u + iv$ is an analytic function, then $|F|^p$ is subharmonic for all $p > 0$. We can show this in the following way: Let $\mathscr{R}$ be the set of points, $x + iy = z$, in the domain of F such that $F(z) \neq 0$; then $s = \log |F|$ is harmonic in $\mathscr{R}$ and, a fortiori, subharmonic. Since $\varphi(x) = e^{px}, p > 0$, defines a convex function, the composition $\varphi \circ s = |F|^p$ is subharmonic in $\mathscr{R}$. The fact that $|F|^p$ is subharmonic in the domain of F now follows from Theorem 4.4′.

(4) If $s_1, s_2, \ldots, s_k$ are subharmonic in a domain $\mathscr{D}$ and $\alpha_1, \alpha_2, \ldots, \alpha_k$

[13] One of the many forms of Jensen's inequality is the following: Suppose μ is a finite measure on the space M, f a μ-integrable function and φ a convex function whose domain includes the range of f, then

$$(*) \qquad \varphi\left(\int_M f(t)\, d\mu(t)/\mu(M)\right) \leqq \int_M \varphi(f(t))\, d\mu(t)/\mu(M).$$

If M is the interval $[x, y]$, $\mu(\{x\}) = 1 = \mu(\{y\})$, while $\mu(E) = 0$ whenever a Borel set E contains neither x nor y, and f is the identity function $f(t) = t$, for $x \leqq t \leqq y$, then this inequality becomes the defining inequality (4.1) for convex functions. It is not hard, however, to obtain (*) from (4.1) by a simple limiting argument.

[14] If u is harmonic in a region $\mathscr{R} \subset E_2$ a harmonic function, v, such that $F = u + iv$ is analytic in $\mathscr{R}$ is called a harmonic conjugate of u. It is an elementary fact that, if $\mathscr{R}$ is simply connected, then such a harmonic conjugate exists and is unique up to an additive constant. In particular, a harmonic conjugate of u exists in a neighborhood of each point of the region $\mathscr{R}$ where u is harmonic.

are nonnegative real numbers, then the linear combination $\alpha_1 s_1 + \alpha_2 s_2 + \cdots + \alpha_k s_k$ is a subharmonic function in $\mathscr{D}$. Similarly $s(x) = \max_{x \in \mathscr{D}} \{s_1(x), s_2(x)\}$ defines a subharmonic function in $\mathscr{D}$ when s_1 and s_2 are subharmonic in this domain.

If a harmonic function, u, majorizes the function s in a domain we say that u is a *harmonic majorant* of s. If $u \leqq h$ whenever h is another harmonic majorant of s, we say that u is the *least harmonic majorant* of s. Theorem 4.5 tells us that in case s is continuous on the closure $\bar{\mathscr{D}}$, of the bounded domain $\mathscr{D}$ and is subharmonic in $\mathscr{D}$, then u is a harmonic majorant of s in $\mathscr{D}$ provided it has a continuous extension to $\bar{\mathscr{D}}$ that majorizes s on $\partial\mathscr{D}$. If $\mathscr{D}$ admits the solution of the Dirichlet problem, it follows immediately from this that the least harmonic majorant of s is the continuous function on $\bar{\mathscr{D}}$ that is harmonic in $\mathscr{D}$ and equals s on $\partial\mathscr{D} = \bar{\mathscr{D}} - \mathscr{D}$. We end this section by proving a related result for certain subharmonic functions defined in upper half-spaces. This result will be used (see Chapter III and VI), in conjunction with Theorem 3.19 or Theorem 3.24, to obtain the existence of boundary values for certain classes of harmonic functions.

THEOREM 4.6. *Suppose s is a nonnegative subharmonic function defined in E_{n+1}^+ satisfying*

$$\int_{E_n} [s(x, y)]^q \, dx \leqq c^q < \infty \tag{4.7}$$

for all $y > 0$, where $1 \leqq q < \infty$ and c is independent of $y > 0$. Then s has a least harmonic majorant in E_{n+1}^+. Moreover, in case $q > 1$ this majorant is the Poisson integral of a function $f \in L^q(E_n)$, with $\|f\|_q \leqq c$; if $q = 1$ it is the Poisson–Stieltjes integral of a finite Borel measure on E_n having total measure not exceeding c.

PROOF. We first observe that s satisfies the hypotheses of Lemma 2.6 except for harmonicity. The proof of that lemma, however, depends only on inequality (4.2) and not on the full force of the mean-value property of harmonic functions. Consequently, there exists a constant $A = A_{n,p} > 0$ such that

$$\|s(\cdot, y)\|_\infty = \sup_{x \in E_n} |s(x, y)| \leqq Acy^{-n/q}. \tag{4.8}$$

We shall show that, for each proper sub-half space $E_{n+1,y_0}^+ = \{(x, y) \in E_{n+1}; y \geqq y_0 > 0\}$, s belongs to the class $C_0(E_{n+1,y_0}^+)$. That is,

$$\lim_{|x|^2 + y^2 \to \infty} s(x, y) = 0 \tag{4.9}$$

provided $y \geqq y_0$. First we note that it follows from (4.8) that $s(x, y)$ is small provided y is large. Thus, it suffices to show that $s(x, y)$ tends to 0 as $|x| \to \infty$ when y lies between y_0 and some fixed $y_1 > y_0$. Let us fix a positive $r < y_0$ and put $B = \{(x, y) \in E_{n+1}; y_0 - r < y < y_1 + r\}$ $(\subset E_{n+1}^+)$. Then

$$\int_B [s(x, y)]^q \, dx \, dy = \int_{y_0 - r}^{y_1 + r} \int_{E_n} [s(x, y)]^q \, dx \, dy \leqq (y_1 + 2r - y_0)c^q < \infty.$$

Consequently, if $B_k = B \cap \{(x, y) \in E_{n+1}; |x| \geqq k\}$, we have

$$\lim_{k \to \infty} \int_{B_k} [s(x, y)]^q \, dx \, dy = 0. \tag{4.10}$$

If $|x| \geqq m$ and $y_0 \leqq y \leqq y_1$, we can find a (solid) sphere S of radius r about the point (x, y) that is contained in the set B_{m-r}. Since $\varphi(t) = t^q$ defines a nondecreasing convex function, $\varphi \circ s = s^q$ is also subharmonic (see example (2) above) and, therefore,

$$[s(x, y)]^q \leqq \frac{1}{|S|} \int_S [s(\xi, \eta)]^q \, d\xi \, d\eta \leqq \frac{1}{|S|} \int_{B_{m-r}} [s(\xi, \eta)]^q \, d\xi \, d\eta. \text{ [15]}$$

But, by (4.10), the last term tends to 0 as $k = m - r$ tends to ∞, and (4.9) is established.

In order to construct the desired harmonic majorant we define, for each $\varepsilon > 0$ and $y > 0$,

$$m_\varepsilon(x, y) = \int_{E_n} s(x - t, \varepsilon) P(t, y) \, dt = \int_{E_n} s(t, \varepsilon) P(x - t, y) \, dt.$$

Then, by inequality (2.2), we have

$$\int_{E_n} [m_\varepsilon(x, y)]^q \, dx \leqq c^q \tag{4.11}$$

for all $\varepsilon > 0$ and $y > 0$. Moreover, it follows from (4.9) and part (b) of Theorem 2.1 that

$$0 = \lim_{y \to 0} \|s(\cdot, \varepsilon) - m_\varepsilon(\cdot, y)\|_\infty = \lim_{y \to 0} \left\{\sup_{x \in E_n} |s(x, \varepsilon) - m_\varepsilon(x, y)|\right\}.$$

[15] Inequality (4.2) asserts that the mean value of s over the surface of a sphere of radius r about x_0 dominates $s(x_0)$. By the same argument that was used in the proof of Lemma 2.6 we can show that $s(x_0)$ is dominated by the mean value of s over the solid sphere of radius r about x_0. It is this inequality, applied to s^q at the point (x, y), that is being used here.

From this we can conclude that, if δ is a positive number, there exists a sufficiently small $y_0 > 0$ such that $s(x, \varepsilon + y_0) - m_\varepsilon(x, y_0) < \delta$ for all $x \in E_n$. By (4.9) we see that $s(x, \varepsilon + y) - m_\varepsilon(x, y) < \delta$ if either $|x|$ or y is large, say $|x| \geqq k$ or $y \geqq y_1$. Thus, the inequality

$$s(x, \varepsilon + y) - m_\varepsilon(x, y) < \delta$$

holds on the boundary of the region $R = \{(x, y) \in E_{n+1}^+; |x| \leqq k, y_0 \leqq y \leqq y_1\}$. An examination of the proof of the maximum property for harmonic functions shows that only the mean value inequality (4.2) was used; therefore, this property is satisfied by subharmonic functions. Applying it to the subharmonic function $s(x, \varepsilon + y) - m_\varepsilon(x, y)$ ($-m_\varepsilon$ being harmonic, is subharmonic) and the region R we see that the last inequality is valid throughout R. Letting $\delta \to 0$ and, accordingly, letting $k, y_1 \to \infty$ and $y_0 \to 0$ we obtain

$$s(x, \varepsilon + y) \leqq m_\varepsilon(x, y) \tag{4.12}$$

for all $(x, y) \in E_{n+1}^+$.

Employing the same argument that was used in the proof of Theorem 2.5 (with u replaced by s and p replaced by q) we can find a null sequence $\{\varepsilon_k\}$ and an L^q function f, when $q > 1$, such that

$$\lim_{k\to\infty} m_{\varepsilon_k}(x, y) = \lim_{k\to\infty} \int_{E_n} s(t, \varepsilon_k) P(x - t, y)\, dt$$

$$= \int_{E_n} f(t) P(x - t, y)\, dt = m(x, y)$$

while, when $q = 1$, instead of f there exists a finite Borel measure μ, such that

$$\lim_{k\to\infty} m_{\varepsilon_k}(x, y) = \lim_{k\to\infty} \int_{E_n} s(t, \varepsilon_k) P(x - t, y)\, dt$$

$$= \int_{E_n} P(x - t, y)\, d\mu(t) = m(x, y).$$

In either case, letting $\varepsilon = \varepsilon_k$ tend to zero in inequality (4.12), we obtain

$$s(x, y) \leqq m(x, y).$$

The function m is the desired Poisson (or Poisson–Stieltjes) integral. The fact that it is the least harmonic majorant of s follows from the obvious fact that $m_\varepsilon(x, y)$ is the least harmonic majorant of $s(x, \varepsilon + y)$: for any harmonic majorant, h, of s satisfies

$$m(x, y) = \lim_{k\to\infty} m_{\varepsilon_k}(x, y) \leqq \lim_{k\to\infty} h(x, \varepsilon_k + y) = h(x, y).$$

5. Further Results

5.1. At the beginning of this chapter we introduced the mean value $\mathscr{M}_{x,u}(r) = \mathscr{M}(r)$ of a function u taken over the surface of the sphere $S_r(x)$ of radius r centered at x. It is just as natural to form the mean value

$$\mathscr{A}_{x,u}(r) = \mathscr{A}(r) = \frac{1}{\Omega_n r^n} \int_{|t| \leqq r} u(x + t)\, dt,$$

of the function u, taken over the volume of the sphere $S_r(x)$. We can, therefore, consider two different mean-value properties: the one introduced at the beginning of this chapter and the analogous one involving $\mathscr{A}(r)$. The argument used at the beginning of the proof of Lemma 2.6 shows that if a function satisfies the first of these mean value properties then it satisfies the second. Similarly, as we have seen in the proof of Theorem 4.6, a function satisfying inequality (4.2) satisfies the inequality obtained by replacing $\mathscr{M}(r)$ by $\mathscr{A}(r)$. Moreover, it is not hard to show that, by replacing $\mathscr{M}(r)$ by $\mathscr{A}(r)$, we obtain another characterization of harmonic and subharmonic functions, similar to the ones in Theorem 1.7 and inequality (4.2) (see Rado [1]).

5.2. There are several results concerning harmonic (and analytic) functions defined on unbounded domains that are extensions of the maximum principle as stated in Corollary 1.3′. These results can be described in a general way as follows: Suppose we are given a continuous function, defined on a closed unbounded region $\mathscr{E} \subset E_n$, that is harmonic in the interior of $\mathscr{E}$. If the function satisfies certain weak restrictions on its growth (as $x \in \mathscr{E}$ tends to ∞) and satisfies certain more stringent conditions on the boundary of $\mathscr{E}$ (often the function is assumed to be bounded on $\partial\mathscr{E} = \bar{\mathscr{E}} - \mathscr{E}$), then it satisfies these more stringent conditions throughout $\mathscr{E}$. Such results are referred to as *Phragmén–Lindelöf* theorems. Corollary 1.15 is a particularly simple example of such a result. The following is an example of a Phragmén–Lindelöf theorem that is considerably more general: Suppose u is harmonic in E^+_{n+1}, continuous on $\overline{E^+_{n+1}} = E^+_{n+1} \cup E_n$ and satisfies $u(x, y) = o(e^{a|x|})$ as $|x| \to \infty$, for all $a > 0$, in each strip of the form $S = \{(x, y) \in E^+_{n+1}; 0 < y \leqq y_0\}$. If, in addition, $u(x, y) = o(y)$ as $y \to \infty$ then

(a) $u(x, 0) \leqq A < \infty$ for all $x \in E_n$ implies $u(x, y) \leqq A$ for all $(x, y) \in E^+_{n+1}$;

(b) $u(x, 0) \geqq B > -\infty$ for all $x \in E_n$ implies $u(x, y) \geqq B$ for all $(x, y) \in E^+_{n+1}$;

(c) $|u(x, 0)| \leqq C < \infty$ for all $x \in E_n$ implies $|u(x, y)| \leqq C$ for all $(x, y) \in E^+_{n+1}$.

To show, say, part (b) we can assume $B = 0$ since the general case can be reduced to this one by considering the function $u(x, y) - B$. Fix $\delta > 0$ and let $g(x, y) = \sin(a\sqrt{n}(y + \delta)) \prod_{k=1}^{n} \cosh ax_k$; then g is a harmonic function in E_{n+1} (see example (6) at the beginning of this chapter). Part (b) would be established if we prove that $u(x, y) + \varepsilon y + \eta g(x, y) \geqq 0$ in $\overline{E_{n+1}^+}$ for each fixed $\varepsilon > 0$ and $\eta > 0$. By the minimum principle we see that it suffices to show that this inequality holds on the boundary of a cylinder of the form $\{(x, y) \in E_{n+1}; |x| \leqq N, 0 \leqq y \leqq y_0\}$ for all N and y_0 that are sufficiently large. Choosing $a > 0$ and $\delta > 0$ so that $a < \pi/\sqrt{n}(y_0 + \delta)$, it follows that $\sin(a\sqrt{n}(y + \delta))$ is positive and bounded away from 0 when $0 \leqq y \leqq y_0$ and, moreover,

$$\prod_{k=1}^{n} \cosh ax_k \geqq 2^{-n} \prod_{k=1}^{n} e^{a|x_k|} = 2^{-n} \exp\left\{a \sum_{1}^{n} |x_k|\right\} \geqq 2^{-n} \exp\{a|x|\}.$$

These estimates, together with the hypotheses $u(x, y) = o(e^{a|x|})$ and $u(x, y) = o(y)$, imply the desired inequality $u(x, y) + \varepsilon y + \eta g(x, y) \geqq 0$ for $|x| \geqq N$ (if N is sufficiently large) and $0 \leqq y \leqq y_0$. By decreasing a appropriately we can choose the number y_0 to be as large as we wish. This, then, implies (b); part (a) is obtained from (b) by considering $-u$, and these two results clearly imply (c).

5.3. We showed (see Lemma 2.6 and inequality (4.8)) that if a (non-negative) subharmonic function s in E_{n+1}^+ satisfied $\|s(\cdot, y)\|_p \leqq c$ for all $y > 0$ and some $p \geqq 1$, then $\|s(\cdot, y)\|_\infty = O(y^{-n/p})$. As a consequence of this result we find, more generally, that $\|s(\cdot, y)\|_q = O(y^{(n/q)-(n/p)})$ when $q \geqq p$. To see this we observe that,

$$\int_{E_n} [s(x, y)]^q \, dx = \int_{E_n} [s(x, y)]^p [s(x, y)]^{q-p} \, dx \leqq \|s(\cdot, y)\|_\infty^{q-p} c^p$$
$$\leqq A[y^{-n/p}]^{(q-p)},$$

and the desired inequality follows by taking q-th roots.

5.4. Theorems 3.19 and 3.24 indicate that the concept of nontangential convergence at the boundary is essential. That this is indeed the fact can be seen from the following two results dealing with functions defined on the unit disk (Analogous results for the upper half-plane can be obtained with the aid of conformal mappings):

(i) Let C_0 be a simple closed curve passing through the point $z = 1$, all of whose other points are contained in the interior of the unit circle $K = \{z = x + iy; |z| = 1\}$. Suppose, further, that C_0 is tangent to the unit circle at 1 and that C_θ is the curve obtained by rotating C_0 about the origin through the angle θ. Then there exists a bounded analytic function,

F, defined in the interior of K such that, for almost all θ, $F(z)$ does not tend to any limit as z tends to $e^{i\theta}$ within C_θ (see Zygmund [1], Vol. I, page 280).

(ii) Let G be a continuous function defined in the interior of the unit circle, K, and let E be a set of the first category in K. Then there exists an analytic function F defined in the interior of K such that

$$\lim_{r\to 1} \{F(re^{i\theta}) - G(re^{i\theta})\} = 0$$

whenever $e^{i\theta} \in E$ (see Zygmund [1], Vol. II, page 204).

Since E can have measure 2π, this last result asserts that the radial behavior, at almost all points of K, of analytic functions is no better than that of continuous functions. Thus, for example, an analytic function F can tend to 0 along almost every radius and, yet, not vanish identically. On the other hand, an analytic function having nontangential limit 0 at the points of a set of positive measure in K must be identically zero (see Zygmund [1], Vol. II, page 203). Moreover, an analytic function can be bounded (though, obviously, not uniformly) on almost every radius and, yet, fail to have radial limits at almost all points of K.

5.5. Theorem 3.19 has several extensions. Carleson (see [1]) has shown that the same result holds if we assume that the (real-valued) harmonic function u is nontangentially bounded from below or from above at each point of S. In the two-dimensional case this result can easily be extended to regions more general than the upper half-plane with the aid of conformal mappings. Hunt and Wheeden (see [1]) have obtained a similar extension in higher dimensions for a domain $\mathscr{D}$ whose boundary satisfies a Lipschitz condition. They have shown that if u is harmonic in such a domain $\mathscr{D}$, $S \subset \partial\mathscr{D}$ has the property that for each $q \in S$ there exists an open cone with vertex q that is exterior to $\mathscr{D}$, and u is bounded from below (or above) in a truncated cone having vertex q, for each $q \in S$, then u has a finite nontangential limit at each point of S except, possibly, for a set of harmonic measure zero.

5.6. By slightly modifying the argument used in the proof of Lemma 3.3 we can obtain the following closely related result: Suppose that to each point x of the bounded set $F \subset E_n$ there corresponds a non-empty open sphere S_x with center x. Then there exist a countable collection $\{S_{x_i}\}$ of these spheres covering F and such that each $x \in E_n$ belongs to at most a_n of these spheres (a_n depends only on the dimension; in fact, it can be shown that a_n can be chosen to be 3^n). Using this result we can generalize Theorem 3.4: Suppose ν_1 and ν_2 are two nonnegative finite Borel measures on E_n and $\varphi(x) = \sup\{\nu_1(S_x)/\nu_2(S_x)\}$, where the supremum is taken over all open spheres S_x centered at $x \in E_n$ (when ν_2 is Lebesgue measure and $\nu_1(S_x) = \int_{S_x} f$, for some integrable f, then φ is the Hardy–Littlewood maximal function m_f). Then, if $F_s = \{x \in E_n; \varphi(x) > s > 0\}$, we have $\nu_2(F_s) \leqq a_n\nu_1(E_n)s^{-1}$. From this one can show that ν_1 is differentiable with respect to ν_2; that is, $\lim \{\nu_1(S_x)/\nu_2(S_x)\} = f(x)$ exists for all $x \in E_n$ outside a set of ν_2-measure zero, the limit being taken over spheres centered at x with radii tending to zero. If we decompose ν_1 into its absolutely continuous and singular parts, with respect to ν_2, then the Radon–Nikodym derivative of the former equals $f(x)$ almost everywhere (again, with respect to ν_2). Finally, we remark that these results still hold if other geometric figures are allowed to replace spheres (see Cotlar [1], pp. 126–127).

5.7. In theorem 3.22 and Corollary 3.23 we assumed $1 < p < \infty$. The case $p = \infty$ was discussed at the end of the proof of Theorem 3.22. There, we pointed out that the result in this case follows from similar arguments. The case $p = 1$ is quite different: one can show that there exist functions $f \in L^1(E_2)$, e.g., such that the iterated Poisson integrals

$$u(x_1, \varepsilon_1, x_2, \varepsilon_2) = \int_{-\infty}^{\infty}\int_{-\infty}^{\infty} f(x_1 - t_1, x_2 - t_2)P(t_1, \varepsilon_1)P(t_2, \varepsilon_2)\, dt_1\, dt_2$$

do not converge to $f(x_1, x_2)$ as $\varepsilon_1, \varepsilon_2 \to 0$ for almost all $x = (x_1, x_2) \in E_2$ (see Jessen, Marcinkiewicz, and Zygmund [1] where the iterated Poisson integral for the unit circle is discussed. The argument used there can be modified to include the case under consideration here). See also (6.6) and (6.8) of Chapter III.

5.8. It follows easily from our development of subharmonic functions that if s has continuous second partial derivatives then the following three conditions are equivalent: (i) s is subharmonic; (ii) $\Delta s \geqq 0$; (iii) if $\mathscr{R}$ is a bounded domain whose closure is contained in the domain of definition of s, u is a continuous function on $\bar{\mathscr{R}}$ that is harmonic in $\mathscr{R}$ and satisfies $u \geqq s$ on the boundary $\partial\mathscr{R} = \bar{\mathscr{R}} - \mathscr{R}$, then $u \geqq s$ on $\mathscr{R}$ (see Theorem 4.5). There are several ways of extending these results. We shall sketch one such extension that is closely connected with the ideas developed in our treatment of tempered distributions (see Section 3 of Chapter I). Let us say that a continuous function s, defined on some domain D, had a nonnegative generalized Laplacian if $\int_D s\Delta\varphi \geqq 0$ whenever φ is a nonnegative C^∞ function with compact support interior to D (integrating by parts, we see that if s has continuous second partial derivatives then this property is equivalent to (ii) above). We first claim that such a function s is subharmonic. In order to do this we show that s satisfies the mean-value inequality (4.2). In view of Theorem 1.25, Chapter

I, it suffices to show that this is the case for each of the (everywhere defined) functions $s_\varepsilon(x) = \int_D s(t)\varphi_\varepsilon(x-t)\,dt$, where φ is a nonnegative C^∞ function with compact support having integral 1 and $\varepsilon > 0$ sufficiently small. By Theorem 4.3 we see that this is true provided $\Delta s_\varepsilon \geqq 0$. As we have observed above, this is equivalent to $\int_D s_\varepsilon \Delta\psi \geqq 0$ for all nonnegative C^∞ functions ψ having compact support within D. Choosing ε small and the support of ψ within an appropriate compact subset of D we have $\int_D s_\varepsilon\, \Delta\psi = \int_D s\, \Delta(\tilde{\varphi}_\varepsilon * \psi) \geqq 0$ ($\tilde{\varphi}$ denotes the reflection of φ, as defined in Chapter I preceding equality (3.12)). This shows that s is subharmonic. Theorem 4.5 tells us that (i), in turn, implies the harmonic majorization property (iii). It is immediate, however, that the latter implies the mean-value inequality (4.2). Assuming that s satisfies this last property we shall show that it has a nonnegative generalized Laplacian. Let η_δ be the function obtained after dividing by $\Omega_n \delta^n$ the characteristic function of the sphere of radius δ about 0. Then (4.2) can be expressed as the inequality $s \leqq s * \eta_\delta$. Thus, $\int_D s\varphi \leqq \int_D (s * \eta_\delta)\varphi$ or, equivalently,

$$\int_D s(\eta_\delta * \varphi - \varphi) \geqq 0$$

for all nonnegative C^∞ functions with compact support interior to D. But $\{\eta_\delta * \varphi - \varphi\}/\delta^2$ tends to a positive constant times $\Delta\varphi$ as δ tends to 0; therefore, $\int_D s\, \Delta\varphi \geqq 0$.

Bibliographical Notes

Kellog [1] is a comprehensive text on the theory of harmonic functions of three variables. For a more modern treatment of the general n-dimensional case see Brelot [1]. The characterization of Poisson integrals is a classical result and the proofs used here are not very different from those used for the characterization of the Cesáro and Abel means of trigonometric series that can be found in Zygmund [1], Chapter III. Theorem 3.10 and inequality (3.18) are an extension to upper half-spaces of the corresponding results for Poisson integrals of functions defined on the unit circle originally obtained by Hardy and Littlewood [1]; for these and other variants see K. T. Smith [1]. Theorems 3.19 and 3.24 are due to Calderón [1]. The maximal function was first introduced, in one dimension, by Hardy and Littlewood [1]. The extension to several variables is due to Wiener [2]. In connection with the covering lemma used here see Wiener [2], Marcinkiewicz and Zygmund [1], Besicovitch [2], Cotlar [1], and de Guzmán [1]. Theorem 3.12 is closely related to a result of Cotlar [1], an earlier and more general result is due to Banach [1]. For a comprehensive treatment of the theory of subharmonic functions of two variables see Rado [1]. Theorem 4.6 can be found in Stein and G. Weiss [1]. For a more complete treatment of strong differentiability of integrals see Saks [1].

CHAPTER III

The Theory of H^p Spaces on Tubes

Comparatively strong conditions have to be imposed on harmonic functions, defined in upper half-spaces, in order to insure the existence of boundary values. In sections two and three of the previous chapter we studied this situation. If, however, we consider a system of harmonic functions satisfying certain partial differential equations (for example, the real and imaginary parts of an analytic function form such a system satisfying the Cauchy–Riemann equations) weaker conditions are sufficient to insure the existence of boundary values. A more detailed discussion of these facts is presented in the first section. The rest of the chapter is devoted to the theory of holomorphic functions defined on tubes. These are, perhaps, the most natural domains where we can study the relation between the theory of holomorphic functions of several complex variables and Fourier analysis. In the second section we study the H^2 space theory of such functions. In the third section we consider an important special case of these spaces: the H^2 spaces over tubes whose bases are cones. An extension of the Paley–Wiener theorem to several variables is obtained in section four as a consequence of these results. In the fifth section we study the more general H^p spaces over tube domains by reducing it to tubes based on certain special cones.

1. Introductory Remarks

We have seen in the previous chapter (see Theorem 2.1) how, by considering the Poisson integral of a function f belonging to $L^p(E_n)$, $1 \leqq p \leqq \infty$, we obtain a harmonic function u, defined on E_{n+1}^+, satisfying

$$\int_{E_n} |u(x, y)|^p \, dx \leqq A^p < \infty \tag{1.1}$$

for all $y > 0$, where $A = \|f\|_p$. Furthermore, $u(x, y)$ converges nontangentially to $f(x)$ for almost all $x \in E_n$ (see Theorem 3.16).

Theorem 2.5 of the previous chapter tells us that inequalities of the type (1.1) characterize those harmonic functions on E_{n+1}^+ that are Poisson integrals, and, thus, imply the existence of nontangential boundary values. It can be shown that these results are best possible in the sense that if $0 < p < 1$ inequality (1.1) does not imply the existence of such boundary values as $y \to 0$; that is, for each such p there exists a harmonic function satisfying (1.1) that does *not* have nontangential limits a.e. at the boundary, E_n, of E_{n+1}^+.

When $n = 1$ the situation changes very much if, instead of harmonic functions, we consider analytic functions. For example, if $F(z) = F(x + iy)$ is holomorphic for $y > 0$ (that is, in the upper half-plane E_2^+) and satisfies, for $p > 0$,

$$\int_{-\infty}^{\infty} |F(x + iy)|^p \, dx \leqq A^p < \infty, \qquad y > 0, \tag{1.2}$$

then the limits $\lim_{z \to x} F(z) = F(x)$ exist at almost every $x \in (-\infty, \infty)$ if z is allowed to approach x nontangentially. The class of all holomorphic functions on E_2^+ satisfying (1.2) for some constant $A < \infty$ is denoted by $H^p = H^p(E_2^+)$. Better known are the corresponding H^p-spaces associated with the unit disk $D = \{z \in E_2; |z| < 1\}$. These consist of those analytic functions, F, in D satisfying

$$\int_{-\pi}^{\pi} |F(re^{i\theta})|^p \, d\theta \leq A^p < \infty, \qquad 0 \leqq r < 1. \tag{1.3}$$

Again, the nontangential limits $F(e^{i\theta})$ exist for almost every $\theta \in [-\pi, \pi]$. Moreover, in both cases these are limits in the "norm"; that is, when $F \in H^p(E_2^+)$ then

$$\lim_{y \to 0} \int_{-\infty}^{\infty} |F(x + iy) - F(x)|^p \, dx = 0, \tag{1.4}$$

while, if $F \in H^p(D)$ then

$$\lim_{r \to 1} \int_{-\pi}^{\pi} |F(re^{i\theta}) - F(e^{i\theta})|^p \, d\theta = 0. \tag{1.4'}$$

It is our intention to extend these notions and results to higher dimensions. In so doing we meet with several essentially different approaches. One, involving the theory of functions of several complex variables, is the following. Let B be an open subset of E_n. Then, the *tube*, T_B, *with base* B is the subset of all $z = (z_1, \ldots, z_n) = (x_1 + iy_1, \ldots, x_n + iy_n) = x + iy \in \mathbf{C}_n$ (= n-dimensional complex Euclidean space) such that $y \in B$. For example, E_2^+ is the tube in $\mathbf{C}_1$ with base $B = \{y \in E_1; y > 0\}$. A function defined and holomorphic[1] on a tube T_B is said to belong to the space

[1] We say that a function F whose domain D is contained in $\mathbf{C}_n$ is *holomorphic* if for each $z_0 = (z_1^0, \ldots, z_n^0) \in D$ there exists a polydisk $\{z = (z_1, \ldots, z_n) \in \mathbf{C}_n;\ |z_1 - z_1^0| < r_1, \ldots, |z_n - z_n^0| < r_n\} \subset D$ in which F is represented as an absolutely convergent power series:

$$F(z) = \sum_{k_1, \ldots, k_n \geqq 0} a_{k_1 \cdots k_n} (z_1 - z_1^0)^{k_1} \cdots (z_n - z_n^0)^{k_n}.$$

$H^p = H^p(T_B), p > 0$, if there exists a constant $A < \infty$ such that

$$\int_{E_n} |F(x + iy)|^p \, dx \leqq A^p, \qquad \text{for all } y \in B. \tag{1.5}$$

This clearly extends the definition of the spaces $H^p(E_2^+)$. The norm of $F \in H^p(T_B)$ is the positive number $\|F\|_p$ defined as the least A for which (1.5) holds.

Another approach arises from the observation that a function $F = u + iv$ of one complex variable is analytic in a simply connected region if and only if (v, u) is the gradient of a harmonic function in this region. Thus, we may consider a vector-valued function $F = (u_1, \ldots, u_n)$, defined in a region of E_n, to be a "generalized analytic function" if it is the gradient of a harmonic function in this region. More generally, we shall consider a vector-valued function $F = (u_1, u_2, \ldots, u_n)$, defined in a region $D \subset E_n$, for which the partial differential equations

$$\sum_{j=1}^{n} \frac{\partial u_j}{\partial x_j} = 0, \qquad \frac{\partial u_i}{\partial x_j} = \frac{\partial u_j}{\partial x_i}, \tag{1.6}$$

$i, j = 1, 2, \ldots, n$, are satisfied. These conditions can also be stated in the form

$$\operatorname{div} F = 0, \qquad \operatorname{curl} F = 0. \tag{1.6'}$$

We note that the second condition implies that, in a simply connected subregion of D, F is the gradient of some function h, while the first condition assures us that h is harmonic. When $n = 2$ Eqs. (1.6) are, simply, the Cauchy–Riemann equations; in this case $u_2 + iu_1$ is an analytic function of $z = x_1 + ix_2$. If F is such a function defined in E_{n+1}^+ and if there exists a constant $A < \infty$ such that

$$\int_{E_n} |F(x, y)|^p \, dx = \int_{E_n} \left(\sum_{j=1}^{n+1} |u_j(x, y)|^2 \right)^{p/2} dx \leqq A^p \tag{1.7}$$

for all $y > 0$,[2] we obtain an extension of the definition of the spaces $H^p(E_2^+)$.

In this chapter we shall study the H^p spaces associated with tube domains. The second extension will be studied in Chapter VI.

2. *The H^2 Theory*

We fix an open connected subset, B, of E_n and shall study the space $H^2(T_B)$. It is not hard to construct functions that belong to this space. In

[2] We warn the reader that in (1.7) we are dealing with harmonic functions of $n + 1$ variables, where $y = x_{n+1}$.

fact, let f be a function satisfying

$$\sup_{y\in B} \int_{E_n} |f(t)|^2 e^{-4\pi y\cdot t}\, dt \leqq A^2 < \infty. \tag{2.1}$$

Let us write $z = x + iy$ and $z\cdot t = z_1t_1 + z_2t_2 + \cdots + z_nt_n$. We shall show that $|e^{2\pi i z\cdot t} f(t)| = e^{-2\pi y\cdot t}|f(t)|$ is majorized by an integrable function when y is restricted to a compact subset of B. We note that this implies

$$F(z) = \int_{E_n} e^{2\pi i z\cdot t} f(t)\, dt \tag{2.2}$$

defines a holomorphic function in T_B.

It suffices to show that $e^{-2\pi y\cdot t}|f(t)|$ is majorized by an integrable function in a neighborhood of an arbitrary point $y_0 \in B$. Since B is open there clearly exists such a neighborhood $N \subset B$; moreover,

$$\int_{E_n} |f(t)|^2 e^{-4\pi y_0\cdot t} e^{-4\pi(y - y_0)\cdot t}\, dt \leqq A^2$$

whenever $y \in N$. Let us decompose E_n into a finite union of nonoverlapping polygonal cones, $\Gamma_1, \Gamma_2, \ldots, \Gamma_k$, having vertex at the origin 0 and such that whenever two points, v and w, belong to one of these cones, then the angle between the segments $0v$ and $0w$ is less than, say, $\pi/4$. Since N is a neighborhood of y_0 there exists $\delta > 0$ such that $\{y : |y - y_0| = \delta\} \subset N$. Let $\varepsilon = 4\pi\delta/\sqrt{2}$ and choose y such that $(y_0 - y) \in \Gamma_j$ and $|y - y_0| = \delta$; then $\varepsilon|t| \leqq -4\pi(y - y_0)\cdot t$ whenever $t \in \Gamma_j$. Consequently,

$$\int_{\Gamma_j} |f(t)|^2 e^{-4\pi y_0\cdot t} e^{\varepsilon|t|}\, dt \leqq \int_{\Gamma_j} |f(t)|^2 e^{-4\pi y_0\cdot t} e^{-4\pi(y - y_0)\cdot t}\, dt \leqq A^2.$$

Hence,

$$\int_{E_n} |f(t)|^2 e^{-4\pi y_0\cdot t} e^{\varepsilon|t|}\, dt = \sum_{j=1}^{k} \int_{\Gamma_j} |f(t)|^2 e^{-4\pi y_0\cdot t} e^{\varepsilon|t|}\, dt \leqq kA^2 < \infty.$$

Therefore,

$$\int_{E_n} |f(t)| e^{-2\pi y_0\cdot t} e^{(\varepsilon/4)|t|}\, dt$$

$$= \int_{E_n} (|f(t)| e^{(\varepsilon/2)|t|} e^{-2\pi y_0\cdot t}) e^{-(\varepsilon/4)|t|}\, dt$$

$$\leqq \left(\int_{E_n} |f(t)|^2 e^{\varepsilon|t|} e^{-4\pi y_0\cdot t}\, dt\right)^{1/2} \left(\int_{E_n} e^{-(\varepsilon/2)|t|}\, dt\right)^{1/2} < \infty.$$

Hence, for y lying in a sphere of radius $\varepsilon/8\pi$ about y_0 we have

$$|f(t)|e^{-2\pi y\cdot t} \leqq |f(t)|e^{-2\pi y_0\cdot t}e^{(\varepsilon/4)|t|},$$

which is an integrable function of t.

An immediate application of the Plancherel theorem then gives us, for $y \in B$,

$$\int_{E_n} |F(x + iy)|^2\, dx = \int_{E_n} |f(t)|^2 e^{-4\pi y\cdot t}\, dt \leqq A^2 < \infty.$$

Thus, the analytic function $F(z)$ defined by (2.2) belongs to $H^2(T_B)$. The basic representation theorem concerning functions in this space asserts that all such functions have this form. That is,

THEOREM 2.3. *F belongs to $H^2(T_B)$ if and only if it has the form* (2.2) *where f is a function satisfying* (2.1).

We shall illustrate the importance of this theorem by first deriving several of its consequences. We postpone its proof until the end of this section.

If $B \subset E_n$ we let B^c denote the *convex hull* of B; that is, B^c is the smallest convex set containing B. It is easy to see that B^c consists of all the finite sums $x = \sum \lambda_i x_i$, where $\sum \lambda_i = 1$, $\lambda_i \geqq 0$ and x_i belongs to B. It follows immediately from this that if B is open so is B^c. The following corollary shows that we might as well restrict ourselves to bases B that are convex.

COROLLARY 2.4. *If $F \in H^2(T_B)$ then the integral* (2.2) *is well-defined for all $z \in T_{B^c}$ and gives us a function in $H^2(T_{B^c})$ having the same norm as F.*

PROOF. We first note that the Plancherel theorem, together with Theorem 2.3, implies that

$$\|F\|_2 = \sup_{y\in B}\left(\int_{E_n} |f(t)|^2 e^{-4\pi y\cdot t}\, dt\right)^{1/2}. \tag{2.5}$$

Let

$$S = \left\{y \in E_n;\ \int_{E_n} |f(t)|^2 e^{-4\pi y\cdot t}\, dt \leqq \|F\|_2^2\right\}.$$

Clearly, $B \subset S$ and it suffices to show that S is convex. Suppose, therefore, that y' and y'' belong to S and $y = \alpha y' + (1 - \alpha)y''$, where $0 \leqq \alpha \leqq 1$. Using the inequality $u^\alpha v^{1-\alpha} \leqq \alpha u + (1 - \alpha)v$ (valid for any two nonnegative numbers u and v) we obtain

$$\begin{aligned}\int_{E_n} |f(t)|^2 e^{-4\pi y\cdot t}\,dt &= \int_{E_n} |f(t)|^2 e^{-4\pi\alpha y'\cdot t} e^{-4\pi(1-\alpha)y''\cdot t}\,dt \\ &= \int_{E_n} (|f(t)|^2 e^{-4\pi y'\cdot t})^{\alpha} (|f(t)|^2 e^{-4\pi y''\cdot t})^{1-\alpha}\,dt \\ &\leqq \alpha \int_{E_n} |f(t)|^2 e^{-4\pi y'\cdot t}\,dt + (1-\alpha)\int_{E_n} |f(t)|^2 e^{-4\pi y''\cdot t}\,dt \\ &\leqq \|F\|_2^2.\end{aligned}$$

Thus $y \in S$, which shows that S is convex. This proves the corollary.

From now on we shall assume the base B to be convex, as well as open.

COROLLARY 2.6. *A necessary and sufficient condition for $H^2(T_B)$ to contain a function that is not identically zero is that no entire straight line lies in B.*

PROOF. Suppose B contains the line consisting of all points y satisfying $y = a\tau + b$, $-\infty < \tau < \infty$. Let $N(t_0)$ be a spherical neighborhood in E_n about t_0 in which $a\cdot t$ is bounded away from 0. Then, for y on this line, $F \in H^2(T_B)$ and f satisfying (2.2) and (2.1) (the existence of such an f is guaranteed by Theorem 2.3)

$$\|F\|_2^2 \geqq \int_{E_n} |f(t)|^2 e^{-4\pi y\cdot t}\,dt \geqq \int_{N(t_0)} |f(t)|^2 e^{-4\pi\tau(a\cdot t)} e^{-4\pi(b\cdot t)}\,dt.$$

But, since we can choose τ arbitrarily on the real line, we can make the term $e^{-4\pi\tau(a\cdot t)}$ as large as we wish in $N(t_0)$. This shows that $f(t) = 0$ for almost every t in $N(t_0)$ and the first part of the corollary follows.

In order to see that if B contains no line then $H^2(T_B)$ contains a function $F \not\equiv 0$ we first observe that such a set B implies the existence of an open convex cone, Γ, containing it and such that Γ contains no entire straight lines.[3] It follows that Γ is regular (see §3), and for such cones, $H^2(T_\Gamma)$ does contain an $F \not\equiv 0$. This, therefore, will establish Corollary 2.6.

In view of the one-dimensional theory we would expect a central question in the theory of H^p spaces over a tube T_B to be the following one: *If y_0 is a boundary point of B does the limit*

$$F(x + iy_0) = \lim_{y\to y_0,\, y\in B} F(x + iy) \tag{2.7}$$

exist and, if so, in what sense?

A positive, but not altogether satisfactory, answer to this question can

[3] We sketch a proof of this geometric fact in (6.15) below.

be given. We first note that if y_0 is such a boundary point of B, Fatou's lemma and (2.5) imply

$$\int_{E_n} |f(t)|^2 e^{-4\pi y_0 \cdot t}\, dt \leqq \|F\|_2^2.$$

Thus, since $f(t)e^{-2\pi y_0 \cdot t}$ defines a function in $L^2(E_n)$, we can extend the definition of the integral (2.2) to $z = x + iy_0$ by taking the inverse Fourier transform of $f(t)e^{-2\pi y_0 \cdot t}$. That is, written formally, we obtain the almost everywhere defined L^2-function of x

$$F(x + iy_0) = \int_{E_n} e^{2\pi i(x + iy_0)\cdot t} f(t)\, dt. \tag{2.8}$$

In this sense, by taking the Fourier transforms of the functions $F_y = F(\cdot + iy)$, passing to the limit and then taking the inverse Fourier transform we can give meaning to the limit in (2.7).

It is not unreasonable, however, to expect, as is the case in one dimension, that $F(x + iy)$ tends to $F(x + iy_0)$ as $y \in B$ tends to y_0, either in the L^2-norm or for almost all x. We shall now show, however, that this is not the case in general.

Suppose that l is a line in E_2 and y_1 a point not on it. The equation of l can be written in the form $y \cdot a = \beta$ for some fixed vector a and real number β. E_2 is then split into two disjoint half-spaces: the set of all y such that $y \cdot a > \beta$ and the set of all y such that $y \cdot a < \beta$. Suppose y_1 belongs to the former. Then the function of two complex variables $z = (z_1, z_2) = (x + iy)$ defined by $G(z) = \exp\{-i\rho(z \cdot a - i\beta)\}$, $\rho > 0$, satisfies $|G(z)| = \exp\{\rho(y \cdot a - \beta)\}$, which is 1 for y on l, less than one in the half-space not containing y_1 and is equal to a number $N > 1$ if $y = y_1$. This N can be made as large as we wish by choosing ρ large enough. Moreover, $|G(z)| \leqq N$ if $z = x + iy$ satisfies $(y_1 - y) \cdot a \geqq 0$.

Now, suppose B is a disk in E_2, having 0 on its boundary and lying in the upper half-plane. Choose a sequence $\{y_k\}$ on the boundary of B that converges to 0 and let $\{\sigma_k\}$ be an accompanying sequence of sectors of B, each of which consists of a region between a line l_k cutting the boundary of B on both sides of y_k and the arc on this boundary that contains y_k and is determined by these two points of intersection. We further assume that l_k is so close to y_k that the σ_k's are mutually disjoint (see figure) and that l_k is parallel to the tangent to B at y_k. For each k we can construct a function, G_k, of the type described in the previous paragraph, satisfying

(i) *G_k is analytic in $\mathbf{C}_2$;*
(ii) *$|G_k(x + iy)|$ depends only on y;*
(iii) *$|G_k(z)| \leqq 1$ if $z \in T_B - T_{\sigma_k}$;*
(iv) *$|G_k(x + iy_k)| = 1 + 2^{k+2} = N_k$ and, for $z \in T_{\sigma_k}$, $|G_k(z)| \leqq N_k$.*

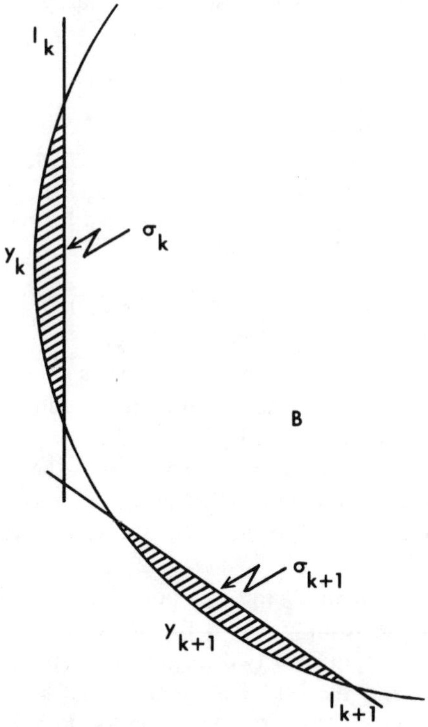

We then define a function F by letting

$$F(z) = \sum_{k=1}^{\infty} 2^{-k} G_k(z).$$

If $z \in T_B$ then it either belongs to precisely one of the tubes T_{σ_k}, say $z \in T_{\sigma_{k_0}}$, or it belongs to none of them; in the first case, by (iii) and in the second part of (iv), we have

$$|F(z)| \leqq \sum_{k=1}^{k_0-1} 2^{-k} + (2^{-k_0} + 4) + \sum_{k=k_0+1}^{\infty} 2^{-k} = 5,$$

while in the second case $|F(z)| \leqq \sum_{k=1}^{\infty} 2^{-k} = 1$. In particular, this shows that $F \in H^{\infty}(T_B)$. By (ii) and (iv) we can find points $y'_k \in \sigma_k$ so close to y_k that $|G_k(x + iy'_k)| > N_k - 1$, for $k = 1, 2, \ldots$. Thus,

$$|F(x + iy'_k)| \geqq 2^{-k}|G_k(x + iy'_k)| - \sum_{j \neq k} 2^{-j}|G_k(x + iy'_k)|$$

$$> 2^{-k}(N_k - 1) - \sum_{j \neq k} 2^{-j} > 4 - 1 = 3.$$

This shows that if we approach the boundary point 0 with a sequence that avoids all sectors σ_k we have $|F(x + iy)| \leqq 1$ for all y in this sequence. On the other hand $\lim_{k\to\infty} y'_k = 0$ and $|F(x + iy'_k)| > 3$. In particular, this shows that the pointwise limit

$$\lim_{y\in B,\, y\to 0} F(x + iy)$$

cannot exist for *any* $x \in E_2$.

To give an example of a function in $H^2(T_B)$ that does not have a limit in the L^2-norm as we approach 0 arbitrarily within B it suffices to find a function $G \in H^2(T_{B'})$, where $\bar{B} \subset B'$, such that $G(x + i0) = G(x) \not\equiv 0$ and multiply it by the F just obtained. Then $\int_{E_n} |F(x + iy)G(x + iy)|^2\, dx \geqq 9 \int_{E_n} |G(x + iy)|^2\, dx$ whenever $y = y'_k$, $k = 1, 2, \ldots,$ while

$$\int_{E_n} |F(x + iy)G(x + iy)|^2\, dx \leqq \int_{E_n} |G(x + iy)|^2\, dx$$

if $y \in B$ belongs to none of the σ_k. Thus, the limit

$$\lim_{y\in B,\, y\to 0} F(x + iy)G(x + iy)$$

cannot exist in the L^2-norm. $G(z) = 1/(z_1 + i)(z_2 + i)$ for $z = (z_1, z_2) = (x_1 + iy_1, x_2 + iy_2)$, $y_1 > -\frac{1}{2}$, $y_2 > -\frac{1}{2}$, is an example of such a function. It is clearly defined and analytic in a tube $T_{B'}$ with $\bar{B} \subset B'$; furthermore,

$$\int_{E_2} |G(x + iy)|^2\, dx = \int_{-\infty}^{\infty}\int_{-\infty}^{\infty} \frac{1}{|x_1 + i(y_1 + 1)|^2|x_2 + i(y_2 + 1)|^2}\, dx_1\, dx_2$$

$$\leqq \left\{ \int_{-\infty}^{\infty} \frac{1}{x^2 + \frac{1}{4}}\, dx \right\}^2 < \infty.$$

Thus, $G \in H^2(T_{B'})$.

If we specialize B, however, we do obtain L^2-limits as we approach a boundary point. We define an *open polyhedron* in E_n to be the interior of the convex hull of a finite subset of E_n. The following result asserts, in particular, that when B is an open polyhedron the limits (2.7) exist in the L^2-sense at all boundary points of B:

COROLLARY 2.9. *Let P be an open polyhedron in E_n and $F \in H^2(T_P)$. If we then extend the definition of F to the set $T_{\bar{P}}$ by (2.8) the mapping $y \to F(x + iy)$, from $\bar{P}$ to $L^2(E_n)$, is continuous.*

PROOF. By the Plancherel theorem it is sufficient to show that the mapping $y \to f(t)e^{-2\pi y\cdot t}$ is continuous. Suppose $\bar{P}$ is the convex hull of the finite set $\{y_1, y_2, \ldots, y_k\} \subset E_n$. Let

$$G(t) = \sum_{j=1}^{k} e^{-4\pi y_j \cdot t} |f(t)|^2.$$

The function G is clearly integrable in E_n. Moreover, it majorizes $e^{-4\pi y \cdot t}|f(t)|^2$ for each y in $\bar{P}$. For, if $y \in \bar{P}$, then $y = \alpha_1 y_1 + \alpha_2 y_2 + \cdots + \alpha_k y_k$, where the α_j's are nonnegative numbers whose sum is 1; but this implies that

$$e^{-4\pi y \cdot t} = \exp\left\{-4\pi \sum_{j=1}^{k} \alpha_j (y_j \cdot t)\right\} = \prod_{j=1}^{k} (e^{-4\pi (y_j \cdot t)})^{\alpha_j} \leqq \sum_{j=1}^{k} \alpha_j e^{-4\pi y_j \cdot t}.$$

Now, for $y, \bar{y} \in \bar{P}$, $y \to \bar{y}$ we have

$$|f(t)e^{-2\pi y \cdot t} - f(t)e^{-2\pi \bar{y} \cdot t}|^2 \to 0;$$

moreover, this convergence is dominated by $4G(t)$. The corollary now follows from the Lebesgue dominated convergence theorem.

COROLLARY 2.10. *Let B be an open convex subset of E_n and y_0 a point on its boundary. Suppose $F \in H^2(T_B)$ and P is an open polyhedron contained in B having y_0 as a boundary point, then $F(x + iy)$ tends to $F(x + iy_0)$ in the L^2-norm as y tends to y_0 within P, where $F(x + iy_0)$ is the function defined by* (2.8).

Since $H^2(T_B) \subset H^2(T_P)$ this is just a special case of Corollary 2.9.

It is not hard to see that the counter example we have just given can be modified in such a way as to include each boundary point that is not a *polygonal boundary point*; by this term we mean a point y_0 that is the point of intersection of two straight line segments making up part of the boundary ∂B of B (we do not exclude the possibility that these two segments lie on the same line; in this case, y_0 would be interior to a single line segment that lies in ∂B). If y_0 is not a polygonal boundary point, it follows from the convexity of B that we can find a sequence $\{y_j\}$ in ∂B converging to y_0 and such that the open segments $\overline{y_j y_0}$ lie in the interior of B. This clearly allows us to find a sequence $\{\sigma_k\}$ of mutually disjoint sectors with accompanying analytic functions $\{G_k\}$ enabling us to construct a function F just as we did in our counter example. This, together with Corollary 2.9, gives us the following necessary and sufficient conditions for the limits (2.7) to exist in the L^2 sense, when the dimension is two:

THEOREM 2.11. *If B is an open convex subset of E_2 and $y_0 \in \partial B$ then*

$$\lim_{y \to y_0,\, y \in B} F(x + iy) = F(x + iy_0)$$

exists in L^2 for every $F \in H^2(T_B)$ if and only if y_0 is a polygonal boundary point of B.

In general, if the limits in (2.7) exist (in the L^2 sense, or for almost every x, or in $L^p, \ldots$) we shall refer to them as *unrestricted limits* (in L^2, or a.e., or in $L^p, \ldots$). If such limits exist whenever y converges to y_0 within a polyhedron in B having y_0 as a boundary point we say that (L^2, or a.e., or $L^p, \ldots$) *restricted* limits exist at y_0. In this terminology, Corollary 2.10 asserts that L^2 restricted limits exist at all boundary points for all H^2 functions; Theorem 2.11, on the other hand, gives necessary and sufficient conditions, when $n = 2$, for the existence of L^2 unrestricted limits.

We now pass to the proof of Theorem 2.3. We begin with a lemma asserting, in particular, that if $F \in H^2(T_B)$ then F is uniformly bounded in "strictly smaller" tubes. We state and prove this lemma for the wider class of spaces $H^p(T_B)$ (see (1.5)) since we will need this more general result later on. More precisely,

LEMMA 2.12. *Suppose $F \in H^p(T_B)$, $p > 0$, and $B_0 \subset B$ satisfies $d(B_0, {}^cB) = \inf\{|y_1 - y_2|; y_1 \in B_0, y_2 \notin B\} \geqq \varepsilon > 0$ then there exists a constant $C = C(\varepsilon, n)$, depending on ε and n but not on F, such that*

$$\sup_{z \in T_{B_0}} |F(z)| \leqq C\|F\|_p.$$

PROOF. Let $z_0 = x_0 + iy_0 \in T_{B_0}$ and $S_\varepsilon = \{z \in \mathbf{C}_n; |z - z_0| < \varepsilon\}$. If $\Sigma_\varepsilon = \{y \in E_n; |y - y_0| < \varepsilon\}$ then $S_\varepsilon \subset T_{\Sigma_\varepsilon} \subset T_B$. Hence, recalling that Ω_m denotes the volume of the unit ball of E_m, $m \geqq 1$,

$$\left(\int_{S_\varepsilon} |F(z)|^p \, dx \, dy\right)^{1/p} \leqq \left(\int_{T_{\Sigma_\varepsilon}} |F(z)|^p \, dx \, dy\right)^{1/p}$$
$$= \left(\int_{\Sigma_\varepsilon}\left\{\int_{E_n} |F(x + iy)|^p \, dx\right\} dy\right)^{1/p}$$
$$\leqq \|F\|_p(\Omega_n \varepsilon^n)^{1/p}.$$

On the other hand, since $|F|^p$ is subharmonic as a function of $2n$ variables (see example (3) in Section 4 and Theorem 4.4′ in Chapter II),

$$|F(z_0)|^p \leqq \Omega_{2n}^{-1}\varepsilon^{-2n}\left(\int_{S_\varepsilon} |F(z)|^p \, dx \, dy\right).$$

Combining these inequalities we obtain $|F(z_0)| \leqq C\|F\|_p$, with $C = (\Omega_n/\Omega_{2n})^{1/p}\varepsilon^{-n/p}$.

In order to establish Theorem 2.3 we need to show that if $F \in H^2(T_B)$ then there exists a function f satisfying (2.1) giving us the representation (2.2) for F (the converse has already been established).

Let f_y, for $y \in B$, be the Fourier transform of $F(x + iy)$ considered as a function of x. It clearly suffices to show that if $y, y' \in B$ then $e^{2\pi y' \cdot t} f_{y'}(t) = e^{2\pi y \cdot t} f_y(t)$, a.e. in t. For, then, $f(t) = e^{2\pi y \cdot t} f_y(t)$ is almost everywhere defined independently of $y \in B$, f satisfies (2.1) with $A = \|F\|$ and (2.2) must hold (see the argument preceding Theorem 2.3). It clearly suffices to show this when y, y' belong to a cube Q whose closure lies in B and whose sides are parallel to the coordinate axes. We further assume, for the moment, that $|F(x + iy)|$, as a function of x, is majorized by a function vanishing sufficiently rapidly at ∞, uniformily in $y \in Q$. Lastly, we remark that, by an iteration argument, it suffices to consider the case where the two points $y, y' \in Q$ have the form $y = (\eta_1, y_2, \ldots, y_n)$ and $y' = (\eta_1', y_2, \ldots, y_n)$. Suppose $\eta_1' \geqq \eta_1$, then by the Cauchy integral theorem,

$$\begin{aligned} 0 = &\int_{-R}^{R} e^{-2\pi i(x_1 + i\eta_1)t_1} F(x_1 + i\eta_1, \ldots, x_n + iy_n)\, dx_1 \\ &+ \int_{\eta_1}^{\eta_1'} e^{-2\pi i(R + i\eta)t_1} F(R + i\eta, \ldots)\, d\eta \\ &+ \int_{R}^{-R} e^{-2\pi i(x_1 + i\eta_1')t_1} F(x_1 + i\eta_1', \ldots)\, dx_1 \\ &+ \int_{\eta_1'}^{\eta_1} e^{-2\pi i(-R + i\eta)t_1} F(-R + i\eta, \ldots)\, d\eta. \end{aligned}$$

Because of our assumptions, the 2nd and 4th terms tend to 0 as $R \to \infty$. Thus, after an integration in $x_2, \ldots, x_n$, we get

$$\begin{aligned} e^{2\pi y \cdot t} f_y(t) &= \int_{E_n} e^{-2\pi i(x + iy)\cdot t} F(x + iy)\, dx \\ &= \int_{E_n} e^{-2\pi i(x + iy')\cdot t} F(x + iy')\, dx \\ &= e^{2\pi y' \cdot t} f_{y'}(t). \end{aligned}$$

We now show how the case of the general $F \in H^2(T_B)$ follows from this one. By Lemma 2.12 F is bounded on T_Q, say $|F(z)| \leqq M$ for $z \in T_Q$. Thus, if we define $F^{(\varepsilon)}$, for $\varepsilon > 0$, by letting

$$F^{(\varepsilon)}(z) = e^{\left\{-\varepsilon \sum_{j=1}^{n} z_j^2\right\}} F(z),$$

for $z = (z_1, \ldots, z_n) \in T_Q$, we have

$$|F^{(\varepsilon)}(x + iy)| \leqq M e^{\varepsilon n a^2} e^{-|x|^2 \varepsilon}$$

for $y \in Q$, where $a = \max_{y \in Q} \{|y_1|, \ldots, |y_n|\}$. The argument just presented, therefore, applies to the Fourier transforms $f_y^{(\varepsilon)}$ of $F^{(\varepsilon)}(x + iy)$ and we have

$$e^{2\pi y\cdot t}f_y^{(\varepsilon)}(t) = e^{2\pi y'\cdot t}f_{y'}^{(\varepsilon)}(t) \tag{2.13}$$

for $y, y' \in Q$. But $\int_{E_n} |F^{(\varepsilon)}(x + iy) - F(x + iy)|^2\, dx \to 0$ as $\varepsilon \to 0$. Thus, the equality (2.13) is preserved in the limit if we let $\varepsilon \to 0$. That is, we can apply the Plancherel theorem to obtain the L^2-convergences $f_y^{(\varepsilon)} \to f_y$ and $f_{y'}^{(\varepsilon)} \to f_{y'}$. In particular

$$e^{2\pi y\cdot t}f_y(t) = e^{2\pi y'\cdot t}f_{y'}(t)$$

for almost every t. This proves the theorem.

3. Tubes Over Cones

It is natural to expect that, if we restrict the type of base, the theory of H^2 spaces (or, more generally, H^p spaces) becomes richer. This is particularly true if we consider tubes over bases that are *open cones*. By the latter we mean nonempty subsets $\Gamma \subset E_n$ satisfying (i) $0 \notin \Gamma$, and (ii) whenever $x, y \in \Gamma$ and $\alpha, \beta > 0$ then $\alpha x + \beta y \in \Gamma$. We note that, in particular, Γ is convex.

A *closed cone* is the closure of an open cone. It is clear that if Γ is an open cone then $\Gamma^* = \{x \in E_n;\, x\cdot t \geqq 0,\, t \in \Gamma\}$ is closed. Moreover, if Γ^* has a nonvoid interior it is a closed cone; in this case we say that Γ is a *regular* cone. Γ^* is called the cone *dual* to Γ. When $n = 1$ the open cones are the half-lines $\{x \in E_1;\, x > 0\}$ and $\{x \in E_1;\, x < 0\}$. When $n = 2$ the open cones are the angular regions between two rays meeting at the origin and forming an angle that is less than or equal to π. Such a cone will be regular if and only if this angle is strictly less than π.

When the base is a cone one obtains the following representation theorem for H^2-functions (a sharper result than Theorem 2.3):

THEOREM 3.1. *Suppose Γ is an open cone. Then $F \in H^2(T_\Gamma)$ if and only if*

$$F(z) = \int_{\Gamma^*} e^{2\pi i z\cdot t}f(t)\, dt \tag{3.2}$$

where f is a measurable function on E_n satisfying

$$\int_{\Gamma^*} |f(t)|^2\, dt < \infty.$$

Furthermore,

$$\|F\|_2 = \left(\int_{\Gamma^*} |f(t)|^2\, dt\right)^{1/2}.$$

Consequently, the correspondence $F \leftrightarrow f$ is a unitary linear mapping of

$H^2(T_\Gamma)$ onto $L^2(\Gamma^*)$. In particular, $H^2(T_\Gamma)$ contains a function that is not identically zero if and only if Γ is regular.

PROOF. Since $y \cdot t \geqq 0$ whenever $y \in \Gamma$ and $t \in \Gamma^*$, the fact that $F \in H^2(T_\Gamma)$ when it has the representation (3.2), with $f \in L^2(\Gamma^*)$, is a consequence of Theorem (2.3).

Conversely, suppose $F \in H^2(T_\Gamma)$. Then, by Theorem 2.3 and (2.5),

$$F(z) = \int_{E_n} e^{2\pi i z \cdot t} f(t)\, dt,$$

where

$$\|F\|_2^2 = \sup_{y \in \Gamma} \int_{E_n} e^{-4\pi y \cdot t} |f(t)|^2\, dt.$$

We shall show that f vanishes a.e. in the complement of Γ^*. If $t_0 \notin \Gamma^*$ then there exists $y_0 \in \Gamma$ such that $y_0 \cdot t_0 < 0$. Thus, there exists a neighborhood of t_0, $N = N(t_0) \subset E_n - \Gamma^*$, and a $\delta > 0$ such that $y_0 \cdot t < -\delta < 0$ for $t \in N$. Hence, $(ky_0) \cdot t < -k\delta$ for $t \in N$ and positive k. Since $ky_0 \in \Gamma$ we also have $\int_N e^{-4\pi k y_0 \cdot t} |f(t)|^2\, dt \leqq \int_{E_n} e^{-4\pi k y_0 \cdot t} |f(t)|^2\, dt \leqq \|F\|_2^2 < \infty$. But this implies that

$$\int_N e^{4\pi k \delta} |f(t)|^2\, dt \leqq \|F\|_2^2 < \infty$$

for all $k > 0$, and this clearly implies that $f(t) = 0$ for almost every t in $N(t_0)$. It follows that $f(t) = 0$ for almost all t outside Γ^*.[4] The other parts of the statement of Theorem 3.1 are now obvious.

In the classical case, when F belongs to the H^2 space associated with the upper half plane (i.e., when the base is the cone of positive real numbers) it is easy to establish the Cauchy integral representation

$$F(z) = \frac{1}{2\pi i} \int_{E_1} \frac{F(\zeta)}{\zeta - z}\, d\zeta \tag{3.3}$$

of F in terms of its boundary values. We shall now show that this representation can be extended to include the case when the base is a cone, Γ, in E_n; moreover, this representation involves only those boundary values obtained as $y \in \Gamma$ tends to the vertex of Γ (i.e., the origin, 0, of E_n). The fact that these boundary values exist in the L^2 sense is an easy consequence of Theorem 3.1:

COROLLARY 3.4. *Suppose Γ is an open cone in E_n and $F(x + iy) \in H^2(T_\Gamma)$ then there exists a function $F(x)$, defined on E_n, such that $F(x + iy) \to F(x)$ in the L^2-norm as $y \in \Gamma$ tends to 0.*

[4] Observe that this part of the proof is very similar to that of Corollary 2.6.

PROOF. Using the notation of Theorem 3.1 we define $F(x)$ to be the inverse Fourier transform of f (which we may think of as defined on E_n and 0 outside of Γ^*); thus, formally,

$$F(x) = \int_{\Gamma^*} e^{2\pi i x \cdot t} f(t)\, dt.$$

But (3.2) asserts that $F(x + iy)$ is the inverse Fourier transform of $e^{-2\pi y \cdot t} f(t)$; therefore, since the latter converges to $f(t)$ in the L^2-norm as $y \to 0$, the desired conclusion follows from the Plancherel theorem.

The extension of (3.3) described above will be obtained by means of the following kernel function associated with a tube domain: For $z = x + iy$ in T_Γ we let

$$K(z) = \int_{\Gamma^*} e^{2\pi i z \cdot t}\, dt = \int_{\Gamma^*} e^{2\pi i x \cdot t} e^{-2\pi y \cdot t}\, dt.$$ [5]

We thus obtain a continuous function K defined on T_Γ, called the *Cauchy kernel associated with the tube* T_Γ. As a function of $x = \text{Re}\,\{z\}$, K belongs to $L^2(E_n)$; in fact, it follows from Plancherel's theorem that

$$(3.5) \qquad \int_{E_n} |K(x + iy)|^2\, dx = \int_{\Gamma^*} e^{-4\pi y \cdot t}\, dt = K(2iy)$$

for all $y \in \Gamma$. (We shall see momentarily that $K(2iy)$ is finite.)

THEOREM 3.6. *If* $F \in H^2(T_\Gamma)$ *then*

$$F(z) = \int_{E_n} K(z - \xi) F(\xi)\, d\xi$$

for all $z \in T_\Gamma$, *where* $F(\xi) = \lim_{\eta \to 0,\, \eta \in \Gamma} F(\xi + i\eta)$ *is the limit function of Corollary* 3.4.

PROOF. From Theorem 3.1 and 3.4 we have

$$F(z) = F(x + iy) = \int_{\Gamma^*} e^{2\pi i z \cdot t} f(t)\, dt$$

where f is the Fourier transform of $F(\xi) = \lim_{\eta \to 0,\, \eta \in \Gamma} F(\xi + i\eta)$. That is, f is the limit in the L^2 norm of the sequence of functions

$$f_k(t) = \int_{|\xi| \leq k} F(\xi) e^{-2\pi i t \cdot \xi}\, d\xi,$$

[5] When Γ is the cone of positive real numbers then $K(z) = -1/2\pi i z$ and (3.3) has the form $F(z) = \int_{E_1} K(z - \xi) F(\xi)\, d\xi$. Theorem 3.6 shows that this formula holds for functions in a general tube T_Γ whose base is a cone.

$k = 1, 2, 3, \ldots$. Consequently, using Fubini's theorem and the fact that $K(x - \xi + iy)$, as a function of ξ, belongs to $L^2(E_n)$ (see (3.5)), we obtain

$$\begin{aligned} F(z) &= \lim_{k \to \infty} \int_{\Gamma^*} e^{2\pi i z \cdot t} f_k(t)\, dt \\ &= \lim_{k \to \infty} \int_{\Gamma^*} e^{2\pi i z \cdot t} \left\{ \int_{|\xi| \leqq k} F(\xi) e^{-2\pi i t \cdot \xi}\, d\xi \right\} dt \\ &= \lim_{k \to \infty} \int_{|\xi| \leqq k} F(\xi) \left\{ \int_{\Gamma^*} e^{2\pi i (z - \xi) \cdot t}\, dt \right\} d\xi \\ &= \int_{E_n} F(\xi) K(z - \xi)\, d\xi, \end{aligned}$$

and the theorem is proved.

In the classical one-dimensional case the Poisson kernel

$$P(x, y) = \frac{1}{\pi} \frac{y}{x^2 + y^2}$$

can be expressed in terms of the Cauchy kernel associated with the upper half-plane in the following simple way: For $z = x + iy$, $y > 0$,

$$P(x, y) = |K(z)|^2 / K(2iy).$$

It is natural, therefore, to extend this definition whenever the Cauchy kernel is defined. Thus, if T_Γ is a tube whose base Γ is a regular cone and K is the corresponding Cauchy kernel we define the *Poisson kernel associated with* T_Γ by putting

$$\mathscr{P}(x, y) = |K(x + iy)|^2 / K(2iy)$$

whenever $z = x + iy \in \Gamma$.

We have shown that $K(x + iy)$, considered as a function of x, belongs to $L^2(E_n)$; hence, $\mathscr{P}(\cdot, y)$ must belong to $L^1(E_n)$ for each y in Γ. Moreover, it is not hard to show that $\mathscr{P}(\cdot, y)$ belongs to $L^\infty(E_n)$. In order to do this it clearly suffices to show that $K(x + iy)$ is bounded independently of x for each y in Γ. But

$$|K(x + iy)| = \left| \int_{\Gamma^*} e^{2\pi i (x + iy) \cdot t}\, dt \right| \leqq \int_{\Gamma^*} e^{-2\pi y \cdot t}\, dt = K(iy).$$

Thus, it suffices to show that $K(iy)$ is finite, if $y \in \Gamma$. In order to do this we first claim that, if $y \in \Gamma$, then there exists a $\delta = \delta_y > 0$ such that $\delta |t| \leqq y \cdot t$ for all t in Γ^*. It is enough to show this when $t \in \Gamma^*$ satisfies

$|t| = 1$. From the definition of Γ^* we have $0 \leqq y \cdot t$. On the other hand, equality is not possible; otherwise, since Γ is open, we could find a $u \in E_n$ of small enough absolute value to guarantee that $y + u \in \Gamma$ and such that $(y + u)\cdot t = u \cdot t < 0$, contradicting the fact that $t \in \Gamma^*$. Since the intersection of Γ^* with the surface, Σ, of the unit sphere of E_n is compact, the existence of $\delta_y > 0$ now follows from the fact that $0 < y \cdot t$ for all $t \in \Gamma^* \cap \Sigma$. Consequently,

$$\int_{\Gamma^*} e^{-2\pi y \cdot t}\, dt \leqq \int_{\Gamma^*} e^{-2\pi\delta|t|}\, dt < \infty$$

for $y \in \Gamma$. Since $L^q(E_n) \supset L^1(E_n) \cap L^\infty(E_n)$ for $1 \leqq q \leqq \infty$,[6] we have the result.

COROLLARY 3.7. *For each $y \in \Gamma$, $\mathscr{P}(\cdot, y) \in L^q(E_n)$ whenever $1 \leqq q \leqq \infty$. It follows that if $f \in L^p(E_n)$, $1 \leqq p \leqq \infty$, then*

$$u(x + iy) = \int_{E_n} f(x - t)\mathscr{P}(t, y)\, dt$$

is defined for all $z = x + iy \in T_\Gamma$. As was the case with the Poisson integrals introduced in Chapter I, it can be shown that $u(x + iy)$ tends to $f(x)$ in the L^p norm. That is,

$$\lim_{y \to 0,\, y \in \Gamma} \int_{E_n} |u(x + iy) - f(x)|^p\, dx = 0. \tag{3.8}$$

This is an immediate consequence of the fact that the kernel $\mathscr{P}$ is an approximation of the identity (see the footnote following the end of the proof of Theorem 2.1 of Chapter II and the last part of the proof of Theorem 5.6 below). We mean by this that the following three properties are satisfied by $\mathscr{P}$:

(i) $\mathscr{P}(x, y) \geqq 0$;

(ii) $\int_{E_n} \mathscr{P}(x, y)\, dx = 1$ *for all* $y \in \Gamma$;

(iii) *if* $\delta > 0$ *then* $\int_{|x| > \delta} \mathscr{P}(x, y)\, dx \to 0$ *as* $y \in \Gamma$ *tends to* 0.

Property (i) is obvious. The second property follows from (3.5) after dividing both sides by $K(2iy)$. In order to prove (iii) it suffices to find a function ψ such that

(a) ψ is continuous on E_n,

(b) $\lim_{y \in \Gamma,\, y \to 0} \int_{E_n} \mathscr{P}(x, y)\psi(x)\, dx = 1$,

(c) $|\psi(x)| < 1$ when $x \neq 0$ and $\psi(x) \to 0$ as $|x| \to \infty$.

[6] If $f \in L^1 \cap L^\infty$ and $1 < q < \infty$ then $\int |f|^q = \int |f|^{q-1}|f| \leqq \|f\|_\infty^{q-1}\|f\|_1 < \infty$.

For, if such a function exists, then, by (b),

$$1 = \lim_{y\in\Gamma,\, y\to 0}\left\{\int_{|x|\leqq\delta} \psi(x)\mathscr{P}(x, y)\, dx + \int_{|x|>\delta} \psi(x)\mathscr{P}(x, y)\, dx\right\}.$$

From (a) and (c) we know that there exists $\varepsilon > 0$ such that $|\psi(x)| \leqq 1 - \varepsilon$ if $|x| > \delta$. Thus (using (i) and (ii), as well),

$$1 \leqq \lim_{y\in\Gamma,\, y\to 0}\left\{\int_{|x|\leqq\delta} \mathscr{P}(x, y)\, dx + (1-\varepsilon)\int_{|x|>\delta} \mathscr{P}(x, y)\, dx\right\}$$

$$= \lim_{y\in\Gamma,\, y\to 0}\left\{\int_{E_n} \mathscr{P}(x, y)\, dx - \varepsilon\int_{|x|>\delta} \mathscr{P}(x, y)\, dx\right\}$$

$$= \lim_{y\in\Gamma,\, y\to 0}\left\{1 - \varepsilon\int_{|x|>\delta} \mathscr{P}(x, y)\, dx\right\}.$$

But this clearly implies (iii).

The existence of such a function ψ will be an easy consequence of the following representation theorem:

THEOREM 3.9. *If $F \in H^2(T_\Gamma)$ then*

$$F(z) = \int_{E_n} \mathscr{P}(x - t, y)F(t)\, dt$$

for all $z = x + iy$ in T_Γ.

PROOF. Let $w = u + iv$ be a point of T_Γ. Then, for $z \in T_\Gamma$,

$$|K(z + w)| = \left|\int_{\Gamma^*} e^{2\pi i(x+u)\cdot t}e^{-2\pi(y+v)\cdot t}\, dt\right| \leqq \int_{\Gamma^*} e^{-2\pi v\cdot t}\, dt = M_v < \infty.$$

Thus, $F(z)K(z + w)$, as a function of z, belongs to $H^2(T_\Gamma)$ and has norm not exceeding $\|F\| M_v$. We can thus apply Theorem 3.6 to this function and obtain

$$F(z)K(z + w) = \int_{E_n} K(z - t)F(t)K(t + w)\, dt, \tag{3.10}$$

for all $z \in T_\Gamma$.

If we let $w = -x + iy$ then $K(z - t)K(t + w) = |K(z - t)|^2$ and $K(z + w) = K(2iy)$; therefore (3.10) in this case is equivalent to

$$F(z) = \int_{E_n} F(t)\{|K(z - t)|^2/K(2iy)\}\, dt = \int_{E_n} F(t)\mathscr{P}(x - t, y)\, dt,$$

which is the result we are seeking.

We now pass to the construction of the function ψ. Choose $\varphi \geqq 0$ that is continuous, has compact support contained in Γ^* and such that $\int_{E_n} \varphi(t)\,dt = 1$ (this can be done since Γ is regular). We claim that

$$\psi(x) = \int_{E_n} e^{2\pi i x\cdot t}\varphi(t)\,dt = \int_{\Gamma^*} e^{2\pi i x\cdot t}\varphi(t)\,dt$$

satisfies (a), (b), and (c). Since $\varphi \in L^1(E_n)$, property (a) is certainly true (see Theorem 1.1, part (b), in the first chapter). The fact that $\psi(x)$ tends to 0 as $|x|$ tends to ∞ is a special case of the Riemann–Lebesgue theorem. If $|\psi(x)| = 1$, say $\psi(x) = e^{2\pi i\theta}$, then

$$1 = \int_{E_n} e^{2\pi i[(x\cdot t)-\theta]}\varphi(t)\,dt = \int_{E_n} \varphi(t)\cos 2\pi[(x\cdot t) - \theta]\,dt$$

(the imaginary part of the first integral must vanish since 1 is a real number). If $x \neq 0$ then $\cos 2\pi[(x\cdot t) - \theta]$, as a function of t, must be strictly less than 1 in a subset of the support of φ having positive measure. This, together with the assumptions that φ is nonnegative and

$$\int_{E_n} \varphi(t)\,dt = 1,$$

implies $|\psi(x)| < 1$ if $x \neq 0$. In order to show that (b) holds we consider

$$F(z) = \int_{\Gamma^*} e^{2\pi i z\cdot t}\varphi(t)\,dt$$

for $z \in T_\Gamma$. By Theorem 3.1 F belongs to $H^2(T_\Gamma)$ and it is clear that

$$F(x) = \lim_{y\in\Gamma,\, y\to 0} F(x+iy) = \int_{\Gamma^*} e^{2\pi i x\cdot t}\varphi(t)\,dt = \psi(x)$$

in the L^2-norm; moreover, by the Lebesgue dominated convergence theorem,

$$F(0) = \lim_{y\in\Gamma,\, y\to 0} F(0+iy) = \psi(0) = 1. \tag{3.11}$$

Applying Theorem 3.9 to this function F we have

$$F(x+iy) = \int_{E_n} \mathscr{P}(x-t, y)F(t)\,dt = \int_{E_n} \mathscr{P}(x-t, y)\psi(t)\,dt. \tag{3.12}$$

Since $\mathscr{P}(-t, y) = \mathscr{P}(t, y)$, (3.12) (with $x = 0$) together with (3.11) gives us property (b).

4. The Paley–Wiener Theorem

This section will be devoted to an application of the theory of H^2 spaces. We shall show how the classical Paley–Wiener theorem can be extended to n-dimensions and proved by using some of the results we have established in this chapter. We begin by discussing the one-dimensional case.

Suppose F is an entire function on the complex plane $\mathbf{C}_1$. We say that F is of *exponential type* $\sigma > 0$ if for each $\varepsilon > 0$ there exists a constant A_ε such that

$$|F(z)| \leqq A_\varepsilon e^{(\sigma+\varepsilon)|z|}$$

for all $z \in \mathbf{C}_1$. An example of such a function can be obtained in the following way: Suppose $f \in L^2(-\tau, \tau)$, we then define an entire function, F, by letting

$$F(z) = \int_{-\tau}^{\tau} f(t) e^{2\pi i z t}\, dt$$

for $z \in \mathbf{C}_1$. Then

$$|F(z)| = |F(x+iy)| \leqq \sqrt{2\tau}\left(\int_{-\tau}^{\tau} |f(t)|^2\, dt\right)^{1/2} e^{2\pi\tau|y|} \leqq A e^{2\pi\tau|z|}.$$

Hence, F is of exponential type $\sigma = 2\pi\tau$. The Paley–Wiener theorem asserts that the converse of this is true, if we know that F restricted to the real axis belongs to $L^2(-\infty, \infty)$. More precisely:

THEOREM 4.1. *Suppose $F \in L^2(-\infty, \infty)$. Then F is the Fourier transform of a function vanishing outside $[-(\sigma/2\pi), \sigma/2\pi] = [-\tau, \tau]$ if and only if F is the restriction to the real axis of an entire function of exponential type σ.*

We have just shown the "only if" part of this theorem.[7] The "if" part will be established with the help of the following *Phragmén–Lindelöf* type result (compare with 5.2 in Chapter II):

LEMMA 4.2. *Let S be the region in $\mathbf{C}_1$ bounded by two rays meeting at the origin at an angle π/α. Suppose f is analytic on $\bar{S}$ and satisfies $|f(z)| \leqq A \exp\{|z|^\beta\}$ for $0 \leqq \beta < \alpha$ and $z \in S$. Then, the condition $|f(z)| \leqq M$ on the two bounding rays implies $|f(z)| \leqq M$ for all $z \in S$.*

[7] According to the notation established in the first chapter, we have shown that if F is the inverse Fourier transform of a function f vanishing outside the interval $[-\tau, \tau]$ then it is the restriction to the real axis of a function of exponential type σ. This is obviously equivalent to having F the Fourier transform of the function g whose values are given by $g(t) = f(-t)$.

PROOF. By rotating, if necessary, we can assume that the rays form angles $\pi/2\alpha$ and $-\pi/2\alpha$ with the real axis. Let $F(z) = f(z)\exp\{-\varepsilon z^{\gamma}\}$ where $\beta < \gamma < \alpha$ and $\varepsilon > 0$. It follows that $|F(z)| \leqq |f(z)| \leqq M$ on the two bounding rays. Moreover, on the arc $R = |z| = |re^{i\theta}|$, $-(\pi/2\alpha) \leqq \theta \leqq \pi/2\alpha$, $|F(z)| \leqq A\exp\{R^{\beta} - \varepsilon R^{\gamma}\cos\gamma\pi/2\alpha\}$. But the last expression tends to 0 as $R \to \infty$. Thus, $|F(z)| \leqq M$ on this arc provided R is large enough. The maximum-modulus principle, thus, implies $|F(z)| \leqq M$ for all $z \in \bar{S}$ with $|z| \leqq R$. Since R can be chosen arbitrarily large this shows that $|F(z)| \leqq M$ for all $z \in \bar{S}$. Hence, $|f(z)| \leqq M\exp\{\varepsilon r^{\gamma}\cos\gamma\theta\}$ for all $z = re^{i\theta} \in \bar{S}$. Letting $\varepsilon \to 0$ we obtain our result.

LEMMA 4.3. *Suppose F is of exponential type σ and $|F(x)| \leqq 1$ for x real then $|F(x + iy)| \leqq \exp\{\sigma|y|\}$ for all complex numbers $z = x + iy$.*

PROOF. For $\varepsilon > 0$ set $F_{\varepsilon}(z) = F(z)e^{i(\sigma+\varepsilon)z}$. Since F is of exponential type σ

$$|F_{\varepsilon}(iy)| = |F(iy)|e^{-(\sigma+\varepsilon)y} \leqq A_{\varepsilon},$$

for all nonnegative y. We also have $|F_{\varepsilon}(x)| \leqq 1$ for all real x. This gives us a bound for F on the positive x and y axes. Moreover, we certainly can find B so that

$$|F_{\varepsilon}(z)| \leqq A_{\varepsilon}e^{(\sigma+\varepsilon)(|z|-y)} \leqq A_{\varepsilon}e^{2(\sigma+\varepsilon)|z|} \leqq Be^{|z|^{3/2}}$$

We can therefore apply Lemma 4.2 with $\beta = \frac{3}{2} < 2 = \alpha$ and obtain

$$|F_{\varepsilon}(z)| \leqq \max\{A_{\varepsilon}, 1\} = A$$

for all $z = x + iy$ such that $x \geqq 0$ and $y \geqq 0$. If we now repeat this argument for the second quadrant, we can then apply Lemma 4.2 to F_{ε} restricted to the upper half-plane and $\beta = 0 < 1 = \alpha$ to obtain $|F_{\varepsilon}(x + iy)| \leqq 1$ for $y \geqq 0$. Letting ε tend to zero we obtain $|F(z)| = |F(x + iy)| \leqq \exp\{\sigma y\}$ for $y \geqq 0$. The lemma is then established by applying this result to $G(z) = F(-z)$.

LEMMA 4.4. *Suppose F is of exponential type σ and that its restriction to the x axis has L^2-norm not exceeding 1. Then*

$$\left(\int_{-\infty}^{\infty} |F(x + iy)|^2\,dx\right)^{1/2} \leqq e^{\sigma|y|}$$

for all real y.

PROOF. Let φ be a bounded function of a real variable with compact support and $\|\varphi\|_2 \leqq 1$. We can then define $G(z) = \int_{-\infty}^{\infty} F(z + t)\varphi(t)\,dt$.

G is, then, analytic and for $\varepsilon > 0$ $|G(z)| \leqq \int_{-\infty}^{\infty} A_\varepsilon e^{(\sigma+\varepsilon)|z|} e^{(\sigma+\varepsilon)|t|} |\varphi(t)|\, dt = B_\varepsilon e^{(\sigma+\varepsilon)|z|}$; thus, G is also of exponential type σ. Furthermore, Schwarz's inequality implies that

$$|G(x)| \leqq \left(\int_{-\infty}^{\infty} |F(x)|^2\, dx \right)^{1/2} \left(\int_{-\infty}^{\infty} |\varphi(x)|^2\, dx \right)^{1/2} \leqq 1.$$

Consequently, we can apply Lemma 4.3 to the function G and obtain

$$\left| \int_{-\infty}^{\infty} F(z+t)\varphi(t)\, dt \right| = |G(z)| = |G(x+iy)| \leqq e^{\sigma|y|} \tag{4.5}$$

for all real y. Thus, making use of the converse of Schwarz's inequality, taking the supremum over all such φ in (4.5) we obtain Lemma 4.4.

We can now finish the proof of the Paley–Wiener theorem. Suppose F is of exponential type σ and its restriction to the real axis belongs to $L^2(-\infty, \infty)$. We shall show that the inverse Fourier transform of this restriction vanishes (a.e.) outside the interval $[-(\sigma/2\pi), \sigma/2\pi] = [-\tau, \tau]$. There is no loss of generality if we assume that $\int_{-\infty}^{\infty} |F(x)|^2\, dx \leqq 1$.

Let $G_+(z) = e^{i\sigma z}F(z)$. Then, using Lemma 4.4,

$$\left(\int_{-\infty}^{\infty} |G_+(x+iy)|^2\, dx \right)^{1/2} = e^{-\sigma y} \left(\int_{-\infty}^{\infty} |F(x+iy)|^2\, dx \right)^{1/2}$$
$$\leqq e^{-\sigma y} e^{\sigma y} = 1$$

for $y \geqq 0$. Thus, G_+ belongs to $H^2(T_{R^+}) = H^2(E_2^+)$, the H^2-space associated with the upper half-plane. It follows from the results of the last section (see, in particular, Theorem 3.1 and Corollary 3.4) that there exists $g \in L^2(-\infty, \infty)$, vanishing on the negative axis, such that

$$G_+(x+iy) = G_+(z) = \int_0^{\infty} g(t) e^{2\pi i z t}\, dt$$

for all $y > 0$. Letting $f(s) = g(\tau - s) = g[(\sigma/2\pi) - s]$, this equality is equivalent to

$$F(z) = F(x+iy) = \int_{-\infty}^{\tau} f(s) e^{-2\pi i z s}\, ds$$

for all $y > 0$. Passing to the limit, $y = 0$, we see that the inverse Fourier transform of $F(x)$ vanishes for almost every $s \geqq \tau$. Applying the same argument to $F(-z)$ we see that it must vanish for almost every $s \leqq -\tau$ as well, and Theorem 4.1 is proved.

We shall now extend this result to n-dimensions. In order to do this we need first to find an appropriate generalization of the notion of exponential type. Toward this end we make the following definitions.

If $\|\cdot\|$ and $|\cdot|$ are two norms on a vector space (over the real or complex numbers) we say that they are *equivalent* if there exist constants c_1 and c_2 such that $0 < c_1 \leqq \|x\|/|x| \leqq c_2 < \infty$ for all nonzero vectors x. It is well known that any norm $\|\cdot\|$ on E_n is equivalent to the Euclidean norm $|\cdot|$. The *unit sphere* with respect to such a norm, $\|\cdot\|$, is the set $K = \{x \in E_n; \|x\| \leqq 1\}$. It is clear that K is *convex*, *compact* and *symmetric* (that is, $x \in K$ implies $-x \in K$). If such a subset of E_n has a non-empty interior it is called a *symmetric body*. It is not hard to show that $K \subset E_n$ is a symmetric body if and only if it is the unit sphere with respect to a norm that is equivalent to the Euclidean norm.

Suppose K is any subset of E_n, then the set $K^* = \{y \in E_n; x \cdot y \leqq 1$ for all $x \in K\}$ is called the *polar set of* K. For example, if, for $p \geqq 1$, $K = \{x = (x_1, x_2) \in E_2; |x_1|^p + |x_2|^p \leqq 1\}$ it is easy to check that $K^* = \{y = (y_1, y_2) \in E_2; |y_1|^q + |y_2|^q \leqq 1\}$, where $1 = 1/p + 1/q$. It is also clear that if K is a symmetric body so is K^*. Thus, K^* is the unit sphere with respect to a norm $\|\cdot\|^*$, called the *dual norm* to the one for which K is the unit sphere. We also have the equivalent way of introducing the dual norm:

$$\|y\|^* = \sup_{x \in K} |x \cdot y|. \tag{4.6}$$

LEMMA 4.7. *Suppose $K \subset E_n$ is convex, closed and that $0 \in K$. Then $K^{**} = (K^*)^* = K$.*

PROOF. It is obvious that $K \subset K^{**}$. Thus, it suffices to show that if $x_0 \notin K$ then $x_0 \notin K^{**}$. Given such a point x_0 choose $y \in K$ so that $|y - x_0|$ is minimal. We now partition E_n into the two subsets $\{x \in E_n; x \cdot (x_0 - y) > y \cdot (x_0 - y)\}$ and $\{x \in E_n; x \cdot (x_0 - y) \leqq y \cdot (x_0 - y)\}$. Since $x_0 \cdot (x_0 - y) - y \cdot (x_0 - y) = |x_0 - y|^2 > 0$ it is clear that x_0 belongs to the former. We claim that K is contained in the latter. If this were not the case, there would exist $y_1 \in K$ such that $(y_1 - y) \cdot (x_0 - y) > 0$. Let us choose $\alpha < 1$ so that $0 < \alpha < 2(y_1 - y) \cdot (x_0 - y)/|y_1 - y|^2$. Since K is convex, $w = (1 - \alpha)y + \alpha y_1 \in K$ and we have

$$|w - x_0|^2 = \alpha\{\alpha|y_1 - y|^2 - 2(y_1 - y) \cdot (x_0 - y)\} + |y - x_0|^2 < |y - x_0|^2,$$

contradicting the fact that $|y - x_0|$ is minimal.

Since $0 \in K$ it follows that $y \cdot (x_0 - y) \geqq 0$. Thus, we can find a positive constant, ε, such that $x_0 \cdot (x_0 - y) > \varepsilon$ while $x \cdot (x_0 - y) \leqq \varepsilon$ for all $x \in K$ (if $y \cdot (x_0 - y) > 0$ we can choose $\varepsilon = y \cdot (x_0 - y)$; while any positive

$\varepsilon < x_0 \cdot (x_0 - y)$ can be chosen if $y \cdot (x_0 - y) = 0$). Letting $v = (x_0 - y)/\varepsilon$ this means

$$K \subset \{x \in E_n; x \cdot v \leqq 1\} \quad \text{and} \quad x_0 \cdot v > 1.$$

But this means that $v \in K^*$ and that x_0 cannot belong to K^{**}; thus, the lemma is proved.

Applying this result and the relation (4.6) we therefore obtain

$$\|x\| = \|x\|^{**} = \sup_{y \in K^*} |x \cdot y|. \tag{4.8}$$

If $z = (z_1, z_2, \ldots, z_n) \in \mathbf{C}_n$ it is then natural to define $\|z\|$ by extending (4.8):

$$\|z\| = \sup_{y \in K^*} |z \cdot y| = \sup_{y \in K^*} |z_1 y_1 + \cdots + z_n y_n|. \tag{4.8'}$$

We now say that an entire function, F, on $\mathbf{C}_n$ is of *exponential type K*, where K is a symmetric body, if for each $\varepsilon > 0$ there exists a constant A_ε such that

$$|F(z)| \leqq A_\varepsilon e^{2\pi(1+\varepsilon)\|z\|}$$

for all z. The class of all functions of exponential type K will be denoted by $\mathscr{E}(K)$.

We can now state an n-dimensional version of the Paley–Wiener theorem:

THEOREM 4.9. *Suppose $F \in L^2(E_n)$. Then F is the Fourier transform of a function vanishing outside a symmetric body K if and only if F is the restriction to E_n of a function in $\mathscr{E}(K^*)$.*

PROOF. If F is the Fourier transform of a function, f, vanishing outside K then it is easy to check that

$$F(z) = \int_{E_n} e^{-2\pi i z \cdot t} f(t)\, dt = \int_K e^{-2\pi i x \cdot t} e^{2\pi y \cdot t} f(t)\, dt \tag{4.10}$$

extends F to a function in $\mathscr{E}(K^*)$. In fact, it follows immediately from (4.10) that

$$|F(z)| = |F(x + iy)| \leqq A e^{2\pi\|y\|^*}.$$

The converse will be deduced from the following n-dimensional extension of Lemma 4.4:

LEMMA 4.11. *Suppose $F \in \mathscr{E}(K^*)$ then*

$$\left(\int_{E_n} |F(x + iy)|^2\, dx\right)^{1/2} \leqq e^{2\pi\|y\|^*} \left(\int_{E_n} |F(x)|^2\, dx\right)^{1/2}.$$

We shall reduce this inequality to the one-dimensional case. Let us fix $y \neq 0$ in E_n and let e_1 be a unit vector in E_n along the direction of y. We now choose an orthonormal basis $\{e_1, e_2, \ldots, e_n\}$ of E_n. We fix $(n-1)$ real numbers $u_2, \ldots, u_n$, put $\alpha = \sum_{j=2}^{n} u_j e_j$ and define

$$\varphi(w_1) = F(w_1 e_1 + \alpha).$$

It is clear that φ is an entire function of the (single) complex variable $w_1 = u_1 + iv_1$. Moreover, it is easy to see that φ is of exponential type $2\pi\|e_1\|^*$. In fact, given $\varepsilon > 0$, the fact that $F \in \mathscr{E}(K^*)$ implies the existence of a constant A_ε such that

$$\begin{aligned}|\varphi(w_1)| &\leqq A_\varepsilon \exp\{2\pi\|w_1 e_1 + \alpha\|^*(1+\varepsilon)\} \\ &\leqq [A_\varepsilon \exp\{2\pi(1+\varepsilon)\|\alpha\|^*\}]e^{2\pi\|e_1\|^*(1+\varepsilon)|w_1|} \\ &= A'_\varepsilon e^{2\pi\|e_1\|^*(1+\varepsilon)|w_1|}.\end{aligned}$$

Thus, by Lemma 4.4,

$$\int_{-\infty}^{\infty} |\varphi(u_1 + iv_1)|^2\, du_1 \leqq e^{4\pi\|e_1\|^*|v_1|} \int_{-\infty}^{\infty} |\varphi(u_1)|^2\, du_1$$

for all $v_1 \in (-\infty, \infty)$.

Choosing v_1 so that $y = v_1 e_1$ this inequality becomes

$$\int_{-\infty}^{\infty} \left|F\left(iy + \sum_{j=1}^{n} u_j e_j\right)\right|^2 du_1 \leqq e^{4\pi\|y\|^*} \int_{-\infty}^{\infty} \left|F\left(\sum_{j=1}^{n} u_j e_j\right)\right|^2 du_1.$$

Integrating both sides with respect to $u_2, u_3, \ldots, u_n$ we obtain Lemma 4.11.

We can now finish the proof of Theorem 4.9. Suppose F is the restriction to E_n of a function in $\mathscr{E}(K^*)$. For simplicity, we shall denote the latter also by F. By Lemma 4.11, we then have $F \in H^2(T_B)$ for all bounded bases B. Thus, by the general representation Theorem 2.3 there exists f such that

$$F(z) = \int_{E_n} e^{2\pi i z\cdot t} f(t)\, dt$$

for all $z = x + iy \in T_B$. We can assume $0 \in B$; then Plancherel's theorem assures us that $f \in L^2(E_n)$ and $\|f\|_2^2 = \int_{E_n} |F(x)|^2\, dx$. Thus, we see that F is the inverse Fourier transform of f (or, equivalently, the Fourier transform of $f(-t)$) and the theorem will be established if we show that f vanishes (a.e.) outside K. In order to do this we first observe that the Plancherel theorem implies that

$$\int_{E_n} |F(x + iy)|^2\, dx = \int_{E_n} e^{-4\pi y\cdot t}|f(t)|^2\, dt$$

for all $y \in E_n$ (since B can be chosen to be arbitrarily large). Thus, by

Lemma 4.11,

$$\int_{E_n} |f(t)|^2 e^{-4\pi y \cdot t}\, dt \leqq e^{4\pi \|y\|^*} \int_{E_n} |f(t)|^2\, dt \tag{4.12}$$

for all $y \in E_n$. We claim that this inequality can hold only if f vanishes almost everywhere outside K. Suppose $t_0 \notin K$; then, by Lemma 4.7, there exists $y_0 \in K^*$ such that $(t_0 \cdot y_0) < -1$ (we are also using the fact that K^* is symmetric). Hence, it is clear that we can find $\delta > 0$ and a neighborhood $N = N(t_0)$ of t_0 such that $(t \cdot y_0) < -(1 + \delta)$ for all $t \in N$. Thus, by (4.12), for $y = \rho y_0$ $(\rho > 0)$

$$\int_N |f(t)|^2 e^{4\pi\rho(1+\delta)}\, dt \leqq \int_N |f(t)|^2 e^{-4\pi y \cdot t}\, dt$$
$$\leqq \|f\|_2^2 e^{4\pi\rho\|y_0\|^*}.$$

Since $y_0 \in K^*$ we must have $\|y_0\|^* \leqq 1$. Thus, we have

$$\left(\int_N |f(t)|^2\, dt\right) e^{4\pi\rho(1+\delta)} \leqq \|f\|_2^2 e^{4\pi\rho}$$

for all $\rho > 0$. If $\int_N |f(t)|^2\, dt$ were not zero, this would imply

$$e^{4\pi\rho\delta} \leqq \|f\|_2^2 \Big/ \int_N |f(t)|^2\, dt,$$

which is clearly impossible for large ρ. Thus $f(t) = 0$ for almost all $t \in N$ and the theorem follows.

5. The H^p Theory

Up to this point we have considered, almost exclusively, properties of analytic functions on a tube that belong to H^2. In particular, we have considered only the existence of L^2 boundary values. In this section we shall derive some results concerning the existence of pointwise limits as we approach a point on the boundary of the base, as well as the existence of L^p limits, $p > 0$.

We begin by examining the special case of the tube T_Γ where Γ is the *first octant*:

$$\Gamma = \{y = (y_1, \ldots, y_n) \in E_n;\, y_1 > 0, \ldots, y_n > 0\}.$$

Γ is, then, a cone and it is clear that $\Gamma^* = \bar{\Gamma}$. Thus, the Cauchy kernel associated with the tube T_Γ is

$$K(z) = \int_0^\infty \int_0^\infty \cdots \int_0^\infty e^{2\pi i(z_1 t_1 + z_2 t_2 + \cdots + z_n t_n)}\, dt_1\, dt_2 \cdots dt_n = \prod_{j=1}^n \frac{-1}{2\pi i z_j},$$

where $z = (z_1, z_2, \ldots, z_n) \in T_\Gamma$. That is, $K(z)$ is simply the product of n one-dimensional Cauchy kernels associated with the upper half-plane. It follows that the Poisson kernel $\mathscr{P}(x, y)$ (see 3.7 and the discussion preceding it) is the product of n one-dimensional Poisson kernels:

$$\mathscr{P}(x, y) = \prod_{j=1}^{n} \frac{1}{\pi} \frac{y_j}{x_j^2 + y_j^2}, \qquad y_1, y_2, \ldots, y_n > 0.$$

This kernel, therefore, is a special case of the iterated Poisson kernels that were studied in the third section of the previous chapter (see, in particular, Theorem 3.22, Corollary 3.23 and the material preceding these results). By identifying the components $z_j = x_j + iy_j$ of a point $z = (z_1, z_2, \ldots, z_n) \in T_\Gamma$ with the ordered pairs (x_j, y_j) we may consider the tube T_Γ to be the Cartesian product $E_2^+ \times E_2^+ \times \cdots \times E_2^+$ of n upper half-planes of two dimensional real Euclidean space. Thus, it is consistent with the definitions introduced in the third section of Chapter II to say that a function u, defined in T_Γ,

(i) has a nontangential limit l in each variable at $x \in E_n$ if $u(\zeta) = u(\xi + i\eta) = u(\xi_1 + i\eta_1, \ldots, \xi_n + i\eta_n)$ tends to l as the point $\zeta = (\xi_1, \eta_1; \xi_2, \eta_2; \ldots; \xi_n, \eta_n)$ tends to $x = (x_1, x_2, \ldots, x_n) = (x_1, 0; x_2, 0; \ldots; x_n, 0)$ within the Cartesian product

$$\gamma_\alpha(x) = \Gamma_{\alpha_1}(x_1) \times \Gamma_{\alpha_2}(x_2) \times \cdots \times \Gamma_{\alpha_n}(x_n) \subset T_\Gamma,$$

for each n-tuple $\alpha = (\alpha_1, \alpha_2, \ldots, \alpha_n)$ of positive real numbers[8];

(ii) is nontangentially bounded in each variable at $x \in E_n$ if u is bounded in the set $\gamma_\alpha(x) \cap \{\xi + i\eta \in T_\Gamma; \eta_1, \eta_2, \ldots, \eta_n \leqq 1\}$ for each n-tuple α of positive real numbers.

The existence of pointwise limits and limits in the norm as we approach the origin within the first octant can be established for all functions in $H^p(T_\Gamma)$, whenever $p > 0$. In fact, we shall prove the following results:

THEOREM 5.1. *Suppose F belongs to $H^p(T_\Gamma)$, $p > 0$, where Γ is the first octant in E_n, then*

(a) *F has a nontangential limit $F(x)$ in each variable at almost every x in E_n. In particular,*

$$\lim_{y \in \Gamma,\, y \to 0} F(x + iy) = F(x)$$

exists for almost every $x \in E_n$;

(b) *$\int_{E_n} |F(x + iy) - F(x)|^p \, dx$ tends to zero as $y \in \Gamma$ tends to $0 \in E_n$.*

[8] We are using the notation developed in Section 3 of Chapter II:

$$\Gamma_{\alpha j}(x_j) = \{(\xi_j, \eta_j) \in E_2^+;\ |\xi_j - x_j| < \alpha_j \eta_j\},\ j = 1, 2, \ldots, n.$$

PROOF. Let us fix $(\zeta_2, \ldots, \zeta_n) = (\xi_2 + i\eta_2, \ldots, \xi_n + i\eta_n)$, $\eta_2, \ldots, \eta_n > 0$, and consider the function g of one complex variable defined by letting $g(\zeta_1) = F(\zeta_1, \zeta_2, \ldots, \zeta_n)$ whenever $\zeta_1 = \xi_1 + i\eta_1$ belongs to the upper half-plane E_2^+. We first show that g belongs to $H^p(E_2^+)$. Since $|F(\zeta_1, \zeta_2, \ldots, \zeta_n)|^p$ is subharmonic as a function of $\zeta_2 = \xi_2 + i\eta_2$ (see example (3) in Section 4 and (5.1) in Chapter II) we have, writing $w_2 = u_2 + iv_2$,

$$|F(\zeta_1, \zeta_2, \ldots, \zeta_n)|^p \leqq \frac{1}{\pi\eta_2^2} \int_{|w_2-\zeta_2|<\eta_2} |F(\zeta_1, w_2, \ldots, \zeta_n)|^p \, du_2 \, dv_2$$
$$\leqq \frac{1}{\pi\eta_2^2} \int_0^{2\eta_2} \int_{-\infty}^{\infty} |F(\zeta_1, w_2, \ldots, \zeta_n)|^p \, du_2 \, dv_2.$$

Repeating this argument for $\zeta_3, \ldots, \zeta_n$ we obtain

$$|F(\zeta)|^p \leqq \pi^{1-n}(\eta_2 \cdots \eta_n)^{-2}$$
$$\times \int_0^{2\eta_n} \cdots \int_0^{2\eta_2} \left(\int_{E_{n-1}} |F(\zeta_1, w_2, \ldots, w_n)|^p \, du_2 \cdots du_n \right) dv_2 \cdots dv_n.$$

Now integrating both sides of this inequality with respect to ξ_1 we have

$$\int_{-\infty}^{\infty} |g(\xi_1 + i\eta_1)|^p \, d\xi_1$$
$$\leqq \pi^{1-n}(\eta_2 \cdots \eta_n)^{-2} \int_0^{2\eta_n} \cdots \int_0^{2\eta_2} \left(\int_{E_n} |F(u + iv)|^p \, du \right) dv_2 \cdots dv_n$$
$$\leqq \pi^{1-n}(\eta_2 \cdots \eta_n)^{-2} \|F\|_p^p (2\eta_2 \cdots 2\eta_n) = A_1 < \infty.$$

This shows that g is a member of $H^p(E_2^+)$.

For $\zeta = (\zeta_1, \zeta_2, \ldots, \zeta_n)$ in T_Γ we let $s(\zeta) = |F(\zeta)|^{p/2}$. The last inequality can then be expressed in terms of the function s:

$$\int_{-\infty}^{\infty} [s(\xi_1 + i\eta_1, \zeta_2, \ldots, \zeta_n)]^2 \, d\xi_1 = \int_{-\infty}^{\infty} |g(\xi_1 + i\eta_1)|^p \, d\xi_1 \leqq A_1 < \infty.$$

Since $s(\zeta_1, \zeta_2, \ldots, \zeta_n)$ is a subharmonic function of ζ_1, the other variables being kept fixed, this inequality together with (4.12) in the second chapter assures us that

$$s(\xi_1 + i[\varepsilon_1 + \eta_1], \zeta_2, \ldots, \zeta_n)$$
$$\leqq \frac{1}{\pi} \int_{-\infty}^{\infty} s(t_1 + i\varepsilon_1, \zeta_2, \ldots, \zeta_n) \frac{\eta_1}{(\xi_1 - t_1)^2 + \eta_1^2} \, dt_1$$

for all $\varepsilon_1 > 0$. Applying the same argument to $\zeta_2, \dots, \zeta_n$ we obtain

$$(5.2) \quad s(\xi + i[\varepsilon + \eta]) = s(\xi_1 + i[\varepsilon_1 + \eta_1], \dots, \xi_n + i[\varepsilon_n + \eta_n]) \leqq \int_{E_n} s(t + i\varepsilon)\mathscr{P}(\xi - t, \eta)\, dt$$

for all $\varepsilon = (\varepsilon_1, \dots, \varepsilon_n)$ in Γ.

The fact that F belongs to $H^p(T_\Gamma)$ implies that the family $\{f_\varepsilon\}$, where $f_\varepsilon(t) = s(t + i\varepsilon)$ for ε in Γ, is uniformly bounded in the L^2 norm (in fact $\|f_\varepsilon\|_2^2 \leqq \|F\|_p^p$). Thus, by the weak compactness of the unit sphere in $L^2(E_n)$, we can find a sequence $\{\varepsilon^{(k)}\}$ tending to $0 \in E_n$ such that $\{f_{\varepsilon^{(k)}}\}$ converges weakly to a function f in $L^2(E_n)$ as $k \to \infty$. Putting $\varepsilon = \varepsilon^{(k)}$ in (5.2) and then letting $k \to \infty$, we obtain

$$(5.3) \qquad s(\xi + i\eta) \leqq \int_{E_n} f(t)\mathscr{P}(\xi - t, \eta)\, dt = m(\xi + i\eta).$$

Since $m(\xi + i\eta)$ is the iterated Poisson integral of a function in $L^2(E_n)$ it is nontangentially bounded in each variable at almost every $x \in E_n$ (see the argument following Corollary 3.23 in the second chapter). Because of (5.3), the same must be true of $s(\xi + i\eta) = |F(\xi + i\eta)|^{p/2}$. We can therefore apply Theorem 3.24 of Chapter II to the real and imaginary parts of F and we obtain part (a) of Theorem 5.1.

Part (b) now follows easily. If we let $Mf = M^{(n)}M^{(n-1)} \cdots M^{(1)}f$ be the n-fold iterate of the one-dimensional maximal functions corresponding to the variables $x_1, x_2, \dots, x_n$ [9] it then follows that $m(x + iy) \leqq A(Mf)(x)$ for all $x + iy \in T_\Gamma$, where A is a constant independent of $y \in \Gamma$ and $f \in L^2(E_n)$ (see the argument preceding Theorem 3.22 in Chapter II). Thus, by (5.3)

$$(5.4) \qquad |F(x + iy)|^p \leqq \{A(Mf)(x)\}^2.$$

By proving part (a) we have established the almost everywhere convergence $\lim_{y \in \Gamma, y \to 0} |F(x + iy) - F(x)| = 0$; it follows from inequality (5.4) that this convergence is dominated almost everywhere by a constant times $\{(Mf)(x)\}^2$:

$$|F(x + iy) - F(x)|^p \leqq 2^p(|F(x + iy)|^p + |F(x)|^p) \leqq 2^{p+1}\{A(Mf)(x)\}^2.$$

Since $f \in L^2(E_n)$ it follows from (3.20), Chapter II, that $Mf \in L^2$. Part (b) is, therefore, an immediate consequence of the Lebesgue dominated convergence theorem.

[9] If g is a function of $x = (x_1, x_2, \dots, x_n)$ then, for $1 \leqq j \leqq n$,

$$(M^{(j)}g)(x) = \sup_{r>0} (2r)^{-1} \int_{|t| \leqq r} |g(x_1, \dots, x_j - t, \dots, x_n)|\, dt.$$

We shall now derive some consequences of this theorem. First, suppose Γ is the interior of the convex hull of n linearly independent rays meeting at the origin. By selecting vectors $a_1, a_2, \ldots, a_n$ in the direction of these rays Γ can be described as the open cone $\{v = v_1a_1 + \cdots + v_na_n \in E_n; v_1 > 0, \ldots, v_n > 0\}$. Theorem 5.1 is then valid for a function F in $H^p(T_\Gamma)$. To see this consider the linear transformation A mapping the standard basis vector e_j onto a_j, $j = 1, 2, \ldots, n$. Then A maps the first octant onto Γ. Extending A linearly to $\mathbf{C}_n$ and putting $G(x + iy) = G(z) = F(Az) = F(Ax + iAy)$ for z in the tube over the first octant, we obtain

$$\int_{E_n} |G(x + iy)|^p \, dx = \frac{1}{|\det A|} \int_{E_n} |F(u + iAy)|^p \, du \leqq \frac{1}{|\det A|} \|F\|_p^p,$$

for all y in the first octant. Thus, G satisfies the hypotheses of Theorem 5.1. But this clearly implies the a.e. existence of the limit

$$F(u) = \lim_{v \in \Gamma,\, v \to 0} F(u + iv)$$

as well as the corresponding limit in the L^p-norm.

More generally, we shall say that Γ is a *polygonal cone* if it is the interior of the convex hull of a finite number of rays meeting in the origin, among which we can find at least n that are linearly independent. A polygonal cone is clearly a finite union of cones of the type just considered in the preceeding paragraph; thus, it follows that the existence of pointwise and norm limits as we approach the origin is valid as well for functions in the class H^p associated with a tube whose base is a polygonal cone.

From these considerations we can easily deduce the following theorem:

THEOREM 5.5. *Suppose* $F \in H^p(T_\Gamma)$, $p > 0$, *where* Γ *is an open convex cone in* E_n, *then if* Γ_1 *is any cone whose closure is contained in* $\Gamma \cup \{0\}$;

(a) $\lim_{y \in \Gamma_1,\, y \to 0} F(x + iy) = F(x)$ *exists for almost every* $x \in E_n$;

(b) $\int_{E_n} |F(x + iy) - F(x)|^p \, dx \to 0$ *as* $y \in \Gamma_1$ *tends to* 0.

PROOF. It clearly suffices to show that we can find a polygonal cone Γ_0 such that $\bar{\Gamma}_1 \subset \Gamma_0 \cup \{0\} \subset \Gamma \cup \{0\}$. In order to do this we consider the intersection of $\bar{\Gamma}_1$ with the surface of the unit sphere $\Sigma_{n-1} \subset E_n$. Let S denote this intersection. For each $x \in S$ we can certainly find an open polygonal cone Γ_x such that $x \in \Gamma_x \subset \Gamma$. Since S is compact, a finite number of such cones, $\Gamma_{x_1}, \ldots, \Gamma_{x_m}$, will cover S. Letting Γ_0 be the convex hull of the union of these m cones we then clearly have the desired inclusions $\bar{\Gamma}_1 \subset \Gamma_0 \cup \{0\} \subset \Gamma \cup \{0\}$.

In the second section of this chapter (immediately following Theorem 2.11) we introduced the notions of restricted and unrestricted limits as we approached a boundary point of the base B of a tube T_B. It follows from this proof that Theorem 5.5 has an equivalent formulation in terms of these notions:

THEOREM 5.5′. *If* $F \in H^p(T_\Gamma), p > 0$, *then* $F(z)$ *has restricted limits almost everywhere and in the* L^p*-norm as we approach* 0 *within* Γ.

The proof also shows that whenever $F \in H^p(T_\Gamma)$ then F has restricted nontangential limits for almost every x^0, in the sense that the limit $F(x + iy)$ exists as $x + iy \to x^0$ when $y \in \Gamma_1$ and $x + iy \in \gamma_\alpha(x^0)$, for any proper subcone Γ_1 of Γ and any Cartesian product cone $\gamma_\alpha(x^0)$.

With the help of the Poisson kernel $\mathscr{P}(x, y)$ associated with the tube T_Γ it can be shown that when $p \geqq 1$ we obtain *unrestricted* limits in the norm. More precisely, we shall show that the following result is true:

THEOREM 5.6. *Suppose* Γ *is a regular open convex cone in* E_n *and that* $F \in H^p(T_\Gamma)$, $1 \leqq p < \infty$, *then*

$$\lim_{y \in \Gamma,\, y \to 0} \int_{E_n} |F(x + iy) - F(x)|^p \, dx = 0,$$

where $F(x)$ *is the limit function whose existence was established in the last theorem.*

PROOF. We first show that the Poisson kernel $\mathscr{P}(x, y)$ has the following "semigroup property"[10]: When $F \in H^p(T_\Gamma)$ and y_1, y_2 belong to Γ then

$$F(x + i(y_1 + y_2)) = \int_{E_n} F(t + iy_2)\mathscr{P}(x - t, y_1)\, dt. \tag{5.7}$$

When $p = 2$ this is an immediate consequence of Theorem 3.9. In general, it follows from Lemma 2.12 that (whenever $p > 0$) $G(x + iy) = F(x + i(y + y_2))$ is uniformly bounded in T_Γ. Let φ be a continuous non-negative function with compact support in Γ^* and satisfying $\int_{E_n} \varphi(t)\, dt = 1$ (such a φ exists since Γ is regular). If $\psi(z) = \int_{E_n} e^{2\pi i z \cdot t}\varphi(t)\, dt$, Theorem 3.1 assures us that $\psi \in H^2(T_\Gamma)$; consequently, $G_\varepsilon(z) = \psi(\varepsilon z)G(z)$ belongs

[10] It is an easy consequence of Corollary 1.28 in Chapter I that $u(x, y_1 + y_2) = \int_{E_n} u(t, y_2)P(x - t, y_1)\, dt$ whenever u is a harmonic function in E^+_{n+1} satisfying $\|u(\cdot, y)\|_p \leqq c < \infty$ for all $y > 0$. In the present situation we have to restrict ourselves to holomorphic functions F as the kernel $\mathscr{P}$ does not in general reproduce harmonic (nor multiply harmonic) functions.

to $H^2(T_\Gamma)$ for all $\varepsilon > 0$. Thus, by Theorem 3.9,

$$G_\varepsilon(x + iy) = \int_{E_n} G_\varepsilon(t)\mathscr{P}(x - t, y)\, dt. \tag{5.8}$$

Moreover, $|\psi(z)| \leq \int_{\Gamma^*} e^{-2\pi y \cdot t}\varphi(t)\, dt \leq 1$ and $\lim_{\varepsilon \to 0} \psi(\varepsilon z) = 1$. These facts, together with the fact that $\mathscr{P}(\cdot, y) \in L^q(E_n)$, where $1/p + 1/q = 1$ (see 3.7), allow us to take the limits, as ε approaches zero, of both sides of equality (5.8) and to conclude that $G(x + iy) = \int_{E_n} G(t)\mathscr{P}(x - t, y)\, dt$. But this is precisely the desired property (5.7) when $y = y_1$.

Part (b) of Theorem 5.5 now assures us that, by letting y_2 tend to 0 restrictedly, we obtain

$$F(x + iy) = \int_{E_n} F(t)\mathscr{P}(x - t, y)\, dt \tag{5.9}$$

for all $y \in \Gamma$.

The rest of the proof is analogous to that of Theorem 1.10 of Chapter II and is based on properties (i), (ii), and (iii) of the kernel $\mathscr{P}$ that were established immediately following (3.8).[11] Letting $\omega(r)$ denote the modulus of continuity of F and using (ii), Minkowski's ordinary inequality followed by his integral inequality and (i), in that order, we have

$$\begin{aligned}
&\left(\int_{E_n} |F(x + iy) - F(x)|^p\, dx\right)^{1/p} \\
&\quad = \left(\int_{E_n}\left|\int_{E_n} \{F(x - t) - F(x)\}\mathscr{P}(t, y)\, dt\right|^p dx\right)^{1/p} \\
&\quad \leq \left(\int_{E_n}\left|\int_{|t| \leq \delta} \{F(x - t) - F(x)\}\mathscr{P}(t, y)\, dt\right|^p dx\right)^{1/p} \\
&\qquad\qquad + \left(\int_{E_n}\left|\int_{|t| > \delta} \{F(x - t) - F(x)\}\mathscr{P}(t, y)\, dt\right|^p dx\right)^{1/p} \\
&\quad \leq \int_{|t| \leq \delta}\left(\int_{E_n} |F(x - t) - F(x)|^p\, dx\right)^{1/p} \mathscr{P}(t, y)\, dt \\
&\qquad\qquad + \int_{|t| > \delta}\left(\int_{E_n} |F(x - t) - F(x)|^p\, dx\right)^{1/p} \mathscr{P}(t, y)\, dt \\
&\quad \leq \left\{\sup_{|t| \leq \delta} \omega(|t|)\right\} \int_{E_n} \mathscr{P}(t, y)\, dt + 2\|F\|_p \int_{|t| > \delta} \mathscr{P}(t, y)\, dt.
\end{aligned}$$

[11] In fact, the proof we now give can be used in order to prove equality (3.8).

The first summand equals $\sup_{|t|\leqq\delta} \omega(|t|)$ since, by (ii), $\int_{E_n} \mathscr{P}(t, y)\, dt = 1$; therefore, it is less than, say, $\varepsilon > 0$ if δ is sufficiently small (see the proof of Theorem (1.18) in Chapter I). Having chosen such a δ, property (iii) then asserts that the last summand tends to 0 as y approaches the origin within Γ. Thus, $(\int_{E_n} |F(x + iy) - F(x)|^p\, dx)^{1/p} < \varepsilon$ if $y \in \Gamma$ is close to 0. This proves the theorem.

6. Further Results

6.1. Throughout most of this chapter we considered only tubes having an open convex base B. In the second section we showed that if $F \in H^2(T_B)$, where B is merely open and connected, then F has an analytic extension, G, to the convex hull T_{B^c} of the tube T_B and $G \in H^2(T_{B^c})$ with $\|F\|_2 = \|G\|_2$ (see Corollary 2.4). This fact, as well as most of the results of the second section of this chapter, is an easy consequence of the basic representation theorem (Theorem 2.3). There are many other reasons for considering only convex bases. It is a classical fact that whenever F is holomorphic in a tube T_B with an open and connected base B, then it has an analytic extension, G, to T_{B^c} (see Theorem 9 of Chapter V in Bochner and Martin [1]). It is easy to see that G assumes only the values assumed by F. If this were not the case and z_0 were in the range of G but not in the range of F we would have a holomorphic function $1/(F - z_0)$ in T_B which, by analytic continuation, has the analytic extension $1/(G - z_0)$ in T_{B^c}. This is clearly impossible.

These facts can be used to obtain another proof of Corollary 2.4 that extends to give us the corresponding result for other H^p spaces. Suppose $F \in H^2(T_B)$ and $h \in L^2(E_n)$, where $\int_{E_n} |h(t)|^2\, dt = 1$. We then obtain a holomorphic function f on T_B by letting

$$f(z) = f(x + iy) = \int_{E_n} F(x + t + iy)h(t)\, dt$$

whenever $z \in T_B$. By Hölder's inequality, $|f(z)| \leqq \|F\|_2 \|h\|_2 = \|F\|_2$. If G is the analytic extension of F to T_{B^c} then $g(z) = \int_{E_n} G(z + t)h(t)\, dt$ clearly is the analytic extension of f to T_{B^c}.[12] Thus, it follows from the above remarks that $|g(z)| \leqq \|F\|_2$ for all $z \in T_{B^c}$. Consequently,

[12] It is not hard to show that the last integral is defined. By first considering a subset $B_0 \subset B$ as in Lemma 2.12 we can assume that F and, therefore, G is bounded; then, by choosing h in $L^1(E_n) \cap L^2(E_n)$ we are assured that $G(z + t)h(t)$ is integrable. This clearly suffices for our purposes.

choosing h so that $g(z) = \int_{E_n} G(z+t)h(t)\,dt = (\int_{E_n} |G(z+t)|^2\,dt)^{1/2} = (\int_{E_n} |G(t+iy)|^2\,dt)^{1/2}$ we see that $G \in H^2(T_{B^c})$ with $\|G\|_2 \leqq \|F\|_2$. If $p \geqq 1$ and $\|h\|_q = 1$, where $(1/p) + (1/q) = 1$, the same argument shows that the analytic extension G of an $F \in H^p(T_B)$ belongs to $H^p(T_{B^c})$ and $\|G\|_p = \|F\|_p$. The result actually holds for all $p > 0$, but the proof is more difficult. It makes use of the fact that if $F \in H^p(T_B), p > 0$, then $\log \|F(\cdot + iy)\|_p$ is a convex function of $y \in B$.

6.2. Another application of the H^2 theory (and, in particular, of Theorem 2.3) is the following:

Let B^+ be an open convex subset of E_n such that 0 is one of its boundary points and $B^- = \{x \in E_n;\ -x \in B^+\}$. Suppose that there exist two functions $F^+ \in H^2(T_{B^+})$ and $F^- \in H^2(T_{B^-})$ such that

$$F^+(x+i0) = \lim_{y\in B^+,\, y\to 0} F^+(x+iy) = \lim_{y\in B^-,\, y\to 0} F^-(x+iy) = F^-(x-i0)$$

a.e. (*these limits are to be interpreted in the sense described in* (2.8)). *Then there exists an analytic function, F, defined on the tube over the convex hull of $B^+ \cup B^-$, such that $F(z) = F^+(z)$ if $z \in T_{B^+}$ and $F(z) = F^-(z)$ if $z \in T_{B^-}$.*

To show this we first note that Theorem 2.3 implies that there exist two functions, f^+ and f^-, such that $F^+(z) = \int_{E_n} e^{2\pi i z\cdot t} f^+(t)\,dt$ and $F^-(z) = \int_{E_n} e^{2\pi i z\cdot t} f^-(t)\,dt$. It follows that f^+ and f^- are the Fourier transforms of $F^+(x+i0)$ $(= F^-(x-i0))$ (see the discussion connected with (2.8)). Consequently, $f^+(t) = f^-(t)$ a.e. Letting $f(t) = f^\pm(t)$ a.e. and using the fact that $S = \{y \in E_n;\ \int_{E_n} |f(t)|^2 e^{-4\pi y\cdot t}\,dt \leqq M^2 < \infty\}$ is convex (see the proof of Corollary 2.4) we see that $F(z) = \int_{E_n} e^{2\pi i z\cdot t} f(t)\,dt$ is analytic on T_S, a set containing the tube over the convex hull of $B^+ \cup B^-$ (see (2.2)), and coincides with $F^+(z)$ on T_{B^+} and with $F^-(z)$ on T_{B^-}.

For related results and various extensions see Streater and Wightman [1].

6.3. Theorem 2.11 can be extended to three dimensions. The result can be stated in the following way (see E. M. Stein, G. Weiss, and M. Weiss [1]):

Suppose y_0 belongs to the boundary ∂B of an open convex set $B \subset E_3$. Then y_0 is a point of unrestricted L^2-convergence (that is, $\lim_{y\to y_0,\, y\in B} F(x+iy)$ exists in the L^2-sense for all $F \in H^2(T_B)$) if and only if there exists a finite set $V = \{y_1, y_2, \ldots, y_n\}$ of "vertices" lying in $\bar{B}$ and a neighborhood N of

y_0, so that, whenever $y \in N \cap \partial B$ and π_y is a plane of support at y, then π_y contains at least one element of V.

This condition is not a local one. For example, if B is a circular cone this result shows that every point of ∂B is a point of unrestricted L^2-convergence; however, truncating the cone by removing a neighborhood of the vertex, the remaining boundary points are no longer points of unrestricted convergence.

It is an open problem to find how this result generalizes to higher dimensions.

6.4. If F is holomorphic in the upper half-plane $E_2^+ = \{z = x + iy; y > 0\} \subset \mathbf{C}_1$ then, except for a set of points in the x axis of linear Lebesgue measure zero, the following is true: Either F has a nontangential limit at $(x_0, 0)$ or F maps each triangular region $\Delta(x_0; \alpha, \beta) = \{x + iy \in E_2^+; |x - x_0| < \alpha y < \beta\}$ densely into the complex plane. This is a special case of a classical result of Plessner [1] and it can be easily derived from the special case $n = 1$ of Theorem 3.19 of Chapter II. Similarly, Theorem 3.24 of Chapter II gives us an extension to n-dimensions of Plessner's result (see Calderón [4]): If F is holomorphic in T_Γ, where Γ is the first octant, then except for a subset of E_n of measure zero either F has a nontangential limit in each variable at $x_0 = (x_1^0, x_2^0, \ldots, x_n^0)$ or it maps densely into the complex plane each Cartesian product $\Delta(x_0; \alpha, \beta) = \Delta(x_1^0; \alpha_1, \beta_1) \times \cdots \times \Delta(x_n^0; \alpha_n, \beta_n)$, where $\alpha = (\alpha_1, \ldots, \alpha_n)$ and $\beta = (\beta_1, \ldots, \beta_n)$ belong to Γ. From this, it follows that if the real part of a holomorphic function $F = u + iv$, defined on T_Γ, has a nontangential limit in each variable at the points of a set $S \subset E_n$ then the imaginary part of F has such limits at almost every point of S. We see this by observing that if u has a nontangential limit in each variable at x then F cannot map $\Delta(x; \alpha, \beta)$ into a dense subset of $\mathbf{C}_1$.

6.5. Other notions of restricted convergence have been studied. For simplicity, suppose u is a function defined in T_Γ, where Γ is the first quadrant in E_2. We shall say that u has a restricted curvilinear limit $u(x_0)$ at $x_0 \in E_2$ if $u(x_0 + iy(t)) \to u(x_0)$ as $t \to 0$, whenever the curve $y(t) = (h(t), k(t))$, $t \geqq 0$, is continuous, the functions h and k are strictly increasing to ∞ and $h(0) = 0 = k(0)$.[13] We have already indicated (see (5.7) in Chapter II) that an iterated Poisson integral $u(x + iy)$ of an L^1-function,

[13] More generally, we can consider nontangential restricted curvilinear limits in each variable by letting $x + iy(t)$ tend to x_0 within the regions $\gamma_\alpha(x_0)$ introduced at the beginning of Section 5. The reader should not confuse this type of nontangential (or tangential) approach to x_0 (within T_Γ) with the nontangential (or tangential) approach of $y = y(t)$ to 0 (within Γ). We have also sacrificed generality for the sake of simplicity by considering only the two-dimensional case; these notions have an obvious extension to n-dimensions.

f, need not have unrestricted limits at almost every $x \in E_2$. One can show, however, that $u(x + iy)$ does converge almost everywhere to $f(x)$ in the restricted curvilinear sense we have just described. This can be done by appropriately modifying the argument given in Zygmund [1], Volume 2. We can extend this notion of restricted convergence to tubes having a general cone for a base and obtain another version of Theorem 5.5′. (In this connection see the comment immediately following Theorem 5.5′.) The counterexample that was constructed in Section 2 shows that the situation is considerably more complicated when the base of the tube is not a cone. A further illustration of the aptness of restricted convergence when the base is a cone is provided by the examples described in 6.8 below.

6.6. Theorem 5.1 states, in particular, that the unrestricted limits $\lim_{y \in \Gamma, y \to 0} F(x + iy) = F(x)$ exist for almost every $x \in E_n$ whenever Γ is the first octant and $F \in H^p(T_\Gamma)$, $p > 0$. Corollary 3.23 of Chapter II assures us that the same is true for the iterated Poisson integral of a function f in $L^p(E_n)$, $1 < p$. We have already observed that this unrestricted convergence may fail when $f \in L^1(E_n)$; however, if we assume in addition that $|f|(\log |f|)^{n-1}$ (this expression being 0 when $|f| = 0$) is locally integrable (which is the case when $f \in L^p(E_n)$, $1 < p$) then the iterated Poisson integral of f does converge unrestrictedly to $f(x)$ for almost every $x \in E_n$ (see Jessen, Marcinkiewicz, and Zygmund [1]).

6.7. A regular cone $\Gamma \subset E_n$ is called *self-dual* if its closure $\bar{\Gamma}$ coincides with its dual Γ^*. An automorphism of a regular cone Γ is a linear operator on E_n that maps Γ onto Γ. The collection of all automorphisms of Γ is a closed subgroup $\Sigma = \Sigma(\Gamma)$ of the general linear group $GL(E_n)$. If Σ is transitive (that is, whenever x and y belong to Γ there exists an automorphism $\rho \in \Sigma$ such that $\rho x = y$) and Γ is self-dual then it is called a (*homogeneous*) *domain of positivity*. Except for an exceptional lower-dimensional cone all domains of positivity can be realized as direct sums of the following four types of domains of positivity, often referred to as the "classical domains" (for details see Koecher [1], Rothaus [1], and Vinberg [1]):

(i) The *forward light cone*. This is the cone $C_n \subset E_n$, $n > 1$, consisting of all $x = (x_1, x_2, \ldots, x_n) \in E_n$ such that $x_1^2 - x_2^2 - \cdots - x_n^2 > 0$ and $x_1 > 0$. The automorphism group $\Sigma = \Sigma(C_n)$ consists of all those linear transformations ρ preserving (up to a positive multiplicative constant) the bilinear form $(x, y) = x_1 y_1 - x_2 y_2 - \cdots - x_n y_n$ and the inequality $x_1 > 0$.

(ii) If, for some positive integer m, $n = m(m + 1)/2$ we can identify E_n with the vector space of all $m \times m$ real symmetric matrices. $S_m \subset E_n$ is then the cone of all positive definite real symmetric matrices. The automorphism group $\Sigma = \Sigma(S_m)$ can be obtained from the real general linear group $GL(E_m)$: to each nonsingular linear transformation g on E_m there corresponds the automorphism $\rho = \rho_g$ of S_m that maps $x \in E_n$ onto $\rho x = gxg^*$ (where g^* is the transpose of g). The tube T_{S_m} over this cone is known as the *Siegel generalized upper half-plane* (when $m = 1 = n$ this tube is the upper half-plane E_2^+).

(iii) When $n = m^2$, E_n can be identified with the real vector space of $m \times m$ complex hermitian matrices (that is, complex $m \times m$ matrices x that are equal to their conjugate transpose x^*). $H_m \subset E_n$ is then the cone of all positive definite $m \times m$ complex hermitian matrices. $\Sigma(H_m)$ can be obtained from the complex general linear group $GL(\mathbf{C}_m)$: To each nonsingular linear transformation g on $\mathbf{C}_m$ there corresponds the automorphism $\rho = \rho_g$ of H_m that maps $x \in E_n$ onto gxg^* where g^* is the conjugate transpose of g.

(iv) When $n = 2m^2 - m$, E_n can be identified with the real vector space of $m \times m$ quaternionic hermitian matrices. $Q_m \subset E_n$ is then the cone of all positive definite $m \times m$ quaternionic hermitian matrices. $\Sigma(Q_m)$ can be obtained from the quaternionic general linear group by letting $\rho_g x = gxg^*$ for $x \in E_n$, where g^* is the quaternionic conjugate transpose of the nonsingular $m \times m$ quaternionic matrix g.

The Cauchy and Poisson kernels associated with these cones can be calculated explicitly. When $x + iy$ is a point of the tube T_{C_n} whose base is the forward light cone then the Poisson kernel evaluated at this point has the value

$$\mathscr{P}(x, y) = c_n \frac{(y, y)^{n/2}}{\{[(x, x) - (y, y)]^2 + 4(x, y)^2\}^{n/2}} = c_n \frac{(y, y)^{n/2}}{|(x + iy, x + iy)|^n},$$

when $x + iy \in T_{S_m}$, $\mathscr{P}(x, y) = a_n\{\det y/|\det (x + iy)|^2\}^{(n+1)/2}$, for $x + iy \in T_{H_m}$, $\mathscr{P}(x, y) = a_n'\{\det y/|\det (x + iy)|^2\}^n$. To each quaternion $a_0 + ia_1 + ja_2 + ka_3 = (a_0 + ja_2) + i(a_1 + ja_3)$ there corresponds a 2×2 matrix over the complex numbers

$$\begin{pmatrix} \alpha & -\bar{\beta} \\ \beta & \bar{\alpha} \end{pmatrix},$$

where $\alpha = (a_0 + ja_2)$ and $\beta = (a_1 + ja_3)$. This correspondence is an isomorphism and can be extended between $m \times m$ quaternionic hermitian matrices x and $(2m) \times (2m)$ matrices x' with complex coefficients. In terms of this correspondence we have, for $x + iy \in T_{Q_m}$, $\mathscr{P}(x, y) = a_n''\{\det y'/|\det (x' + iy')|^2\}^{(2n-1)/2}$.

6.8. The result described in 6.6 when Γ is the first octant is not true for general cones. In fact, unrestricted convergence of Poisson integrals may fail even for functions in $L^p(E_n)$, $p > 1$. For example, when $\Gamma = C_n$ (the forward light cone), $n \geqq 3$, and $1 \leqq p < \infty$ then there exists an $f \in L^p(E_n)$ such that for almost every $x \in E_n$

$$\limsup u(x, y) = \limsup \int_{E_n} \mathscr{P}(x - t, y) f(t)\, dt = \infty$$

as $y \in \Gamma$ tends to 0 unrestrictedly (see Stein and N. J. Weiss [1]). We do obtain a positive result, however, when we limit ourselves to restricted convergence and when the base Γ of the tube is one of the domains described in 6.7. In fact, if $f \in L^p(E_n)$, $1 \leqq p$, then, for almost every $x \in E_n$, the Poisson integral $u(x, y)$ converges to $f(x)$ as $y \in \Gamma$ tends to 0 restrictedly see Stein and N. J. Weiss [1]).

6.9. The classical *Cayley transform* is the linear fractional transformation $w = (z - i)/(z + i)$ mapping the upper half-plane onto the unit disk $|w| < 1$. The Siegel upper half-plane corresponding to S_m described in 6.7 (ii) can be mapped by a generalized Cayley transform onto the "disc" D of $m \times m$ complex symmetric matrices w satisfying $w^*w < I$, where I is the $m \times m$ identity matrix. This mapping is given by $w = (z - iI) \times (z + iI)^{-1}$ when $z = x + iy \in T_{S_m}$. The "distinguished boundary" or "spine" $\{z = x + iy \in \mathbf{C}_n; y = 0\}$ then corresponds to the symmetric unitary matrices. There are similar transformations of the tubes over the other classical domains introduced in 6.7 (see Bochner [1], Pyatetskii-Shapiro [1] and Korányi and Wolf [1]). The images of these tubes form an important subclass of the *bounded symmetric domains* of Cartan (see Hua [1] and Helgason [1]).

6.10. Let D be the domain of $m \times m$ complex matrices w satisfying $w^*w < I$. We can consider D as a domain in $\mathbf{C}_{m^2}$. We say that a complex valued function F belongs to $H^p(D)$, $p > 0$, if it is holomorphic in D and $\int_U |F(\rho u)|^p\, du \leqq M < \infty$ for $0 \leqq \rho < 1$, where u ranges over the unitary group U and du is the element of Haar measure on this group. It can be shown that for $F \in H^p(D)$, $p > 0$, (i) $\lim_{\rho \to 1} F(\rho u) = F(u)$ exists for almost every $u \in U$ and (ii) $\int_U |F(\rho u) - F(u)|^p\, du \to 0$ as $\rho \to 1$. A similar result holds for any Reinhardt circular region and, thus, for the bounded symmetric domains of Cartan (see Bochner [5]). When D is the *polydisc* $\{z = (z_1, z_2, \ldots, z_n) \in \mathbf{C}_n; |z_j| < 1, j = 1, 2, \ldots, n\}$ the space $H^p(D)$. $p > 0$, consists of those holomorphic functions F defined in D that satisfy $\int_0^{2\pi} \cdots \int_0^{2\pi} |F(r_1 e^{i\theta_1}, \ldots, r_n e^{i\theta_n})|^p\, d\theta_1 \cdots d\theta_n \leqq M < \infty$ for $0 \leqq r_j < 1$, $j = 1, 2, \ldots, n$. Similarly, when D is the open unit sphere $\{z = (z_1, \ldots, z_n) \in \mathbf{C}_n; |z_1|^2 + \cdots + |z_n|^2 < 1\}$ a holomorphic function F defined in D is said to belong to $H^p(D)$ if $\int_{\partial D} |F(\rho z')|^2\, dz' \leqq M < \infty$ for $0 \leqq \rho < 1$, where the boundary ∂D of D may be identified with the surface Σ_{2n-1} of the unit sphere in E_{2n} and dz' is then the element of surface area. In these cases we can strengthen the two results (i) and (ii) to include nontangential convergence to the boundary. When D is the polydisc the approach can be "unrestricted"; that is, the radii $r_1, r_2, \ldots, r_n$ can approach 1 independently (for details see Zygmund [1] and [3]). See also Korányi [1].

6.11. Suppose D is the domain $\{z = (z_1, z_2, \ldots, z_n); \mathscr{I} z_1 > \sum_{j=2}^{n} |z_j|^2\}$. The mapping $z \to w = (w_1, w_2, \ldots, w_n)$ given by $w_1 = (z_1 - i)/(z_1 + i)$ and $w_j = 2z_j/(z_j + i), j = 2, \ldots, n$, transforms D into the open unit sphere $\{w \in \mathbf{C}_n; \sum_{j=1}^{n} |w_j|^2 < 1\}$ (when $n = 1$ this mapping is the Cayley transform). A general class of domains that contains D as well as the tube domains over convex cones has been introduced by Pyatetskii–Shapiro [1]; these domains are called by him *Siegel domains of the second kind* (see also Korányi [2]). An H^p theory for such domains generalizing Theorem 5.1 can be found in Stein [6].

6.12. Tubes over the n-dimensional cube $Q = Q_n = \{y \in E_n; 0 < y_j < 1, j = 1, 2, \ldots, n\}$ share many properties with tubes over the first octant. For example, a result analogous to Theorem 5.1 holds for the former domains: If F belongs to $H^p(T_Q), p > 0$, and y_0 is a vertex of Q, then F has a nontangential limit $F(x + iy_0)$ in each variable at almost every x in E_n. In particular $\lim_{y \in Q,\, y \to y_0} F(x + iy) = F(x + iy_0)$ exists for almost every $x \in E_n$. Moreover,

$$\int_{E_n} |F(x + iy) - F(x + iy_0)|^p \, dx$$

tends to zero as $y \in Q$ tends to y_0. The proof is similar to that of Theorem 5.1. The rôle of the upper half-plane is assumed by the band $\{z = x + iy \in \mathbf{C}_1; x \in E_1, 0 < y < 1\}$ and the Poisson integral of a function f defined on the boundary of this region is given by the formula

$$u(x + iy) = \tfrac{1}{2} \sin \pi y \int_{-\infty}^{\infty} \left[\frac{f(x - t)}{\cosh \pi t - \cos \pi y} + \frac{f(x - t + i)}{\cosh \pi t + \cos \pi y}\right] dt$$

(see comments following the proof of Lemma 4.2 in Chapter V). With the aid of an appropriate linear transformation we can extend this result to H^p spaces of functions defined on tubes with polyhedral bases. We then obtain the following extension of Theorem 5.5 (and Theorem 5.5′):

If F belongs to $H^p(T_B), p > 0$, where B is an open convex subset of E_n, and y_0 is a point on the boundary of B then the restricted limits

$$\lim_{y \to y_0,\, y \in B} F(x + iy) = F(x + iy_0)$$

exist for almost every $x \in E_n$ *and* $\int_{E_n} |F(x + iy) - F(x + iy_0)|^p \, dx \to 0$ *as* $y \in B$ *tends restrictedly to* y_0.

6.13. In the third section of this chapter we introduced a Poisson kernel associated with each regular cone $\Gamma \subset E_n$. This was done by extending to n-dimensions a relationship between the classical Poisson and Cauchy kernels associated with the upper half-plane (i.e., the tube whose base $\Gamma \subset E_1$ consists of the positive reals). Other relations of this kind exist. Perhaps the simplest is expressed by the following equality:

$$K(x + iy) = -\frac{1}{2\pi i(x + iy)} = \frac{1}{2}\left\{\frac{y}{\pi(x^2 + y^2)} + i\,\frac{x}{\pi(x^2 + y^2)}\right\}$$
$$= \frac{1}{2}\{P(x, y) + iQ(x, y)\}.$$

Consequently, $P(x, y) = 2 \operatorname{Re}\{K(x + iy)\}$. The kernel $Q(x, y) = 2\mathscr{I}\{K(x + iy)\}$ is called the *conjugate Poisson kernel.* Motivated by this relationship, therefore, one might be led to consider

$$P(x, y) = 2 \operatorname{Re}\{K(x + iy)\} = 2 \operatorname{Re}\left\{\int_{\Gamma^*} e^{2\pi i(x+iy)\cdot t}\, dt\right\}$$

and

$$Q(x, y) = 2\mathscr{I}\{K(x + iy)\} = 2\mathscr{I}\left\{\int_{\Gamma^*} e^{2\pi i(x+iy)\cdot t}\, dt\right\}$$

as generalizations of the Poisson and conjugate Poisson kernels associated with a regular cone $\Gamma \subset E_n$. For example, when Γ is the first quadrant of E_2,

$$P(x_1, x_2; y_1, y_2) = P(x, y) = \frac{1}{\pi^2}\frac{y_1 y_2 - x_1 x_2}{(x_1^2 + y_1^2)(x_2^2 + y_2^2)}\cdot$$

In particular, we see that $P(x, y)$ is not necessarily a positive kernel. For this and as well as other reasons this kernel is not a satisfactory approximation to the identity. Nevertheless, it can be used together with $Q(x, y)$ in order to see how the real and imaginary parts of an H^2 function are related to each other and to their boundary values.

If $\Gamma \subset E_n$ is a regular convex open cone we let χ^* denote the characteristic function of the dual cone Γ^* and sgn* the *signum function* associated with Γ^*: $\operatorname{sgn}^* t = 1$ if $t \in \Gamma^*$, $\operatorname{sgn}^* t = -1$ if $-t \in \Gamma^*$ and $\operatorname{sgn}^* t = 0$ otherwise. It then follows immediately from the definitions above that

$$P(x, y) = \int_{E_n} e^{2\pi i x\cdot t} e^{-2\pi|y\cdot t|}[\chi^*(t) + \chi^*(-t)]\, dt$$

and

$$Q(x, y) = -i \int_{E_n} e^{2\pi i x\cdot t} e^{-2\pi|y\cdot t|}[\chi^*(t) - \chi^*(-t)]\, dt$$

$$= -i \int_{E_n} e^{2\pi i x\cdot t} e^{-2\pi|y\cdot t|} \operatorname{sgn}^* t\, dt.$$

Thus, for each $y \in \Gamma$, $\hat{P}_y(t) = [\chi^*(t) + \chi^*(-t)]e^{-2\pi|y\cdot t|}$ and $\hat{Q}_y(t) = (-i \operatorname{sgn}^* t)e^{-2\pi|y\cdot t|}$ are the values of the Fourier transforms of $P_y = P(\cdot, y)$ and $Q_y = Q(\cdot, y)$.

Now suppose $F = u + iv \in H^2(T)$. By Corollary 3.4 the L^2 boundary values $\lim_{y\in\Gamma, y\to 0} F(x + iy) = u(x) + iv(x)$ exist. It is not hard to establish the following results:

(i) *For all* $z \in T_\Gamma$, $F(z) = \int_{E_n} 2K(z - \xi)u(\xi)\, d\xi$;

(ii) $v(z) = v(x + iy) = \int_{E_n} Q(x - \xi, y)u(\xi)\, d\xi$, *when* $z \in T_\Gamma$, *and the boundary values* $v(x)$ *and* $u(x)$ *satisfy the relationship* $\hat{v}(t) = (-i \operatorname{sgn}^* t)\hat{u}(t)$;

(iii) *A real-valued function* $u \in L^2(E_n)$ *is the real part of the* L^2 *boundary value* $F(x) = \lim_{y\in\Gamma, y\to 0} F(x + iy)$ *of an* $F \in H^2(T_\Gamma)$ *if and only if* $u(t)$ *vanishes for almost every* t *outside* $\Gamma^* \cup (-\Gamma^*)$, *where* $-\Gamma^* = \{t \in E_n;\ -t \in \Gamma^*\}$.

The first part of (ii) follows immediately from (i) by equating imaginary parts. The second part is then obtained by taking the Fourier transform of both sides and passing to the limit as $y \in \Gamma$ tends to 0. If $u \in L^2(E_n)$ is the real part of the boundary value of an $F \in H^2(T_\Gamma)$ then from (i) and the definition of the Cauchy kernel we obtain

$$F(z) = 2 \int_{E_n} K(z - \xi)u(\xi)\, d\xi = 2 \int_{\Gamma^*} \hat{u}(t)e^{2\pi i z\cdot t}\, dt$$

for all $z \in T_\Gamma$. We note that, in any case, an F defined in this way belongs to $H^2(T_\Gamma)$ (this follows from Theorem 3.1) and its real part $u_y(x) = u(x + iy)$ equals $\int_{E_n} P(x - \xi, y)u(\xi)\, d\xi$. Thus, taking Fourier transforms

$$\hat{u}_y(t) = e^{-2\pi|y\cdot t|}(\chi^*(t) + \chi^*(-t))\hat{u}(t).$$

Part (iii) is an immediate consequence of this equality. Part (i) is obtained by adding the left and right sides of the equalities

$$F(z) = \int_{E_n} K(z - \xi)F(\xi)\, d\xi$$

and $0 = \int_{E_n} K(z - \xi)\overline{F(\xi)}\, d\xi$. The first of these identities follows from

Theorem 3.6. The second is an easy consequence of Theorem 3.1 and the definition of the Cauchy kernel.

6.14. In the case when $n = 1$ and Γ is the positive real axis then every real-valued $u \in L^2(E_1) = L^2(-\infty, \infty)$ is the real part of the boundary value of an $F = u + iv \in H^2(E_2^+)$. This is an immediate consequence of 6.13, (iii) and the fact that $\Gamma^* \cup (-\Gamma^*) = (-\infty, \infty) = E_1$. From (ii) and Plancherel's theorem we see that the mapping $u \to v$ is well defined and is an isometry of $L^2(E_1)$ (in fact, it is a unitary transformation). This mapping is called the *Hilbert transform*. We shall show later in Chapter V, Section 2, that the Hilbert transform can be defined on $L^p(E_1)$ and is a bounded linear transformation into $L^p(E_1)$, $1 < p < \infty$. This fact is equivalent to saying that there exists a constant $A_p < \infty$ such that, whenever $u + iv$ is the boundary value of an $F \in H^p(E_2^+)$,

$$\|v\|_p = \left(\int_{-\infty}^{\infty} |v(x)|^p \, dx\right)^{1/p} \leqq A_p\left(\int_{-\infty}^{\infty} |u(x)|^p \, dx\right)^{1/p} = A_p\|u\|_p.$$

This inequality is known as the M. Riesz inequality. This one-dimensional result can be extended to n dimensions:

If $F = u + iv \in H^p(T_\Gamma)$, $1 < p < \infty$, where Γ is an open convex cone in E_n, then there exists a constant $A_p < \infty$ such that for $y \in \Gamma$, and for $y = 0$ (this limiting value can be taken in the sense of Theorem 5.6)

$$\left(\int_{E_n} |v(x + iy)|^p \, dx\right)^{1/p} \leqq A_p\left(\int_{E_n} |u(x + iy)|^p \, dx\right)^{1/p}.$$

By a change of coordinates we can reduce this to the case where the ray $(\eta_1, 0, \ldots, 0)$, $\eta_1 > 0$, belongs to Γ. If $y \in \Gamma$ is fixed and $\eta = (\eta_1, 0, \ldots, 0)$, $\eta_1 > 0$, then $y + \eta \in \Gamma$ and, by the one-dimensional result,

$$\int_{-\infty}^{\infty} |v(x_1 + iy_1 + i\eta_1, \ldots, x_n + iy_n)|^p \, dx_1$$
$$\leqq A_p^p \int_{-\infty}^{\infty} |u(x_1 + iy_1 + i\eta_1, \ldots, x_n + iy_n)|^p \, dx_1.$$

Integrating both sides with respect to $x_2, \ldots, x_n$ and then letting $\eta_1 \to 0$ we obtain the desired inequality. The case $y = 0$ follows from this one by letting $y \in \Gamma$ tend to 0.

6.15. Let K be any open convex set containing no (entire straight) lines. Suppose $p \notin K$ and let Γ be the cone generated by p and K. Then Γ does not contain any (entire straight) lines. To see this assume first that $n = 2$.

Consider the supporting lines to K. Either they are all parallel (which would imply that K contains such a line) or there are two properly intersecting ones. Their intersection determines a cone Γ_0 containing K such that no such line lies within Γ_0. The required property for Γ is now obvious. Assume now that $n > 2$, and that l is a line lying in Γ. Let π be a plane containing l and p and let $K' = K \cap \pi$, $\Gamma' = \Gamma \cap \pi$. Then K' is open, convex and does not contain any line; moreover, Γ' is the cone generated by K' and p, and contains the line l, which contradicts the case $n = 2$, already proved.

6.16. If the convex open cone Γ contains no straight lines, then Γ is regular. It follows from Theorem 3.1 that $H^2(T_\Gamma)$ contains a function which is not identically zero. To prove the assertion about Γ one uses the fact (see Rockafellar [1], Theorem 14.1) that $\bar{\Gamma} = (\Gamma^*)^*$. If Γ^* had no interior it would be contained in a hyperplane, which would imply that $\bar{\Gamma}$ contained a straight line, and hence Γ would contain a straight line.

Bibliographical Notes

The basic results of the classical theory of H^p spaces can be found in Zygmund [1], Chapter VII. The H^2 theory over tubes was first studied by Bochner to whom the basic result, Theorem 2.3, is due (see Bochner [8]). Many of the results included in Sections 2, 3 and 5 are presented in Stein, G. Weiss and M. Weiss [1]. The Cauchy kernel associated with a convex cone was introduced by Bochner (see [1]), he also calculated this kernel explicitly for cones corresponding to the classical domains. For other extensions of the Paley–Wiener theorem to n-dimensions see Plancherel and Polya [1] and Stein [1]. The notion of curvilinear restricted convergence described in (6.5) was developed by Calderón and Zygmund [1]. The basic results concerning the Hilbert transform are presented in Titchmarsh [2] and Zygmund [1]. For earlier papers dealing with H^p theory of several complex variables see Bochner [2], Zygmund [3] and Bochner [5]. For an introduction to the theory of convex sets see Valentine [1], and Rockafellar [1].

CHAPTER IV

Symmetry Properties of the Fourier Transform

Fourier analysis is intimately connected with the action of the group of translations on Euclidean space. If we were to restrict our study only to questions related to this translation structure we would present little that cannot be found in a treatment of abstract harmonic analysis on locally compact Abelian groups and, at the same time, lose much of its elegant generality. Harmonic analysis in Euclidean spaces, however, is richer because of its connection with several classes of transformations: the dilations and the rotations as well as the translations. This connection has already appeared implicitly in our discussion of harmonic and holomorphic functions. We now wish to make quite explicit the further connection between the Fourier transform, dilations, and rotations.

The starting point for this is the observation that the Fourier transform has a very simple transformation law under dilations and that, in addition, it commutes with the action of rotations. This second property leads to a direct sum decomposition of $L^2(E_n)$ into various parts, each transforming in a specific manner with respect to the action of rotations. The Fourier transform preserves this direct sum and its restriction to each of these parts can be identified with a classical Bessel transform. This decomposition is obtained by a passage to polar coordinates enabling us first to decompose the L^2 spaces of functions on the surface of the unit sphere of E_n. This decomposition is obtained in terms of the restriction to the surface of the unit sphere of homogeneous harmonic polynomials. The study of these functions, the spherical harmonics, is very natural in view of the considerations mentioned above because the Laplacian is both translation and rotation invariant while the homogeneity of these polynomials reflects the action of dilations.

This chapter is organized in the following way: the first section treats the special two-dimensional case, which is particularly simple since the decomposition of L^2 of the surface of the unit sphere is obtained in terms of the usual one-dimensional Fourier series expansions of square integrable functions. The second section generalizes these expansions to n dimensions by developing the theory of spherical harmonics. The third section begins with the study of the space of radial functions, the simplest of the summands occurring in the above mentioned decomposition of $L^2(E_n)$, and then treats the rest of the direct sum decomposition. The fourth section deals with some applications of this theory involving several important identities in the theory of potentials and singular integrals, where the interplay of dilations and rotations is evident.

1. The Decomposition of $L^2(E_2)$ into Subspaces Invariant Under the Fourier Transform

Our principal goal in this chapter is to study in much greater detail the action of the Fourier transform on functions defined on E_n. This will be done by finding a natural direct sum decomposition of $L^2(E_n)$ that is preserved by the Fourier transform and, then, by making a detailed analysis of how the latter acts on each of the summands. In this connection we will be led to a development of the basic properties of Bessel functions and spherical harmonics. These are of fundamental importance, not only in other aspects of harmonic analysis, but in many other branches of analysis as well.

In one dimension we have a simple natural decomposition of $L^2(E_1)$. The subspace of even functions and the subspace of odd functions are easily seen to be invariant under the Fourier transform, they are orthogonal to each other and their direct sum is all of $L^2(E_1)$. These properties, obtained by decomposing a function f into the sum $f = f_e + f_0$, where the *even part* f_e is defined by $f_e(x) = [f(x) + f(-x)]/2$ and the *odd part* f_0 by $f_0(x) = [f(x) - f(-x)]/2$, are clearly also valid in n dimensions. A much more interesting, though more complicated, situation arises when we consider the following higher dimensional generalization of the even part of a function. Given a locally integrable function f on E_n, its *radial part* is the function φ defined by

$$\varphi(x) = \frac{1}{\omega_{n-1}} \int_{\Sigma_{n-1}} f(rx')\,dx',$$

where, we recall, $r = |x|$, $x' = x/r$ (when $x \neq 0$) and ω_{n-1} is the area of the surface of the unit sphere Σ_{n-1} of E_n. It is clear that φ is a *radial function*; that is, φ depends only on $r = |x|$. When $n = 1$ we see that $\varphi = f_e$.

This notion, which is certainly natural when considering polar coordinates, is also appropriate when studying the action of the Fourier transform. The following remarks, from which we will deduce that the Fourier transform of a radial function is radial, will make this clear. We first observe that a function f, defined on E_n, is radial if and only if $f(\rho x) = f(x)$ for all orthogonal transformations[1] ρ of E_n and all x in E_n. The basic property

[1] We recall that ρ is *orthogonal* if it is a linear operator on E_n that preserves the inner product: $\rho x \cdot \rho y = x \cdot y$ for all x, y in E_n. If $\det \rho = 1$, ρ is called a *rotation*. It is also true that f is radial if and only if $f(\rho x) = f(x)$ for all rotations ρ and all x in E_n, when $n > 1$.

of the Fourier transform with respect to orthogonal transformations is the following result:

THEOREM 1.1. *The Fourier transform, $\mathscr{F}$, commutes with orthogonal transformations. That is, if ρ is an orthogonal transformation, let R_ρ be the mapping taking a function f on E_n into a funtion g whose values are $g(x) = (R_\rho f)(x) = f(\rho x)$ for $x \in E_n$, then, whenever $f \in L^1(E_n)$,*

$$\hat{g}(t) = (\mathscr{F}g)(t) = (\mathscr{F}R_\rho f)(t) = (R_\rho \mathscr{F} f)(t) = (\mathscr{F}f)(\rho t) = \hat{f}(\rho t);$$

that is, the operators $\mathscr{F}$ and R_ρ commute: $\mathscr{F}R_\rho = R_\rho\mathscr{F}$.

PROOF. Since the adjoint of ρ is also its inverse, and the Jacobian in the change of variables $w = \rho x$ is one, we have

$$\hat{g}(t) = \int_{E_n} e^{-2\pi i t \cdot x} f(\rho x)\, dx = \int_{E_n} e^{-2\pi i t \cdot \rho^{-1} w} f(w)\, dw$$
$$= \int_{E_n} e^{-2\pi i \rho t \cdot w} f(w)\, dw = \hat{f}(\rho t).$$

Since it is true that whenever $|x_1| = |x_2|$ for two points of E_n there is an orthogonal transformation ρ such that $\rho x_1 = x_2$, we immediately obtain the above mentioned property of the Fourier transform:

COROLLARY 1.2. *If f is a radial function in $L^1(E_n)$ then $\hat{f}$ is also radial.*

This result extends in an obvious way to $L^2(E_n)$. We thus obtain a direct sum decomposition of this space which is left invariant by the Fourier transform simply by considering the subspace $\mathfrak{H}^0$ of those functions equal almost everywhere to radial functions, and the orthogonal complement of $\mathfrak{H}^0$ (this is an immediate consequence of the Plancherel theorem). Consequently, it is natural to ask whether there exists a simple description of the space of functions orthogonal to the radial ones and, also, if we can obtain more detailed information on the action of the Fourier transform on this subspace. We first consider these questions in the two dimensional case.

Let us choose an f in $L^2(E_2)$ and let us use the standard identification of the points (x, y) in E_2 with the complex numbers $x + iy = z = re^{i\theta}$. Fubini's theorem implies that, for almost every r, $f(re^{i\theta})$ defines a square integrable function of θ. Consequently, we have a Fourier series expansion

$$f(re^{i\theta}) \sim \sum_{k=-\infty}^{\infty} f_k(r) e^{ik\theta} \tag{1.3}$$

which converges to $f(re^{i\theta})$ in the L^2-norm for almost every r. Moreover, $\sum_{-\infty}^{\infty} |f_k(r)|^2 = (2\pi)^{-1} \int_0^{2\pi} |f(re^{i\theta})|^2\, d\theta$ a.e.; hence, by the Lebesgue monotone convergence theorem,

$$\lim_{n\to\infty}\int_0^\infty \sum_{-n}^{n} |f_k(r)|^2 r\,dr = \frac{1}{2\pi}\int_0^\infty \left\{\int_0^{2\pi} |f(re^{i\theta})|^2\,d\theta\right\} r\,dr = \frac{1}{2\pi}\,\|f\|_2^2.$$

Putting $g_k(z) = f_k(r)e^{ik\theta}$, $k = 0, \pm 1, \pm 2, \ldots$, we thus obtain

$$\lim_{n\to\infty}\int_{E_2} |f(z) - \sum_{-n}^{n} g_k(z)|^2\,dz$$

$$= \lim_{n\to\infty} 2\pi \int_0^\infty \left[\frac{1}{2\pi}\left\{\int_0^{2\pi} |f(re^{i\theta})|^2\,d\theta\right\} - \sum_{-n}^{n} |f_k(r)|^2\right] r\,dr = 0.$$

This, together with the orthogonality relations satisfied by the exponential functions $e^{ik\theta}$, $k = 0, \pm 1, \pm 2, \ldots$, shows that we have the direct sum decomposition

$$L^2(E_2) = \sum_{k=-\infty}^{\infty} \oplus\, \mathfrak{H}^k, \tag{1.4}$$ [2]

where $\mathfrak{H}^k = \{g \in L^2(E_2);\ g(z) = f(r)e^{ik\theta}$ a.e. for some measurable function satisfying $\int_0^\infty |f(r)|^2 r\,dr < \infty\}$. This notation is consistent with that used above since $\mathfrak{H}^0$, in either case, is precisely the subspace of those square integrable functions that are equal almost everywhere to radial functions. Moreover, we see from (1.4) that the orthogonal complement of $\mathfrak{H}^0$ is the direct sum of the subspaces $\mathfrak{H}^k$, $k \neq 0$.

We know that the Fourier transform maps $\mathfrak{H}^0$ into itself. It is easy to see that it also maps each of the other spaces $\mathfrak{H}^k$, $k = \pm 1, \pm 2, \ldots$, into themselves: First suppose g belongs to $\mathfrak{H}^k \cap L^1(E_2)$ and set $h(z) = g(e^{i\phi}z)$ for ϕ fixed. We then have $h(z) = e^{ik\phi}g(z)$ for almost every z. On the other hand, multiplication by $e^{i\phi}$ is a rotation of E_2 and, since the Fourier transform commutes with rotations (Theorem 1.1), we have

$$\hat{g}(e^{i\phi}w) = \hat{h}(w) = e^{ik\phi}\hat{g}(w)$$

for all w and ϕ. Putting $w = r \geqq 0$ we see that $\hat{g}$ also belongs to $\mathfrak{H}^k$. Since $\mathfrak{H}^k \cap L^1(E_2)$ is dense in the closed subspace $\mathfrak{H}^k$ this shows that all of this last space is mapped into itself by the Fourier transform. Because of the Plancherel theorem we can deduce that all the spaces $\mathfrak{H}^k$ are mapped onto each other.

Therefore, we have found a direct sum decomposition of $L^2(E_2)$, namely (1.4), that is left invariant by the Fourier transform. This answers our first

[2] By such a direct sum we mean that each of the subspaces $\mathfrak{H}^k$ are closed in $L^2(E_2)$, are mutually orthogonal and the closure of their linear span consists of all of $L^2(E_2)$.

question. In order to answer the second one we now engage in a more detailed study of how the Fourier transform acts on each of the summands $\mathfrak{H}^k$.

Let us choose an f belonging to $\mathfrak{H}^k$; this f must have the form $f(z) = f_0(r)e^{ik\theta}$ for almost every $z = re^{i\theta}$. Since $\hat{f}$ must also belong to $\mathfrak{H}^k$ it has a similar form; that is, $\hat{f}(w) = F_0(R)e^{ik\phi}$ for almost every $w = Re^{i\phi}$. Assuming f is integrable we shall now compute F_0 explicitly (our results extend in an obvious way to square integrable f). Putting $w = Re^{i0} = R$, we have

$$F_0(R) = \hat{f}(Re^{i0}) = \int_0^\infty f_0(r)\left\{\int_0^{2\pi} e^{-2\pi iRr\cos\theta}e^{ik\theta}\,d\theta\right\}r\,dr$$

$$= (-i)^k 2\pi \int_0^\infty f_0(r)\left\{\frac{1}{2\pi}\int_0^{2\pi} e^{2\pi iRr\sin\theta}e^{-ik\theta}\,d\theta\right\}r\,dr.$$

Thus, we see that an explicit expression for $F_0(R)$ in terms of f_0 involves the Fourier coefficients of the function of θ given by the expression $e^{it\sin\theta}$. These coefficients, which depend on t, are known as the *Bessel functions* $J_k(t)$. That is, the expression

$$J_k(t) = \frac{1}{2\pi}\int_0^{2\pi} e^{it\sin\theta}e^{-ik\theta}\,d\theta$$

defines the Bessel function J_k, $k = 0, \pm 1, \pm 2, \ldots$. By a simple change of variables we obtain the relation, valid for all integers k,

$$J_k(t) = (-1)^k J_{-k}(t). \tag{1.5}$$

Hence, we can state our main result on the action of the Fourier transform on the spaces $\mathfrak{H}^k$ in the following way:

THEOREM 1.6. *Suppose f belongs to $L^1(E_2)$ and $f(z) = f_0(r)e^{ik\theta}$, where $z = re^{i\theta}$; then $\hat{f}(w) = F_0(R)e^{ik\phi}$, where $w = Re^{i\phi}$ and $F_0(R) = 2\pi i^k \int_0^\infty f_0(r)J_{-k}(2\pi Rr)r\,dr = 2\pi(-i)^k\int_0^\infty f_0(r)J_k(2\pi Rr)r\,dr$.*

2. Spherical Harmonics

In order to extend this decomposition to n dimensions it would be desirable to be able to represent functions defined on the surface of the unit sphere by expansions similar to Fourier series. This would enable us to obtain a development for $f(x) = f(rx')$ analogous to (1.3). We shall show that this can be done by letting the rôle of the exponentials $e^{ik\theta}$, $k = 0, \pm 1, \pm 2, \ldots$, be assumed by a class of functions called the *spherical harmonics*. Before defining these functions we make some simple observations about trigonometric series $s(\theta) = \sum_{k=-\infty}^{\infty} c_k e^{ik\theta}$.

We consider such a series to be the sequence of *symmetric partial sums* $s_n(\theta) = \sum_{k=-n}^{n} c_k e^{ik\theta} = c_0 + \sum_{k=1}^{n} (c_k e^{ik\theta} + c_{-k} e^{-ik\theta})$. These sums are real valued provided $c_k = \bar{c}_{-k}$. In this case $s_n(\theta)$ is the restriction to the unit circle Σ_1 of the real part $u_n(z)$ of the polynomial $p_n(z) = c_0 + 2\sum_{k=1}^{n} c_k z^k$, $z = re^{i\theta} \in \mathbf{C}_1$. Since there is no restriction imposed on the complex coefficients $c_1, c_2, \ldots, c_n$ and c_0 can be any real number, it follows that $u_n(z) = u_n(x + iy)$ is an arbitrary real harmonic polynomial of degree n in the variables x and y. The summands c_0, $\operatorname{Re}\{c_k z^k\} = q_k(z)$, $k = 1, 2, \ldots$, are then the general homogeneous[3] real harmonic polynomials of degree k. Thus, the linear combinations $Y^{(k)}(e^{i\theta}) = c_k e^{ik\theta} + \bar{c}_k e^{-ik\theta} = a_k \cos k\theta + b_k \sin k\theta$, with $c_k = (a_k - ib_k)/2$, are precisely the restrictions to the unit circle $|z| = 1$ of such polynomials $q_k(z) = q_k(re^{i\theta})$. In particular, since each real valued L^2-function on the unit circle can be represented (in the L^2 norm) by its Fourier series, whose terms are such restrictions, we see that $L^2(\Sigma_1)$ is the closure of the span of these functions $Y^{(k)}$, $k = 0, 1, 2, \ldots$.

In general, the restriction to the surface of the unit sphere Σ_{n-1} of a homogeneous harmonic polynomial of degree k is called a *spherical harmonic of degree k*. This section will be devoted to the development of important properties of spherical harmonics. This will enable us, in particular, to show how $L^2(E_n)$ can be decomposed into a direct sum analogous to (1.4), where the summands $\mathfrak{H}_k$ consist of the subspaces of $L^2(E_n)$ that are spanned by functions that are products of radial functions times spherical harmonics of degree k.[4] The Fourier transform leaves these spaces $\mathfrak{H}_k$ invariant and we shall describe its behavior on them in detail, thus obtaining an extension of Theorem 1.6.

We let $\mathscr{P}_k$ be the set of all homogeneous polynomials of degree k on E_n with complex coefficients. Thus, if $P \in \mathscr{P}_k$ then

$$P(x) = \sum_{|\alpha| = k} c_\alpha x^\alpha,$$

where (as in Chapter I) α denotes an n-tuple $(\alpha_1, \alpha_2, \ldots, \alpha_n)$ of nonnegative integers, $|\alpha| = \alpha_1 + \alpha_2 + \cdots + \alpha_n$ and $x^\alpha = x_1^{\alpha_1} x_2^{\alpha_2} \cdots x_n^{\alpha_n}$. It is clear that the set of monomials x^α, $|\alpha| = k$, is a basis for this space. The number of such monomials, on the other hand, is precisely the number, d_k, of ways an n-tuple $\alpha = (\alpha_1, \alpha_2, \ldots, \alpha_n)$ of nonnegative integers can be chosen so

[3] A function f defined on E_n is said to be *homogeneous of degree k* if $f(ax) = a^k f(x)$ for all $x \in E_n$ and $a > 0$.

[4] When $n = 2$ the space $\mathfrak{H}_k$ is spanned by $\mathfrak{H}^k$ and $\mathfrak{H}^{-k}$, $k = 0, 1, 2, \ldots$.

that $\alpha_1 + \alpha_2 + \cdots + \alpha_n = k$. It is not hard to determine d_k. Let us select $n - 1$ boxes out of a linearly ordered array of $n + k - 1$ boxes and let us fill each of the remaining k boxes with a ball. Then, there will be α_1 balls preceding the first box we chose, α_2 balls in the boxes between the first and second one we chose and so on until we reach the α_n balls that follow the last box we chose. In this way we obtain n nonnegative integers $\alpha_1, \alpha_2, \ldots, \alpha_n$ satisfying $\alpha_1 + \alpha_2 + \cdots + \alpha_n = k$ and all such n-tuples are obtainable in this manner. Thus, there are precisely as many such n-tuples as there are ways of selecting $n - 1$ boxes out of a collection of $n + k - 1$. Consequently, the dimension of $\mathscr{P}_k$ is

$$d_k = \binom{n+k-1}{n-1} = \binom{n+k-1}{k} = \frac{(n+k-1)!}{(n-1)!\,k!}.$$

We introduce an inner product $\langle P, Q\rangle$ on $\mathscr{P}_k$ by letting $\langle P, Q\rangle = P(D)\bar{Q}$ for all P, Q in $\mathscr{P}_k$, where $P(D)$ is the differential operator introduced in Chapter I (see (1.9)). Since P and Q are homogeneous polynomials of the same degree, $\langle P, Q\rangle$ is scalar valued; furthermore it is clearly linear in the first variable, conjugate linear in the second variable and hermitian symmetric. In order to verify that we, indeed, have an inner product, we must, therefore, show that $\langle P, P\rangle \geqq 0$, with equality only if $P = 0$. But if $\alpha = (\alpha_1, \ldots, \alpha_n) \neq (\beta_1, \ldots, \beta_n) = \beta$ then

$$\left(\frac{\partial^{\alpha_1}}{\partial x_1^{\alpha_1}} \frac{\partial^{\alpha_2}}{\partial x_2^{\alpha_2}} \cdots \frac{\partial^{\alpha_n}}{\partial x_n^{\alpha_n}}\right) x_1^{\beta_1} x_2^{\beta_2} \cdots x_n^{\beta_n} = 0,$$

while this derivative equals $\alpha_1!\,\alpha_2! \cdots \alpha_n! = \alpha!$ when $\alpha = \beta$. Consequently, if $P(x) = \sum_{|\alpha|=k} c_\alpha x^\alpha$ then $\langle P, P\rangle = \sum_{|\alpha|=k} |c_\alpha|^2 \alpha!$. But this last expression vanishes if and only if all the coefficients c_α are 0.

We shall use this inner product in order to establish the following basic result:

THEOREM 2.1. *If $P \in \mathscr{P}_k$ then*

$$P(x) = P_0(x) + |x|^2 P_1(x) + \cdots + |x|^{2l} P_l(x),$$

where P_j is a homogeneous harmonic polynomial of degree $k - 2j$, $j = 0, 1, \ldots, l$.

PROOF. Any polynomial of degree less than 2 is harmonic; thus, we may assume that $k \geqq 2$. Consider the linear mapping $\varphi\colon \mathscr{P}_k \to \mathscr{P}_{k-2}$ defined by letting $\varphi(P) = \Delta P$ for $P \in \mathscr{P}_k$, where Δ is the Laplace operator. We first show that φ maps $\mathscr{P}_k$ onto $\mathscr{P}_{k-2}$. If this were not the case we could find a nonzero $Q \in \mathscr{P}_{k-2}$ that is orthogonal to $\mathscr{R}(\varphi)$, the range of φ. That is

$$\overline{\langle \Delta P, Q\rangle} = \langle Q, \Delta P\rangle = 0$$

for all $P \in \mathscr{P}_k$. In particular, this must be true for $P(x) = |x|^2 Q(x)$. Thus

$$0 = \langle Q, \Delta P \rangle = Q(D)\,\overline{\Delta P} = \Delta Q(D)\bar{P} = P(D)\bar{P} = \langle P, P \rangle.$$

But this is impossible since $P \not\equiv 0$.

Let $\mathscr{A}_j \subset \mathscr{P}_j$, $j \geqq 2$, be the class of all polynomials in $\mathscr{P}_j$ that are harmonic. We claim that $\mathscr{P}_j$ is the orthogonal direct sum of $\mathscr{A}_j$ and $\mathscr{B}_j = |x|^2 \mathscr{P}_{j-2} = \{P(x) \in \mathscr{P}_j; P(x) = |x|^2 Q(x) \text{ with } Q \in \mathscr{P}_{j-2}\}$. If $R(x) = |x|^2 Q(x)$ with $Q \in \mathscr{P}_{j-2}$, then $\langle R, P \rangle = 0$ for all $Q \in \mathscr{P}_{j-2}$ if and only if $Q(D)\,\Delta\bar{P} = 0$ for all $Q \in \mathscr{P}_{j-2}$, which is true if and only if $\langle Q, \Delta P \rangle = 0$ for all $Q \in \mathscr{P}_{j-2}$, which in turn is true if and only if $\Delta P = 0$.

In particular, for $j = k$ and $P \in \mathscr{P}_k$ we have $P(x) = P_0(x) + |x|^2 Q(x)$ with P_0 harmonic and $Q \in \mathscr{P}_{k-2}$. Applying our result again for $j = k - 2$ we obtain the decomposition $Q(x) = P_1(x) + |x|^2 Q_1(x)$ with P_1 harmonic and Q_1 in $\mathscr{P}_{k-4}$. Thus, $P(x) = P_0(x) + |x|^2 P_1(x) + |x|^4 Q_1(x)$ and it is clear that the theorem follows by induction.

An immediate consequence of Theorem 2.1 is:

COROLLARY 2.2. *The restriction to the surface of the unit sphere Σ_{n-1} of any polynomial of n-variables is a sum of restrictions to Σ_{n-1} of harmonic polynomials.*[5]

We let $\mathscr{H}_k$ denote the space of spherical harmonics of degree k. $\mathscr{H}_k$ coincides with the collection of all restrictions to Σ_{n-1} of the members of $\mathscr{A}_k$. If we consider the restriction Y of P in $\mathscr{A}_k$ (for $x' \in \Sigma_{n-1}$, $Y(x') = P(x')$) then, because of the homogeneity of P, $P(x) = |x|^k Y(x/|x|)$ for $x \neq 0$. Thus, the restriction map $P \to Y$ has a trivial kernel and must, therefore, be an isomorphism of $\mathscr{A}_k$ onto $\mathscr{H}_k$. In particular,

$$\begin{aligned}\dim \mathscr{H}_k &= \dim \mathscr{A}_k = \dim \mathscr{P}_k - \dim \mathscr{P}_{k-2} = d_k - d_{k-2} \\ &= \binom{n+k-1}{k} - \binom{n+k-3}{k-2}\end{aligned}$$

for $k \geqq 2$. Moreover, $\dim \mathscr{H}_0 = d_0 = 1$ and $\dim \mathscr{H}_1 = d_1 = n$. At the beginning of this section we found that when $n = 2$ the space $\mathscr{H}_k$ is spanned by the two functions $\cos k\theta$ and $\sin k\theta$; thus $\dim \mathscr{H}_k = 2$ for all $k \geqq 1$. This is consistent with the result just obtained:

$$\binom{2+k-1}{k} - \binom{k-1}{k-2} = (k+1) - (k-1) = 2.$$

When $n = 3$ we obtain $\dim \mathscr{H}_k = 2k + 1$ for $k \geqq 0$.

[5] The reader should note that, when $n = 2$, this fact follows easily from the observations we made at the beginning of this section.

The space $\mathscr{A}_k$ is called the space of *solid spherical harmonics*. Sometimes, in order to emphasize the distinction between $\mathscr{H}_k$ and $\mathscr{A}_k$, the members of $\mathscr{H}_k$ are referred to as the *surface spherical harmonics.*

COROLLARY 2.3. *The collection of all finite linear combinations of elements of* $\bigcup_{k=0}^{\infty} \mathscr{H}_k$ *is*

(i) *dense in the space of all continuous functions on* Σ_{n-1} *with respect to the* L^∞ *norm;*

(ii) *dense in* $L^2(\Sigma_{n-1})$.

PROOF. Since the space of continuous functions is dense in $L^2(\Sigma_{n-1})$ it is easy to see that (i) implies (ii): given $\varepsilon > 0$ and $f \in L^2(\Sigma_{n-1})$ choose a continuous g such that $\|f - g\|_2 < \varepsilon/2$. If (i) is true we can then find a finite linear combination, h, of elements of $\bigcup_{k=0}^{\infty} \mathscr{H}_k$ such that $\|g - h\|_\infty < \varepsilon/(2\sqrt{\omega_{n-1}})$. Thus, $\|f - h\|_2 \leqq \|f - g\|_2 + \|g - h\|_2 < \varepsilon/2 + \sqrt{\omega_{n-1}}\, \|g - h\|_\infty < \varepsilon/2 + \varepsilon/2 = \varepsilon$.

On the other hand, (i) follows from the Weierstrass approximation theorem: If g is continuous on Σ_{n-1} we can approximate it uniformly by polynomials restricted to Σ_{n-1}. But, by (2.2), these restrictions are finite linear combinations of elements of $\bigcup_{k=0}^{\infty} \mathscr{H}_k$.

COROLLARY 2.4. *If* $Y^{(k)}$ *and* $Y^{(l)}$ *are spherical harmonics of degrees* k *and* l, *with* $k \neq l$, *then*

$$\int_{\Sigma_{n-1}} Y^{(k)}(x')\, Y^{(l)}(x')\, dx' = 0.$$

PROOF. For $x \neq 0$ in E_n let $r = |x|$ and $x' = x/r$. We then define $u(x) = r^k Y^{(k)}(x')$ and $v(x) = r^l Y^{(l)}(x')$ for $x \neq 0$ and we put $u(0) = 0 = v(0)$ if neither k nor l is 0, while if, say, $k = 0$ then $Y^{(k)}$ is a constant function and we let $u(0)$ be this constant. The derivatives of u and v at $x' \in \Sigma_{n-1}$ in the direction of the outward directed normal to Σ_{n-1} are $(dr^k/dr)\, Y^{(k)}(x') = k\, Y^{(k)}(x')$ and $(dr^l/dr)\, Y^{(l)}(x') = l\, Y^{(l)}(x')$. Moreover, since $Y^{(k)}$ and $Y^{(l)}$ are (surface) spherical harmonics, u and v are solid spherical harmonics. Thus, by Green's theorem,

$$\begin{aligned} 0 = \int_{|x| \leqq 1} (u\, \Delta v - v\, \Delta u)\, dx &= \int_{\Sigma_{n-1}} \left(u \frac{\partial v}{\partial r} - v \frac{\partial u}{\partial r}\right) dx' \\ &= \int_{\Sigma_{n-1}} (l\, Y^{(k)}(x')\, Y^{(l)}(x') - k\, Y^{(k)}(x')\, Y^{(l)}(x'))\, dx' \\ &= (l - k) \int_{\Sigma_{n-1}} Y^{(k)}(x')\, Y^{(l)}(x')\, dx'. \end{aligned}$$

The result now follows by dividing by $(l - k)$, which we can do since $l \neq k$.

Let us consider $\mathcal{H}_k$ as a subspace of $L^2(\Sigma_{n-1})$ with the inner product $(f, g) = \int_{\Sigma_{n-1}} f(x')\overline{g(x')}\, dx'$. If $\{Y_1^{(k)}, \ldots, Y_{a_k}^{(k)}\}$, $a_k = d_k - d_{k-2}$, is an orthonormal basis of $\mathcal{H}_k$ then it follows from Corollary 2.4 that the collection $\bigcup_{k=0}^{\infty} \{Y_1^{(k)}, \ldots, Y_{a_k}^{(k)}\}$ is an orthonormal basis of $L^2(\Sigma_{n-1})$. For, if $f \neq 0$ is orthogonal to all these elements then

$$\begin{aligned} \|f - h\|_2^2 &= \|f\|_2^2 - (f, h) - (h, f) + \|h\|_2^2 \\ &= \|f\|_2^2 + \|h\|_2^2 \geqq \|f\|_2^2 > 0 \end{aligned}$$

whenever h is a finite linear combination of members of $\bigcup_{k=0}^{\infty} \mathcal{H}_k$; but, by Corollary 2.3 part (ii), this is impossible. Thus, if $f \in L^2(\Sigma_{n-1})$ then there exists a unique representation,

$$f = \sum_{k=0}^{\infty} Y^{(k)}, \tag{2.5}$$

where the series on the right converges to f in the L^2 norm and $Y^{(k)} \in \mathcal{H}_k$; in fact, $Y^{(k)} = b_1^{(k)} Y_1^{(k)} + b_2^{(k)} Y_2^{(k)} + \cdots + b_{a_k}^{(k)} Y_{a_k}^{(k)}$ with $b_j^{(k)} = (f, Y_j^{(k)})$, $j = 1, 2, \ldots, a_k$. When $n = 2$ the series in (2.5) is the Fourier series of f. In this case

$$Y_1^{(k)}(e^{i\theta}) = \frac{1}{\sqrt{\pi}} \cos k\theta$$

and

$$Y_2^{(k)}(e^{i\theta}) = \frac{1}{\sqrt{\pi}} \sin k\theta$$

form an orthonormal basis for $\mathcal{H}_k$.

The functions $Y_1^{(k)}$ play a special rôle in the theory of Fourier series. For example, the Abel means $u(r, \theta) = \sum_{k=-\infty}^{\infty} c_k r^{|k|} e^{ik\theta}$, $0 \leqq r < 1$, of the Fourier series of an integrable and periodic function f have simple expressions in terms of the $Y_1^{(k)}$'s and the function f. More precisely, using the basic trigonometric identity

$$\cos(\theta - \phi) = \cos\phi \cos\theta + \sin\phi \sin\theta \tag{2.6}$$

we find that

$$u(r, \theta) = \int_0^{2\pi} f(\phi) p(r, \theta - \phi)\, d\phi,$$

where

$$p(r, \theta) = \frac{1}{\pi}\left\{\frac{1}{2} + \sum_{k=1}^{\infty} r^k \cos k\theta\right\} = \frac{1}{2\pi} \frac{1 - r^2}{1 - 2r\cos\theta + r^2} \tag{2.7}$$

is the Poisson kernel for the unit circle (see Theorem 1.9 in Chapter II). We shall now show that we have a similar situation in higher dimensions. In order to avoid the somewhat exceptional cases of $n = 1$ and $n = 2$ (which were anyway treated above), we shall assume that $n > 2$ throughout the rest of this section.

Let us fix a point x' in Σ_{n-1} and consider the linear functional L on $\mathscr{H}_k$ that assigns to each Y in $\mathscr{H}_k$ the value $Y(x')$. By the self-duality of the finite dimensional inner product space $\mathscr{H}_k$ there exists a unique spherical harmonic $Z_{x'}^{(k)}$ such that

$$L(Y) = Y(x') = \int_{\Sigma_{n-1}} Y(t')Z_{x'}^{(k)}(t')\,dt'$$

for all $Y \in \mathscr{H}_k$. This function $Z_{x'}^{(k)}$ is called the *zonal harmonic of degree k with pole x'*. We now establish some elementary but basic properties of zonal harmonics.

LEMMA 2.8. (a) *If* $\{Y_1, Y_2, \ldots, Y_{a_k}\}$ *is an orthonormal basis of* $\mathscr{H}_k$ *then* $Z_{x'}^{(k)}(t') = \sum_{m=1}^{a_k} \overline{Y_m(x')}\,Y_m(t')$;

(b) $Z_{x'}^{(k)}$ *is real-valued and* $Z_{x'}^{(k)}(t') = Z_{t'}^{(k)}(x')$;

(c) *If* ρ *is a rotation then* $Z_{\rho x'}^{(k)}(\rho t') = Z_{x'}^{(k)}(t')$.

PROOF. Since $\{Y_1, Y_2, \ldots, Y_{a_k}\}$ is an orthonormal basis of $\mathscr{H}_k$ we must have

$$Z_{x'}^{(k)} = \sum_{m=1}^{a_k} (Z_{x'}^{(k)}, Y_m)Y_m.$$

But, by the defining property of zonal harmonics,

$$(Z_{x'}^{(k)}, Y_m) = \int_{\Sigma_{n-1}} \overline{Y_m(t')}Z_{x'}^{(k)}(t')\,dt' = \overline{Y_m(x')}$$

and part (a) is proved. Our derivation of the formula for the dimension of $\mathscr{H}_k$ was independent of whether we considered the space $\mathscr{H}_k$ to consist of real- or complex-valued functions. Consequently, we can choose the orthonormal basis in (a) to consist of real valued functions. If we do this, we see immediately that $Z_{x'}^{(k)}$ must be real-valued. Part (b) is now an immediate consequence of part (a).

Letting $w' = \rho t'$ we have, for $Y \in \mathscr{H}_k$,

$$\int_{\Sigma_{n-1}} Z_{\rho x'}^{(k)}(\rho t')Y(t')\,dt' = \int_{\Sigma_{n-1}} Z_{\rho x'}^{(k)}(w')Y(\rho^{-1}w')\,dw'$$
$$= Y(\rho^{-1}(\rho x')) = Y(x').$$

Thus, by the uniqueness of the representation of linear functionals we must have $Z^{(k)}_{\rho x'}(\rho t') = Z^{(k)}_{x'}(t')$ and part (c) is proved.

COROLLARY 2.9. (a) *$Z^{(k)}_{x'}(x') = a_k \omega_{n-1}^{-1}$ for all $x' \in \Sigma_{n-1}$, where a_k is the dimension of $\mathscr{H}_k$ and ω_{n-1} is the surface area of Σ_{n-1};*

(b) *$\sum_{m=1}^{a_k} |Y_m(x')|^2 = a_k \omega_{n-1}^{-1}$ for all $x' \in \Sigma_{n-1}$, independently of the orthonormal basis $\{Y_1, Y_2, \ldots, Y_{a_k}\}$ of $\mathscr{H}_k$;*

(c) *$|Z^{(k)}_{t'}(x')| \leqq a_k \omega_{n-1}^{-1}$ for all x' and t' in Σ_{n-1}.*

PROOF. Suppose x_1' and x_2' are two points of Σ_{n-1}. We can then find a rotation ρ such that $\rho x_1' = x_2'$. By part (c) of Lemma 2.8, therefore, we must have

$$Z^{(k)}_{x_2'}(x_2') = Z^{(k)}_{x_1'}(x_1').$$

Consequently, $Z^{(k)}_{x'}(x')$ is a constant, independent of $x' \in \Sigma_{n-1}$. From part (a) of Lemma 2.8 we see that this constant, c, must equal $\sum_{m=1}^{a_k} |Y_m(x')|^2$, where $\{Y_1, Y_2, \ldots, Y_{a_k}\}$ is an orthonormal basis of $\mathscr{H}_k$. But then

$$a_k = \sum_{m=1}^{a_k} \int_{\Sigma_{n-1}} |Y_m(x')|^2\, dx' = \int_{\Sigma_{n-1}} \sum_{m=1}^{a_k} |Y_m(x')|^2\, dx'$$

$$= \int_{\Sigma_{n-1}} c\, dx' = c\omega_{n-1}.$$

Thus, $c = a_k \omega_{n-1}^{-1}$ which proves both parts (a) and (b).

In order to show part (c) we first observe that it follows from the defining property of zonal harmonics that for t' and x' in Σ_{n-1}

(i) $$Z^{(k)}_{t'}(x') = \int_{\Sigma_{n-1}} Z^{(k)}_{t'}(w') Z^{(k)}_{x'}(w')\, dw'.$$

On the other hand, if $\{Y_1, Y_2, \ldots, Y_{a_k}\}$ is an orthonormal basis of $\mathscr{H}_k$ it follows from Lemma 2.8, part (a), and the result we have just established that for all $u' \in \Sigma_{n-1}$

$$\|Z^{(k)}_{u'}\|_2^2 = \int_{\Sigma_{n-1}} |Z^{(k)}_{u'}(w')|^2\, dw' = \sum_{m=1}^{a_k} |Y_m(u')|^2 = a_k \omega_{n-1}^{-1}.$$

Thus, (i), Schwarz's inequality and this last equality give us

$$|Z^{(k)}_{t'}(x')| \leqq \|Z^{(k)}_{t'}\|_2 \|Z^{(k)}_{x'}\|_2 = \sqrt{a_k \omega_{n-1}^{-1}}\,\sqrt{a_k \omega_{n-1}^{-1}} = a_k \omega_{n-1}^{-1}$$

and the last part of the corollary is established.[6]

[6] The reader should observe that when $n = 2$ Lemma 2.8 and Corollary 2.9 reduce to familiar elementary properties of the trigonometric functions. For example, Lemma 2.8 (a) is then essentially the identity (2.6) and Corollary 2.9 (b) extends the equality $\cos^2 \theta + \sin^2 \theta = 1$.

From formula (2.7) we see how the Poisson kernel for the unit circle can be expressed in terms of zonal harmonics. We now show that such an expression is valid in n-dimensions. We recall (see Theorem 1.9 of Chapter II) that the Poisson kernel for the unit sphere in E_n is given by the formula

$$p(t', x) = \frac{1}{\omega_{n-1}} \frac{1 - |x|^2}{|x - t'|^n}$$

for $0 \leqq |x| < 1 = |t'|$.

THEOREM 2.10. *Let $x = rx'$ with $r = |x| < 1$ then*

$$p(t', x) = \sum_{k=0}^{\infty} r^k Z_{x'}^{(k)}(t') = \sum_{k=0}^{\infty} r^k Z_{t'}^{(k)}(x')$$

for all $t' \in \Sigma_{n-1}$.

PROOF. It follows from Corollary 2.9 (c) and the fact that

$$a_k = d_k - d_{k-2} = \{(n + 2k - 2)/k\}\binom{n + k - 3}{k - 1}$$

is less than or equal to a constant times $k^{(n-2)}$ that the series

$$q(t', x) = \sum_{k=0}^{\infty} r^k Z_{t'}^{(k)}(x')$$

converges uniformly in each closed region $\{x \in E_n;\ |x| \leqq r_0 < 1\}$. Now suppose $u(t') = \sum_{j=0}^{m} Y_j(t')$ is a finite linear combination of spherical harmonics (where $Y_j \in \mathscr{H}_j$). Then by Theorem 1.10 of Chapter II, the definition of spherical harmonics, and the uniqueness of the solution of the Dirichlet problem

$$\sum_{j=0}^{m} |x|^j Y_j(x') = u(x) = \int_{\Sigma_{n-1}} u(t')p(t', x)\, dt'$$

must be the continuous function on $|x| \leqq 1$ that is harmonic for $|x| < 1$ and equal to $u(x')$ for $|x'| = 1$. But by Lemma 2.8 (b), the defining property of zonal harmonics, and the orthogonality relation Corollary 2.4,

$$\begin{aligned} \int_{\Sigma_{n-1}} u(t')q(t', x)\, dt' &= \sum_{j=0}^{m} \int_{\Sigma_{n-1}} Y_j(t')q(t', x)\, dt' \\ &= \sum_{j=0}^{m} \left\{ \sum_{k=0}^{\infty} |x|^k \int_{\Sigma_{n-1}} Y_j(t') Z_{t'}^{(k)}(x')\, dt' \right\} \\ &= \sum_{j=1}^{m} |x|^j Y_j(x') = u(x). \end{aligned}$$

Thus, $\int_{\Sigma_{n-1}} [p(t', x) - q(t', x)]u(t')\,dt' = 0$ for all finite linear combinations of spherical harmonics. Since the latter are dense in $L^2(\Sigma_{n-1})$ (Corollary 2.3, part (ii)) and both $p(t', x)$ and $q(t', x)$ are continuous functions of t' it follows that $p(t', x) = q(t', x)$ and our theorem is established.

The zonal harmonics can be characterized by a simple geometrical property. In order to describe this property we define a *parallel of* Σ_{n-1} *orthogonal to a point e in* Σ_{n-1} to be the intersection of the unit sphere with a hyperplane that is perpendicular to the line determined by the origin and e. If ρ is a rotation leaving e fixed it then follows from Lemma 2.8 (c) that $Z_e^{(k)}(x') = Z_e^{(k)}(\rho x')$ for all x' in Σ_{n-1}. But such a rotation ρ must clearly map a parallel orthogonal to e into itself; moreover, given two points, x_1' and x_2', on this parallel it follows from the two-dimensional case that there exists a rotation ρ that leaves e fixed and such that $\rho x_1' = x_2'$ (for ρ can be chosen so that it leaves the orthogonal complement of the plane spanned by x_1' and x_2' fixed). Consequently, the zonal harmonic $Z_e^{(k)}$ with pole e is constant on the parallels of Σ_{n-1} orthogonal to e. We shall show that this property characterizes the zonal harmonics up to constant multiples. We observe that such functions are invariant under the rotations which leave e fixed. To utilize this remark we also need the following result.

LEMMA 2.11. *Suppose P is a polynomial on E_n, $n \geqq 2$, such that $P(\rho x) = P(x)$ for all rotations ρ and points $x \in E_n$, then there exist constants $c_0, c_1, c_2, \ldots, c_m$ such that*

$$P(x) = \sum_{k=0}^{m} c_k(x_1^2 + x_2^2 + \cdots + x_n^2)^k.$$

PROOF. We can always write $P(x) = \sum_{l=0}^{j} P_l(x)$, where P_l is homogeneous of degree l. Then, for each $\varepsilon > 0$ and each rotation ρ,

$$\sum_{l=0}^{j} \varepsilon^l P_l(x) = P(\varepsilon x) = P(\varepsilon\rho x) = \sum_{l=0}^{j} \varepsilon^l P_l(\rho x).$$

Consequently, we must have $P_l(\rho x) = P_l(x)$ for $l = 0, 1, \ldots, j$. If we let $F(x) = |x|^{-l}P_l(x)$ then F is a homogeneous function of degree 0 that is invariant under the rotation group (that is, $F(\rho x) = F(x)$ for all rotations ρ). But this implies that F is a constant function: $F(x) = c_l'$ for all x in E_n. Thus, $P_l(x) = c_l'|x|^l$. Since P_l is a polynomial, however, l must be even when $c_l' \neq 0$. Thus, $P(x) = \sum_{k=0}^{m} c_k|x|^{2k}$, where $c_k = c_{2k}'$ for $k = 0, 1, 2, \ldots, m$ and m is the largest integer less than or equal to $j/2$.

THEOREM 2.12. *Suppose e is a point of Σ_{n-1}, then $Y \in \mathscr{H}_k$ is constant on parallels of Σ_{n-1} orthogonal to e if and only if there exists a constant c such that $Y = cZ_e^{(k)}$.*

PROOF. We have already shown that zonal harmonics have this property. Suppose, therefore, that Y is constant on parallels of Σ_{n-1} orthogonal to e. If $e_1 = (1, 0, \ldots, 0) \in \Sigma_{n-1}$ and τ is a rotation satisfying $e = \tau e_1$, then the spherical harmonic W whose values are $W(x') = Y(\tau x')$ is constant on parallels of Σ_{n-1} orthogonal to e_1. If we show that $W = cZ_{e_1}^{(k)}$ then, by Lemma 2.8 (c),

$$Y(y') = W(\tau^{-1}y') = cZ_{e_1}^{(k)}(\tau^{-1}y') = cZ_{\tau^{-1}e}^{(k)}(\tau^{-1}y') = cZ_e^{(k)}(y')$$

for all $y' \in \Sigma_{n-1}$. Consequently, it suffices to show that $W = cZ_{e_1}^{(k)}$.

Let $P(x) = |x|^k W(x/|x|)$ for $x \neq 0$ and $P(0) = 0$. Then, if ρ is a rotation leaving e_1 fixed then $P(\rho x) = P(x)$ for all $x \in E_n$ and, also, all polynomials of the form x_1^m, $m = 0, 1, 2, \ldots$, are invariant under the action of ρ. Consequently, if we write

$$P(x) = \sum_{j=0}^{k} x_1^{k-j} P_j(x_2, x_3, \ldots, x_n)$$

it follows that $P_0, P_1, \ldots, P_k$ are invariant under the action of ρ (observe that $\rho(x_1, x_2, \ldots, x_n) = (x_1, x_2', \ldots, x_n')$ and the mapping $(x_2, \ldots, x_n) \to (x_2', \ldots, x_n')$ is a rotation of the $(n-1)$-dimensional space $x_1 = 0$; moreover, by varying ρ over all rotations leaving e_1 fixed we obtain all rotations of this $(n-1)$-dimensional space in this way). By Lemma 2.11, therefore, P_j is zero if j is odd while $P_j(x_2, \ldots, x_n) = c_j(x_2^2 + \cdots + x_n^2)^{j/2}$ if j is even. Letting $R = (x_2^2 + \cdots + x_n^2)^{1/2}$, therefore, we have shown that

(i) $$P(x) = c_0 x_1^k + c_2 x_1^{k-2} R^2 + \cdots + c_{2l} x_1^{k-2l} R^{2l}.$$

Because W is a surface spherical harmonic the polynomial P must be a solid spherical harmonic. Thus,

$$0 = \Delta P(x) = \sum_{j=0}^{l-1} [c_{2j}\alpha_j + c_{2(j+1)}\beta_j] x_1^{k-2(j+1)} R^{2j},$$

where $\alpha_j = (k-2j)(k-2j-1)$ and $\beta_j = 2(j+1)(n+2j-1)$. Hence, $c_{2(j+1)} = -(\alpha_j/\beta_j)c_{2j}$ for $j = 0, 1, 2, \ldots, l-1$. But this asserts that all the coefficients $c_0, c_2, \ldots, c_{2l}$ are determined by c_0; consequently, any two nonzero harmonic polynomials having form (i) must be constant multiples of each other. On the other hand, we have shown that any homogeneous polynomial of degree k whose restriction to Σ_{n-1} is constant on parallels of Σ_{n-1} orthogonal to e_1 must have form (i). Since, $Z_{e_1}^{(k)}(x/|x|)|x|^k$ has this property and, moreover, is harmonic it follows that $W(x') = P(x') = cZ_{e_1}^{(k)}(x')$ for all $x' \in \Sigma_{n-1}$, and the theorem is proved.

COROLLARY 2.13. *Suppose $F_{y'}(x')$ is defined for all $x', y' \in \Sigma_{n-1}$ and*

(a) *$F_{y'}$ is a spherical harmonic of degree k for each $y' \in \Sigma_{n-1}$,*

(b) *if* ρ *is a rotation then* $F_{\rho y'}(\rho x') = F_{y'}(x')$, *then there exists a constant c such that* $F_{y'}(x') = cZ_{y'}^{(k)}(x')$ *for all* x', y' *in* Σ_{n-1}.

PROOF. Fix y' in Σ_{n-1} and let ρ be a rotation leaving y' fixed. Then, using (b) we have

$$F_{y'}(x') = F_{\rho y'}(\rho x') = F_{y'}(\rho x')$$

for all x' in Σ_{n-1}. But this means, because of assumption (a), that $F_{y'}$ is a spherical harmonic that is constant on parallels of Σ_{n-1} orthogonal to y'. By theorem 2.12, there exists $c(y')$ such that $F_{y'} = c(y')Z_{y'}^{(k)}$. The corollary will be proved, therefore, if we can show that $c(y_1') = c(y_2')$ whenever y_1' and y_2' belong to Σ_{n-1}. In order to do this we consider a rotation σ such that $\sigma y_1' = y_2'$. Then, using assumption (b), $c(y_2')Z_{y_2'}^{(k)}(\sigma x') = F_{y_2'}(\sigma x') = F_{\sigma y_1'}(\sigma x') = F_{y_1'}(x') = c(y_1')Z_{y_1'}^{(k)}(x')$. On the other hand, by Lemma 2.8 (c),

$$Z_{y_1'}^{(k)}(x') = Z_{\sigma y_1'}^{(k)}(\sigma x') = Z_{y_2'}^{(k)}(\sigma x').$$

Consequently, $c(y_1') = c(y_2')$.

The zonal harmonics have a particularly simple expression in terms of the ultraspherical (or Gegenbauer) polynomials P_k^λ. The latter can be defined in terms of a generating function. If we write

$$(1 - 2rt + r^2)^{-\lambda} = \sum_{k=0}^{\infty} P_k^\lambda(t) r^k,$$

where $0 \leqq |r| < 1$, $|t| \leqq 1$ and $\lambda > 0$, then the coefficient $P_k^\lambda(t)$ is called the *ultraspherical polynomial of degree k associated with* λ. We shall derive some of the most elementary properties of the functions P_k^λ. Putting $r = 0$ we see that

(i) $$P_0^\lambda(t) \equiv 1.$$

Since

$$\begin{aligned} 2r\lambda \sum_{k=0}^{\infty} P_k^{\lambda+1}(t) r^k &= 2r\lambda(1 - 2rt + r^2)^{-\lambda-1} \\ &= \frac{d}{dt}(1 - 2rt + r^2)^{-\lambda} \\ &= \sum_{k=0}^{\infty} r^k \frac{d}{dt} P_k^\lambda(t) \end{aligned}$$

we have

(ii) $$\frac{d}{dt} P_k^\lambda(t) = 2\lambda P_{k-1}^{\lambda+1}(t) \text{ for } k \geqq 1.$$

From (i) and (ii) we have $(d/dt)P_1^\lambda(t) = 2\lambda$; thus, $P_1^\lambda(t)$ is a polynomial of degree 1. An induction argument using property (ii) gives us

(iii) *$P_k^\lambda(t)$ is a polynomial in t of (precisely) degree k.* It follows from this last property that the polynomials $1, t, t^2, \ldots, t^k, \ldots$ are obtainable as finite linear combinations of $P_0^\lambda, P_1^\lambda, \ldots, P_k^\lambda, \ldots$. Thus, by the Weierstrass approximation theorem,

(iv) *The finite linear combinations of the polynomials $P_k^\lambda(t)$, $k = 0, 1, \ldots,$ are uniformly dense in the space of continuous functions on the closed interval* $[-1, 1]$.

Since

$$\sum_{k=0}^{\infty} P_k^\lambda(-t)r^k = (1 + 2rt + r^2)^{-\lambda} = \sum_{k=0}^{\infty} P_k^\lambda(t)(-r)^k$$

we must have

(v) $$P_k^\lambda(-t) = (-1)^k P_k^\lambda(t) \text{ for } k \geqq 0.$$

The following representation of the zonal harmonics is an easy consequence of these properties:

THEOREM 2.14. *If $n > 2$ is an integer, $\lambda = (n-2)/2$ and $k = 0, 1, 2, \ldots$ then there exists a constant $c_{k,n}$ such that*

$$Z_{y'}^{(k)}(x') = c_{k,n}P_k^\lambda(x' \cdot y')$$

for all $x', y' \in \Sigma_{n-1}$.

PROOF. For y' in Σ_{n-1} we let $F_{y'}(x) = |x|^k P_k^\lambda(x \cdot y'/|x|)$ for x in E_n. If we can show that $F_{y'}(x')$ satisfies the hypotheses of Corollary 2.13 the theorem would then be immediate. By properties (iii) and (v), if k is even, say $k = 2m$, then $P_k^\lambda(t)$ must have the form $P_k^\lambda(t) = \sum_{j=0}^m d_{2j}t^{2j}$ while, if k is odd, say $k = 2m + 1$, $P_k^\lambda(t) = \sum_{j=0}^m d_{2j+1}t^{2j+1}$. Thus, in the former case,

$$|x|^k P_k^\lambda\left(\frac{x \cdot y'}{|x|}\right) = \sum_{j=0}^{m} d_{2j}(x_1^2 + \cdots + x_n^2)^{m-j}(x_1y_1' + \cdots + x_ny_n')^{2j},$$

and, in the second case

$$|x|^k P_k^\lambda\left(\frac{x \cdot y'}{|x|}\right) = \sum_{j=0}^{m} d_{2j+1}(x_1^2 + \cdots + x_n^2)^{m-j}(x_1y_1' + \cdots + x_ny_n')^{2j+1}.$$

In either case, $F_{y'}(x)$ is a (nonzero) homogeneous polynomial of degree k.

If ρ is a rotation then, since it preserves the inner product,

$$F_{\rho y'}(\rho x') = P_k^\lambda(\rho x' \cdot \rho y') = P_k^\lambda(x' \cdot y') = F_{y'}(x')$$

for all points x', y' of Σ_{n-1}. Thus, property (b) of Corollary 2.13 is satisfied; consequently, we only need to show that $F_{y'}$ is harmonic. We have already observed (see Examples 3 and 5 at the beginning of Section 1 in Chapter II) that $|x - x_0|^{2-n}$ is a harmonic function of x in the region $E_n - \{x_0\}$. In particular,

$$(2.15)\qquad s^{2-n}\left|x - \frac{y'}{s}\right|^{2-n} = \left[1 - 2s|x|\left(\frac{x}{|x|}\cdot y'\right) + (s|x|)^2\right]^{-\lambda} = \sum_{k=0}^{\infty} s^k|x|^k P_k^\lambda\left(\frac{x}{|x|}\cdot y'\right)$$

is harmonic in the region $\mathscr{R} = \mathscr{R}_s = \{x \in E_n; 0 < |x| < 1/s\}$, for $s \neq 0$ and y' in Σ_{n-1} fixed. But this implies that each of the coefficients $F_{y'}(x) = |x|^k P_k^\lambda(x\cdot y'/|x|)$ must be a harmonic function of x. This can be seen, for example, by taking the mean value of both the left and right sides of equality (2.15) on spheres about x that lie within the region $\mathscr{R}_s$. Since the left side satisfies the mean-value property (Theorem 1.1 of Chapter II) for $0 < s \leqq s_0 < \infty$, the same must be true for each of the coefficients $F_{y'}(x) = |x|^k P_k^\lambda(x\cdot y'/|x|)$. From this and Theorem 1.7 of Chapter II we can now deduce the desired result that $F_{y'}$ is harmonic.

The ultraspherical polynomials can be used to obtain some important orthogonal expansions of functions. The basic orthogonality property needed for carrying out such expansions is an easy consequence of the last theorem.

COROLLARY 2.16. *The polynomials $P_k^{(n-2)/2}(t)$, $k = 0, 1, 2, \ldots$, are mutually orthogonal with respect to the inner product*

$$(f, g) = \int_{-1}^{1} f(t)g(t)(1 - t^2)^{(n-3)/2}\, dt.$$

PROOF. Let $e = (1, 0, \ldots, 0)$ and, for $x' \in \Sigma_{n-1}$, define θ, $0 \leqq \theta \leqq \pi$, by letting $e\cdot x' = \cos\theta$. We can evaluate integrals over Σ_{n-1} by first integrating over the parallel $L_\theta = \{x' \in \Sigma_{n-1}; e\cdot x' = \cos\theta\}$ orthogonal to e, thus obtaining a function of θ, $0 \leqq \theta \leqq \pi$, which can then be integrated over the interval $[0, \pi]$. The measure of L_θ is $\omega_{n-2}(\sin\theta)^{n-2}$ (the surface area of a sphere in E_{n-2} of radius $\sin\theta$). Using this fact together with Corollary 2.4, Theorems 2.12 and 2.14 we thus have

$$\begin{aligned} 0 &= \int_{\Sigma_{n-1}} Z_e^l(x')Z_e^k(x')\, dx' \\ &= c_{l,n}c_{k,n}\omega_{n-2}\int_0^\pi P_l^{(n-2)/2}(\cos\theta)P_k^{(n-2)/2}(\cos\theta)(\sin\theta)^{n-2}\, d\theta \\ &= \omega_{n-2}c_{l,n}c_{k,n}\int_{-1}^{1} P_l^{(n-2)/2}(t)P_k^{(n-2)/2}(t)(1 - t^2)^{(n-3)/2}\, dt. \end{aligned}$$

From this result and property (iv) of the ultraspherical polynomials we obtain

COROLLARY 2.17. *The polynomials* $P_k^{(n-2)/2}$, $k = 0, 1, 2, \ldots,$ *form an orthogonal basis for the space* $L^2([-1, 1]; (1 - t^2)^{(n-3)/2}\, dt)$.[7]

Having completed our study of the spherical harmonics, mainly from the point of view of functions on the sphere, we shall now show how they can be used in the Fourier analysis of functions on E_n. More precisely, we shall study the subspaces $\mathfrak{H}_k$ of $L^2(E_n)$. We recall that $\mathfrak{H}_k$ was defined to be the space of all linear combinations of functions of the form $f(r)P(x)$, where f ranges over the radial functions and P over the solid spherical harmonics of degree k, in such a way that $f(r)P(x)$ belongs to $L^2(E_n)$.

We now state and prove the following lemma as preparatory to the development in Section 3.

LEMMA 2.18. *The direct sum decomposition* $L^2(E_n) = \sum_{k=0}^{\infty} \oplus \mathfrak{H}_k$ *holds in the sense that*

(a) *each subspace* $\mathfrak{H}_k$ *is closed;*

(b) *the* $\mathfrak{H}_k$ *are mutually orthogonal;*

(c) *every element of* $L^2(E_n)$ *is a limit of finite linear combinations of elements belonging to the spaces* $\mathfrak{H}_k$. *In addition, the Fourier transform maps* $\mathfrak{H}_k$ *into itself.*

PROOF. The space of solid spherical harmonics of degree k, being isomorphic to $\mathscr{H}_k$, has (finite) dimension $a_k = d_k - d_{k-2}$. Let $P_1, P_2, \ldots, P_{a_k}$ be an orthonormal basis of this space (where the inner product is that which is inherited from $L^2(\Sigma_{n-1})$). Then, clearly, every element of $\mathfrak{H}_k$ can be written in the form $\sum_{j=1}^{a_k} f_j(r)P_j(x)$. Moreover,

$$\int_{E_n} |f(x)|^2\, dx = \sum_{j=1}^{a_k} \int_0^{\infty} |f_j(r)|^2 r^{2k+n-1}\, dr.$$

From this conclusion (a) is obvious. The mutual orthogonality of the $\mathfrak{H}_k$'s follows immediately from the orthogonality of the spherical harmonics (Corollary 2.4) and by an integration involving polar coordinates. To prove the completeness property (c), it suffices to show that, if a function of $L^2(E_n)$ is orthogonal to all the $\mathfrak{H}_k$'s then it vanishes almost everywhere. By the completeness of the spherical harmonics on the sphere, however, such a function must vanish a.e. on almost every sphere centered at the origin, which gives the desired conclusion.

The action of the Fourier transform on $\mathfrak{H}_k$ will be the subject matter of the following section. An elementary argument, however, enables us to

[7] Corollaries 2.16 and 2.17 are valid for all $\lambda > 0$ and not just $\lambda = (n-2)/2$. Different arguments, however, are needed to establish them.

prove at this point that the Fourier transform preserves $\mathfrak{H}_k$. It suffices, for this purpose, to consider $f \in L^1(E_n) \cap L^2(E_n)$ having the form $f(u) = f_0(\rho)P(u) = \rho^k f_0(\rho) Y(u')$ with $Y \in \mathscr{H}_k$, where $\rho = |u|$ and $u = \rho u'$. Since the finite linear combinations of such functions are dense in $\mathfrak{H}_k$, this last space will be invariant under the action of the Fourier transform provided $\hat{f} \in \mathfrak{H}_k$ whenever f has the above form. Letting $r = |x|$ and $x = rx'$ we then obtain

$$\hat{f}(x) = \int_{E_n} e^{-2\pi i x \cdot u} f(u)\, du = \int_0^\infty f_0(\rho)\rho^{k+n-1}\left\{\int_{\Sigma_{n-1}} e^{-2\pi i r\rho x' \cdot u'}\, Y(u')\, du'\right\} d\rho.$$

If we can show that there exists a function φ, defined on $[0, \infty)$, such that

$$\int_{\Sigma_{n-1}} e^{-2\pi i s x' \cdot u'}\, Y(u')\, du' = \varphi(s)\, Y(x') \tag{2.19}$$

for $s \geqq 0$ then

$$\hat{f}(x) = \left\{\int_0^\infty f_0(\rho)\varphi(r\rho)\rho^{k+n-1}\, d\rho\right\} Y(x')$$

and, consequently, $\hat{f} \in \mathfrak{H}_k$.

By the defining property of the zonal harmonic $Z_{u'}^{(k)}(v')$ and the property $Z_{u'}^{(k)}(v') = Z_{v'}^{(k)}(u')$ we must have

$$\begin{aligned}\int_{\Sigma_{n-1}} e^{-2\pi i s x' \cdot u'}\, Y(u')\, du' &= \int_{\Sigma_{n-1}} e^{-2\pi i s x' \cdot u'}\left\{\int_{\Sigma_{n-1}} Y(v') Z_{u'}^{(k)}(v')\, dv'\right\} du' \\ &= \int_{\Sigma_{n-1}} Y(v')\left\{\int_{\Sigma_{n-1}} e^{-2\pi i s x' \cdot u'} Z_{v'}^{(k)}(u')\, du'\right\} dv'.\end{aligned}$$

If we denote the expression in the last brackets by $F_{x'}(v')$ an immediate application of Fubini's theorem and Corollary 2.4 gives us the fact that $F_{x'}$, as a function of $v' \in \Sigma_{n-1}$, is orthogonal to all the spaces $\mathscr{H}_j$, $j \neq k$. But, by (2.5) this implies that $F_{x'} \in \mathscr{H}_k$. From Lemma 2.8, part (c) and the change of variables $u' = \sigma w'$, we obtain the property $F_{\sigma x'}(\sigma v') = F_{x'}(v')$ for all rotations σ. Thus, by Corollary 2.13 there exists a number $c(= \varphi(s))$ such that $F_{x'}(v') = cZ_{x'}^{(k)}(v')$ for all x', v' in Σ_{n-1}. Consequently,

$$\begin{aligned}\int_{\Sigma_{n-1}} e^{-2\pi i s x' \cdot u'}\, Y(u')\, du' &= \int_{\Sigma_{n-1}} Y(v') F_{x'}(v')\, dv' = \int_{\Sigma_{n-1}} Y(v')\varphi(s) Z_{x'}^{(k)}(v')\, dv' \\ &= \varphi(s)\, Y(x')\end{aligned}$$

and (2.19) is established.

3. The Action of the Fourier Transform on the Spaces $\mathfrak{H}_k$

We have shown that the Fourier transform of a radial function is a radial function (see Corollary 1.2). In the two dimensional case we have obtained an explicit formula relating these two radial functions (the special case $k = 0$ of Theorem 1.6). This formula has a natural extension to $n > 2$ dimensions and, as is the case when $n = 2$, it involves a Bessel function. We shall show that this Bessel function is $J_{(n-2)/2}$. When n is odd $(n - 2)/2$ is not an integer; consequently, in this case $J_{(n-2)/2}$ is not one of the functions introduced in the first section. We shall show, however, that there is a natural way of extending the class of Bessel functions we introduced so as to include the ones we will encounter. In order to do this we establish the following identity (known as the *Poisson representation of Bessel functions*):

LEMMA 3.1. *If k is a nonnegative integer then*

$$J_k(t) = \frac{(t/2)^k}{\Gamma[(2k+1)/2]\Gamma(\frac{1}{2})} \int_{-1}^{1} e^{its}(1 - s^2)^{(2k-1)/2}\, ds.$$

PROOF. We define J_k^* by letting

$$J_k^*(t) = \frac{(t/2)^k}{\Gamma[(2k+1)/2]\Gamma(\frac{1}{2})} \int_{-1}^{1} e^{its}(1 - s^2)^{(2k-1)/2}\, ds.$$

From the definition of J_0 (see Section 1) and the change of variables $s = \sin\theta$ we immediately obtain the fact that $J_0^* = J_0$. The lemma will be established, therefore, if we can show that the recursion relation

$$\frac{d}{dt}(t^{-k}G_k(t)) = -t^{-k}G_{k+1}(t),\ t \neq 0, \qquad k = 0, 1, 2, \ldots, \tag{3.2}$$

is satisfied by both of the sequences $\{J_k\}$ and $\{J_k^*\}$. But,

$$\begin{aligned}
\frac{d}{dt}(t^{-k}J_k(t)) &= -t^{-k}\left\{\frac{k}{t}J_k(t) - J_k'(t)\right\} \\
&= -t^{-k}\left\{\frac{k}{2\pi t}\int_0^{2\pi} e^{it\sin\theta}e^{-ik\theta}\,d\theta - \frac{1}{2\pi}\int_0^{2\pi}\left(\frac{d}{dt}e^{it\sin\theta}\right)e^{-ik\theta}\,d\theta\right\} \\
&= -\frac{t^{-k}}{2\pi}\int_0^{2\pi}\left\{i\frac{d}{d\theta}\left[\frac{e^{it\sin\theta}e^{-ik\theta}}{t}\right] + \cos\theta e^{it\sin\theta}e^{-ik\theta} - i\sin\theta e^{it\sin\theta}e^{-ik\theta}\right\}d\theta \\
&= -\frac{t^{-k}}{2\pi}\int_0^{2\pi} e^{it\sin\theta}e^{-i(k+1)\theta}\,d\theta = -t^{-k}J_{k+1}(t).
\end{aligned}$$

While, integrating by parts and using the fact that

$$\left(\frac{2k+1}{2}\right)\Gamma\left(\frac{2k+1}{2}\right) = \Gamma\left(\frac{2k+3}{2}\right),$$

we have

$$\begin{aligned}\frac{d}{dt}(t^{-k}J_k^*(t)) &= \frac{2^{-k}i}{\Gamma[(2k+1)/2]\Gamma(\frac{1}{2})}\int_{-1}^{1} e^{its}s(1-s^2)^{(2k-1)/2}\,ds \\ &= \frac{2^{-k}i}{\Gamma[(2k+1)/2]\Gamma(\frac{1}{2})}\int_{-1}^{1} \frac{2it}{(2k+1)}e^{its}\frac{(1-s^2)^{(2k+1)/2}}{2}\,ds \\ &= -t^{-k}J_{k+1}^*(t).\end{aligned}$$

The integral occurring in Lemma 3.1 is well-defined for all real $k > -\frac{1}{2}$. We therefore obtain a wider class of functions than the one considered in Section 1 if we define the *Bessel function* J_k, where k is a real number greater than $-\frac{1}{2}$, by letting

$$J_k(t) = \frac{(t/2)^k}{\Gamma[(2k+1)/2]\Gamma(\frac{1}{2})}\int_{-1}^{1} e^{its}(1-s^2)^{(2k-1)/2}\,ds$$

for $t > 0$.

Now suppose f is a radial function in $L^1(E_n)$; thus $f(x) = f_0(|x|)$ for a.e. x in E_n, where $\int_0^\infty |f_0(r)|r^{n-1}\,dr < \infty$. The Fourier transform $\hat{f}$ is then also radial; consequently, $\hat{f}(x) = F_0(|x|)$ for all $x \in E_n$. Thus, if $r = |x|$, $x = rx'$, $s = |u|$ and $u = su'$, we have $F_0(r) = \hat{f}(x) = \int_{E_n} f(u)e^{-2\pi i x\cdot u}\,du = \int_0^\infty f_0(s)\{\int_{\Sigma_{n-1}} e^{-2\pi irs(x'\cdot u')}\,du'\}s^{n-1}\,ds$. We can evaluate the inner integral by the method used in the proof of Corollary 2.16: First we integrate over the parallel $L_\theta = \{u' \in \Sigma_{n-1};\, x'\cdot u' = \cos\theta\}$ orthogonal to x' obtaining a function of θ, $0 \leqq \theta \leqq \pi$, which we then integrate over the interval $[0, \pi]$. Thus, since $e^{-2\pi irs(x'\cdot u')} = e^{-2\pi irs\cos\theta}$ is constant on L_θ and the measure of this parallel is $\omega_{n-2}(\sin\theta)^{n-2} = (2\pi^{(n-1)/2}/\Gamma[(n-1)/2])\,(\sin\theta)^{n-2}$,

$$\begin{aligned}\int_{\Sigma_{n-1}} e^{-2\pi irs(x'\cdot u')}\,du' &= \int_0^\pi e^{-2\pi irs\cos\theta}\omega_{n-2}(\sin\theta)^{n-2}\,d\theta \\ &= \omega_{n-2}\int_{-1}^{1} e^{2\pi irs\xi}(1-\xi^2)^{(n-3)/2}\,d\xi \\ &= \frac{2\pi^{(n-1)/2}\Gamma[(n-1)/2]\Gamma(\frac{1}{2})}{\Gamma[(n-1)/2](\pi rs)^{(n-2)/2}}J_{(n-2)/2}(2\pi rs) \\ &= 2\pi(rs)^{-(n-2)/2}J_{(n-2)/2}(2\pi rs).\end{aligned}$$

We therefore obtain the following result:

THEOREM 3.3. *Suppose f is a radial function in $L^1(E_n)$, $n \geqq 2$; thus, $f(x) = f_0(|x|)$ for a.e. $x \in E_n$. Then the Fourier transform $\hat{f}$ is also radial and has the form $\hat{f}(x) = F_0(|x|)$ for all $x \in E_n$, where*

$$F_0(|x|) = F_0(r) = 2\pi r^{-[(n-2)/2]} \int_0^\infty f_0(s) J_{(n-2)/2}(2\pi r s) s^{n/2}\, ds.^{8}$$

This theorem shows how the Fourier transform acts on the space $\mathfrak{H}_0$. We now turn our attention to the spaces $\mathfrak{H}_k$, $k \geqq 1$. We begin by first examining the action of the Fourier transform on an important class of functions:

THEOREM 3.4. *Suppose $f(u) = e^{-\pi|u|^2}P(u)$ for u in E_n, where $P(u)$ is a solid spherical harmonic of degree k, then $\hat{f}(v) = i^{-k}f(v)$ for all v in E_n.*

PROOF. For $t \in E_n$ fixed we have

$$\int_{E_n} e^{-\pi|u|^2} P(u+t)\, du = \int_0^\infty r^{n-1} e^{-\pi r^2} \left\{ \int_{\Sigma_{n-1}} P(t + ru')\, du' \right\} dr.$$

Since P is harmonic it satisfies the mean-value property (Theorem 1.1 of Chapter II) and, thus, $\int_{\Sigma_{n-1}} P(t + ru')\, du' = \omega_{n-1}P(t)$. Consequently,

$$\begin{aligned} \int_{E_n} e^{-\pi|u|^2} P(u+t)\, du &= P(t) \int_0^\infty r^{n-1} e^{-\pi r^2} \omega_{n-1}\, dr \\ &= P(t) \int_0^\infty r^{n-1} e^{-\pi r^2} \left\{ \int_{\Sigma_{n-1}} du' \right\} dr \\ &= P(t) \int_{E_n} e^{-\pi|x|^2}\, dx = P(t). \end{aligned}$$

$P(t) = P(t_1, \ldots, t_n)$, being a polynomial, has an obvious analytic continuation $P(z) = P(z_1, z_2, \ldots, z_n)$ to all of $\mathbf{C}_n$. It then follows from the

[8] The reason why both here and in Theorem 1.6 we consider the action of the Fourier transform on $L^1(E_n)$, rather than on $L^2(E_n)$, is that in this way all integrals encountered are defined in the L^1 sense. These results, however, have obvious extensions to $L^2(E_n)$.

last equality that

$$\int_{E_n} e^{-\pi|u|^2}P(u+z)\,du = P(z)$$

for all $z = x + iy = (x_1 + iy_1, x_2 + iy_2, \ldots, x_n + iy_n) \in \mathbf{C}_n$. In particular for $z = -iv$ we have

$$\int_{E_n} e^{-\pi|u|^2}P(u-iv)\,du = P(-iv) = (-i)^k P(v),$$

the last equality being a consequence of the homogeneity of P. But an n-fold application of Cauchy's integral theorem shows that

$$\int_{E_n} e^{-\pi(u+iv)\cdot(u+iv)}P(u)\,du = \int_{E_n} e^{-\pi u\cdot u}P(u-iv)\,du = (-i)^k P(v),$$

where $(u + iv)\cdot(u + iv) = \sum_{j=1}^{n}(u_j + iv_j)^2$. If we now multiply the left and right sides by $e^{-\pi|v|^2}$ we obtain the desired equality $\hat{f}(v) = (-i)^k f(v) = i^{-k}f(v)$.

Theorem 3.4 enables us to study the behavior of the Fourier transform on a still larger class of functions. For $\alpha > 0$ let $g(x) = e^{-\pi|\alpha x|^2}P(x) = \alpha^{-k}f(\alpha x)$ for x in E_n. If δ_α denotes dilation by α it then follows from (1.6) of Chapter I, Theorem 3.4 and the homogeneity of $P(x)$ that

(3.5) $$\hat{g}(x) = \alpha^{-k}(\delta_\alpha f)^\wedge(x) = \alpha^{-k}\alpha^{-n}\hat{f}(\alpha^{-1}x) = i^{-k}\alpha^{-n-2k}e^{-\pi|x|^2/\alpha^2}P(x).$$

On the other hand, if h is the radial function whose values are $h(x) = e^{-\pi|\alpha x|^2}$ for $x \in E_{n+2k}$, then an application of Theorem 1.13 of Chapter I gives us

(3.6) $$\hat{h}(x) = \alpha^{-n-2k}e^{-\pi|x|^2/\alpha^2}.$$

We can interpret these two identities in the following way: Let $\mathscr{D}_m$ be the Hilbert space of functions φ on $(0, \infty)$ satisfying

$$\|\varphi\| = \left(\int_0^\infty |\varphi(r)|^2 r^{m-1}\,dr\right)^{1/2} < \infty.$$

If $m = n + 2k$ and $P(x)$ is a nonzero solid spherical harmonic of degree k we consider the function g whose values are $g(x) = \varphi(|x|)P(x)$ for x in E_n. Then, if $\|P\|_\Sigma = (\int_{\Sigma_{n-1}} |P(x')|^2\,dx')^{1/2}$,

$$\|g\|_2^2 = \int_{E_n} |\varphi(|x|)P(x)|^2\,dx = \left(\int_0^\infty |\varphi(r)|^2 r^{n+2k-1}\,dr\right)\|P\|_\Sigma^2 < \infty.$$

Thus, $g \in L^2(E_n)$ and, by the Plancherel theorem, $\hat{g} \in L^2(E_n)$ with $\|\hat{g}\|_2 = \|g\|_2$. It follows from (2.19) and this equality between the norms of g and $\hat{g}$ that $\hat{g}(x) = \psi(|x|)P(x)$ for almost every x in E_n and that $\|\psi\| = \|\varphi\|$. We therefore obtain a bounded linear operator, T_k^n, on $\mathscr{D}_{n+2k}$ by letting $T_k^n\varphi = \psi$ (in fact, T_k^n is an isometry). In a similar way, by considering the radial function h defined by letting $h(x) = \varphi(|x|)$ for x in E_{n+2k} we obtain another bounded linear operator, T_0^{n+2k}, on $\mathscr{D}_{n+2k}$. In order to define this operator we first observe that it follows from Corollary 1.2 that $\hat{h}$ is also radial, say $\hat{h}(x) = \theta(|x|)$ for almost every x in E_{n+2k}. Putting $T_0^{n+2k}\varphi = \theta$ we obtain the desired bounded linear operator on $\mathscr{D}_{n+2k}$ (again, it is an immediate consequence of the Plancherel theorem that T_0^{n+2k} is an isometry).

Equalities (3.5) and (3.6) show that $T_0^{n+2k}\varphi = i^kT_k^n\varphi$ whenever $\varphi(r) = e^{-\varepsilon r^2}$, $\varepsilon > 0$. Thus, T_0^{n+2k} and $i^kT_k^n$ agree when they are restricted to the space $\mathscr{W}$ of finite linear combinations of all such functions φ obtained by varying ε throughout the positive real numbers. But it is not hard to show that $\mathscr{W}$ is dense in the Hilbert space $\mathscr{D}_{n+2k}$. If this were not the case, we could find a $b \in \mathscr{D}_{n+2k}$, not equal to 0 almost everywhere, such that $\int_0^\infty \varphi(r)b(r)r^{n+2k-1}\,dr = 0$ for all φ in $\mathscr{W}$. In particular,

$$\int_0^\infty e^{-\varepsilon r^2}b(r)r^{n+2k-1}\,dr = 0 \tag{3.8}$$

for all $\varepsilon > 0$. Let Φ be the function whose value at $s \geqq 0$ is $\Phi(s) = \int_0^s e^{-r^2}b(r)r^{n+2k-1}\,dr$. Then, putting $\varepsilon = (m + 1)$, where m is a positive integer, and integrating the integrand in (3.8) by parts we have

$$0 = \int_0^\infty e^{-mr^2}\Phi'(r)\,dr = 2m\int_0^\infty e^{-mr^2}\Phi(r)r\,dr.$$

By the change of variable $u = e^{-r^2}$ this equality is equivalent to

$$0 = \int_0^1 u^{m-1}\Phi\left(\sqrt{\log\frac{1}{u}}\right)du \qquad \text{for } m = 1, 2, 3, \ldots.$$

Since the polynomials are uniformly dense in the space of continuous functions on the closed interval [0, 1], this can only be the case when $\Phi(\sqrt{\log 1/u}) = 0$ for all u in [0, 1]. Thus $\Phi'(r) = e^{-r^2}b(r)r^{n+2k-1} = 0$

for almost every r in $(0, \infty)$, contradicting the hypothesis that $b(r)$ is not equal to 0 almost everywhere.

Since the operators T_0^{n+2k} and $i^k T_k^n$ are bounded and agree on the dense subspace $\mathscr{W}$ they must be equal. Thus, we have shown

(3.9) $$T_0^{n+2k}\varphi = i^k T_k^n \varphi \qquad \text{for all } \varphi \in \mathscr{D}_{n+2k}.$$

In Theorem 3.3 we expressed the Fourier transform of a radial function in terms of an integral involving a Bessel function. Equalities (2.19) and (3.9) show that the Fourier transform of a function in $\mathfrak{H}_k$ of the form $f_0(|x|)P(x)$, where $P(x)$ is a solid harmonic of degree k on E_n, can be expressed in terms of the Fourier transform of the radial function whose values are $h(y) = f_0(|y|)$ for $y \in E_{n+2k}$. Combining these results we obtain the following theorem (which has an obvious extension to all of $L^1(E_n)$ or $L^2(E_n)$):

THEOREM 3.10. *Suppose $n \geqq 2$ and $f \in L^2(E_n) \cap L^1(E_n)$ has the form $f(x) = f_0(|x|)P(x)$, where $P(x)$ is a solid spherical harmonic of degree k, then $\hat{f}$ has the form $\hat{f}(x) = F_0(|x|)P(x)$ where*

$$F_0(r) = 2\pi i^{-k} r^{-[(n+2k-2)/2]} \int_0^\infty f_0(s) J_{(n+2k-2)/2}(2\pi r s) s^{(n+2k)/2}\, ds.$$

This theorem, together with the fact that the spaces $\mathfrak{H}_k$ are spanned by functions of the form $f_0(|x|)P(x)$, gives the promised description of the action of the Fourier transform on $\mathfrak{H}_k$.

The identity describing the action of the Fourier transform on $\mathfrak{H}_k$ naturally leads us to the study of the behavior of the Bessel functions $J_m(r)$. Near the origin we have the trivial estimate $J_m(r) \sim cr^m$ as $r \to 0$. The behavior for large r is less obvious and is given in the following lemma.

LEMMA 3.11. $J_m(r) = \sqrt{2/\pi r}\cos(r - \pi m/2 - \pi/4) + O(r^{-3/2})$ *as* $r \to \infty$. *In particular,*

(3.12) $$J_m(r) = O(r^{-1/2}) \qquad \textit{as} \qquad r \to \infty.$$ [9]

PROOF. In view of the definition of $J_m(r)$ for $m > -\frac{1}{2}$, we need to estimate the integral

$$I = \int_{-1}^{1} e^{irs}(1 - s^2)^{m-1/2}\, ds.$$

In order to do this we consider the simply connected region in the complex plane obtained by excluding the rays $(-\infty, -1)$ and $(1, \infty)$. We then

[9] For later purposes it is useful to observe that the proof given below shows that the bounds arising in the terms $O(r^{-1/2})$ and $O(r^{-3/2})$ are uniform in m, as long as m is restricted to any closed, bounded subinterval of $(-\frac{1}{2}, \infty)$.

choose the branch of $f(z) = (1 - z^2)^{m-1/2}$ in this region which is real valued and nonnegative on the interval $[-1, 1]$. Integrating $e^{irz}(1 - z^2)^{m-1/2} = e^{ir(x+iy)}(1 - [x + iy]^2)^{m-1/2}$ around the rectangle whose lower side is $[-1, 1]$ and whose height is $a > 0$, we obtain

$$0 = \int_0^a e^{ir(1+iy)}(y^2 - 2iy)^{m-1/2}\,diy + \int_{-1}^1 e^{irs}(1 - s^2)^{m-1/2}\,ds + \int_a^0 e^{ir(iy-1)}(y^2 + 2iy)^{m-1/2}\,diy + \varepsilon(a),$$

where $\varepsilon(a) \to 0$ as $a \to \infty$. Thus,

$$-i\,I = \int_0^\infty e^{ir(iy-1)}(y^2 + 2iy)^{m-1/2}\,dy - \int_0^\infty e^{ir(1+iy)}(y^2 - 2iy)^{m-1/2}\,dy = I_1 - I_2.$$

Now

$$(y^2 \pm 2iy)^{m-1/2} = \begin{cases} y^{m-1/2}(\pm 2i)^{m-1/2} + O(y^{m+1/2}), & \text{for } 0 \leqq y \leqq 1 \\ y^{m-1/2}(\pm 2i)^{m-1/2} + O(y^{2m-1}), & \text{for } 1 \leqq y < \infty. \end{cases}$$

Therefore,

$$I_1 = e^{-ir}\int_0^\infty e^{-ry}(2i)^{m-1/2}y^{m-1/2}\,dy + O\left(\int_0^1 e^{-ry}y^{m+1/2}\,dy\right) + O\left(\int_1^\infty e^{-ry}y^{2m-1}\,dy\right).$$

The first term is $(2i)^{m-1/2}\Gamma(m + \frac{1}{2})\,(e^{-ir}/r^{m+1/2})$. The second is $O(r^{-m-3/2})$ and the third is $O(e^{-r})$ as $r \to \infty$. Similarly, $I_2 = (-2i)^{m-1/2}\Gamma(m + \frac{1}{2}) \times (e^{ir}/r^{m+1/2}) + O(r^{-m-3/2})$, as $r \to \infty$. Since $J_m(r) = [(r/2)^m/\Gamma(m + \frac{1}{2})\Gamma(\frac{1}{2})] \times i(I_1 - I_2)$, we obtain the desired result by collecting terms.

4. Some Applications

Many important operators in analysis are convolution operators. That is, they map functions f, ranging throughout some function space, into convolutions $K * f$, where K is a fixed function (or, more generally, a tempered distribution). We have encountered several such operators. One example is given by the Poisson integral, where $K(x) = K_y(x) = c_n y/(|x|^2 + y^2)^{(n+1)/2}$ for some $y > 0$ (see Section 1 in Chapter I). If f

belongs to $L^1(E_n)$ (or to $L^2(E_n)$) it follows from Theorems 1.4 and 1.14 of Chapter I that

$$(K * f)^\wedge(x) = e^{-2\pi y|x|}\hat{f}(x)$$

for x in E_n (or a.e. x in E_n). We see, therefore, that if we follow this convolution operator with the Fourier transform we obtain a transformation of a very simple form. For example, if $f \in L^2(E_n)$ it is an immediate consequence of the Plancherel theorem and the convergence of $e^{-2\pi y|x|}$ to 1, as y tends to 0, that $\lim_{y \to 0} \|f - (K_y * f)\|_2 = 0$. This is a particularly simple example of a general situation: After taking the Fourier transform of a convolution operator we obtain a multiplication operator, many of whose important properties are easily derived. The Poisson kernel is a radial function. The other kernels we shall consider are either radial or are products of radial functions with spherical harmonics. In view of Theorem 3.10 it is not unreasonable to expect that the Fourier transform of such a kernel has the same form (even when the kernel is not an L^1 or L^2 function).

Our first application concerns a class of kernels that give rise to convolution transforms that are important in Fourier analysis and the theory of partial differential equations. These are the functions of the form $P(x)/|x|^\beta$, where P is a solid spherical harmonic of degree k and β is a complex number. If K is such a function and $\operatorname{Re}(\beta) < n + k$, then it is locally integrable and it defines a tempered distribution in the sense described in example (3) of the third section in Chapter I. Thus, its Fourier transform $\hat{K}$ is well-defined and is also a tempered distribution. The following result asserts that K is a tempered function for a certain range of β and gives its explicit form.

THEOREM 4.1. *Suppose α is a complex number such that $0 < \operatorname{Re}(\alpha) < n$ and $P(x)$ is a harmonic polynomial on E_n that is homogeneous of degree k. If $K(x) = P(x)/|x|^{n+k-\alpha}$ then $\hat{K}(t) = \gamma P(t)/|t|^{k+\alpha}$, where*

$$\gamma = \gamma_{k,\alpha} = i^{-k}\pi^{(n/2)-\alpha}\Gamma\left(\frac{k+\alpha}{2}\right)\Big/\Gamma\left(\frac{n+k-\alpha}{2}\right).$$

PROOF. We define K_1 by letting $K_1(x) = K(x)$ if $|x| \leqq 1$ and $K_1(x) = 0$ if $|x| > 1$; we then define K_2 to be the difference $K - K_1$. K_1 then belongs to $L^1(E_n)$ and, if we impose the additional restriction that $\operatorname{Re}(\alpha) < n/2$, we can then conclude that K_2 belongs to $L^2(E_n)$. Thus it follows from the definition of the Fourier transform of a distribution (see (3.15), Chapter I) that $\hat{K}$ is the function $\hat{K}_1 + \hat{K}_2$. By approximating K_1 and K_2 with functions in $L^1(E_n) \cap L^2(E_n)$ and using (2.19) we see that $\hat{K}_1$ and $\hat{K}_2$ must have the forms $\hat{K}_1(x) = f_1(|x|)P(x)$ and $\hat{K}_2(x) = f_2(|x|)P(x)$. Thus, $\hat{K}(x) = f(|x|)P(x)$, where $f = f_1 + f_2$. If K were in $L^1(E_n) \cap L^2(E_n)$ it would be an immediate consequence of Theorem 3.10 that f is a homogeneous function of degree $-(k + \alpha)$. Unfortunately this is not the case; thus, we cannot apply that theorem directly in order to obtain this homogeneity property. Although we could use Theorem 3.10 and an approximation argument to show that for a.e. $r > 0$

$$f(\delta r) = \delta^{-k-\alpha} f(r) \tag{4.2}$$

whenever $\delta > 0$, it is easier to do this by exploiting the relation between the Fourier transform and the action of dilations on E_n.

Let φ be a testing function (a member of $\mathscr{S}$). By the multiplication formula (in fact, by the definition of the Fourier transform of tempered distributions) we have

$$\int_{E_n} K(x)\hat{\varphi}(x)\,dx = \int_{E_n} \hat{K}(x)\varphi(x)\,dx.$$

Putting $x = \delta u$ on the left and $x = \delta^{-1}v$ on the right, this equality becomes

$$\delta^n \int_{E_n} K(\delta u)\hat{\varphi}(\delta u)\,du = \delta^{-n} \int_{E_n} \hat{K}(\delta^{-1}v)\varphi(\delta^{-1}v)\,dv.$$

Since $K(\delta u) = \delta^{\alpha-n}K(u)$

$$\delta^\alpha \int_{E_n} K(u)\hat{\varphi}(\delta u)\,du = \delta^{-n} \int_{E_n} \hat{K}(\delta^{-1}v)\varphi(\delta^{-1}v)\,dv.$$

On the other hand, the Fourier transform of the function ψ whose values are $\varphi(\delta^{-1}v)$ is the function defined by $\hat{\psi}(u) = \delta^n\hat{\varphi}(\delta u)$ for $u \in E_n$. Thus,

$$\begin{aligned}\delta^{-\alpha} \int_{E_n} \hat{K}(\delta^{-1}v)\varphi(\delta^{-1}v)\,dv &= \delta^n \int_{E_n} K(u)\hat{\varphi}(\delta u)\,du = \int_{E_n} K(u)\,\hat{\psi}(u)\,du \\ &= \int_{E_n} \hat{K}(u)\psi(u)\,du = \int_{E_n} \hat{K}(v)\varphi(\delta^{-1}v)\,dv\end{aligned}$$

for all testing functions φ. Consequently, $\hat{K}(v) = \delta^{-\alpha}\hat{K}(\delta^{-1}v)$ for a.e. $v \in E_n$ and (4.2) follows immediately. Thus,

$$F(\xi) = \int_0^\xi f(r)\,dr = \int_0^1 f(\xi s)\xi\,ds = F(1)\xi^{1-k-\alpha}$$

for all $\xi > 0$ and, consequently, $f(r) = F'(r) = \gamma r^{-k-\alpha}$. This shows that $\hat{K}(t) = \gamma P(t)/|t|^{k+\alpha}$ for a.e. $t \in E_n$, provided $0 < \operatorname{Re}(\alpha) < n/2$. In order to evaluate γ, we let $\varphi(x) = e^{-\pi|x|^2}P(x)$. By Theorem 3.4 we then have

$\hat{\varphi}(t) = i^{-k}\varphi(t)$; hence,

$$i^{-k}\int_{E_n} e^{-\pi|t|^2}P(t)\{P(t)/|t|^{k+n-\alpha}\}\,dt = \gamma\int_{E_n} e^{-\pi|t|^2}P(t)\{P(t)/|t|^{k+\alpha}\}\,dt.$$

Expressing these two integrals in polar coordinates and then cancelling the common term $\int_{\Sigma_{n-1}}[P(t')]^2\,dt'$, we obtain

$$i^{-k}\int_0^\infty r^{\alpha+k-1}e^{-\pi r^2}\,dr = \gamma\int_0^\infty r^{k+n-\alpha-1}e^{-\pi r^2}\,dr.$$

The value of γ is easily deduced from this equality and the relation

$$\Gamma\left(\frac{\beta+1}{2}\right)\pi^{-(\beta+1)/2} = 2\int_0^\infty e^{-\pi t^2}t^\beta\,dt.$$

Finally, we must extend our result to the entire range $0 < \mathrm{Re}(\alpha) < n$. We have established the equality

$$\int_{E_n}\{P(x)/|x|^{k+n-\alpha}\}\hat{\varphi}(x)\,dx = \gamma_{k,\alpha}\int_{E_n}\{P(x)/|x|^{k+\alpha}\}\varphi(x)\,dx \tag{4.3}$$

for $0 < \mathrm{Re}(\alpha) < n/2$ when φ is a testing function. But each of these two integrals defines an analytic function of α for $0 < \mathrm{Re}(\alpha) < n$. Since $\gamma_{k,\alpha}$ also defines such an analytic function, equality (4.3) must hold for $0 < \mathrm{Re}(\alpha) < n$. The theorem now follows immediately from the definition of the Fourier transform of a distribution.

Kernels of the form $K(x) = P(x)/|x|^{n+k}$, $k \geqq 1$, give rise to an important class of operators in analysis.[10] In view of Theorem 4.1 it is natural to try to justify taking limits, as α tends to 0, and claim that

$$\hat{K}(t) = i^{-k}\pi^{n/2}\left\{\Gamma\left(\frac{k}{2}\right)\Big/\Gamma\left(\frac{n+k}{2}\right)\right\}P(t)/|t|^k.$$

The first difficulty we encounter is that K is not integrable in a neighborhood of the origin. The function K, however, can be used in order to define a tempered distribution L. This is done by letting

$$L(\varphi) = \lim_{\varepsilon\to 0}\int_{0<\varepsilon\leqq|t|} K(t)\varphi(t)\,dt \tag{4.4}$$

[10] These operators are special cases of the singular integral operators that will be discussed in Chapter VI.

for each testing function φ. To see that this limit exists and defines a continuous linear functional on the space $\mathscr{S}$ of testing functions we first observe that it suffices to show that this is true for $\lim_{\varepsilon\to 0}\int_{\varepsilon\leqq|t|\leqq 1} K(t)\varphi(t)\,dt$ (for the function K_1, defined by letting $K_1(t) = 0$ if $|t| \leqq 1$ and $K_1(t) = K(t)$ if $|t| > 1$, is a tempered function). Since $k \geqq 1$ we must have

$$\int_{\varepsilon\leqq|t|\leqq 1} K(t)\,dt = \int_{\varepsilon}^{1} r^{-1}\left\{\int_{\Sigma_{n-1}} P(t')\,dt'\right\} dr = 0.$$

Thus,

$$\int_{\varepsilon\leqq|t|\leqq 1} K(t)\,\varphi(t)\,dt = \int_{\varepsilon\leqq|t|\leqq 1} K(t)[\varphi(t) - \varphi(0)]\,dt.$$

But

$$|\varphi(t) - \varphi(0)| \leqq \left\{\sup_{|s|\leqq 1} |(\nabla\varphi)(s)|\right\}|t|.$$

Thus, if

$$t = |t|t',\ |K(t)[\varphi(t) - \varphi(0)]| \leqq \left\{\sup_{s\in E_n} |(\nabla\varphi)(s)|\right\} |P(t')|/|t|^{n-1};$$

therefore, $K(t)[\varphi(t) - \varphi(0)]$ is locally integrable and

$$\left|\lim_{\varepsilon\to 0}\int_{\varepsilon\leqq|t|\leqq 1} K(t)\varphi(t)\,dt\right| = \left|\int_{|t|\leqq 1} K(t)[\varphi(t) - \varphi(0)]\,dt\right|$$
$$\leqq \left\{\sup_{s\in E_n} |(\nabla\varphi)(s)|\right\} \int_{|t|\leqq 1} \{|P(t')|/|t|^{n-1}\}\,dt.$$

The fact that L is a tempered distribution now follows from Theorem 3.11 of Chapter I.

Distributions obtained by the limiting process (4.4) are called *principal value* distributions; the notation usually used to indicate that L is such a linear functional, defined in terms of the function K, is

$$L(\varphi) = \text{P.V.} \int_{E_n} K(t)\varphi(t)\,dt$$

for $\varphi \in \mathscr{S}$. In such a case we shall denote the distribution also by K.

Since $K(t) = P(t)/|t|^{n+k}$, as we have just seen, defines a (principal value) distribution, the Fourier transform $\hat{K}$ is defined as a tempered distribution. The following theorem justifies the passage to the limit, as α approaches 0, that was discussed above:

THEOREM 4.5. *Suppose $P(x)$ is a harmonic polynomial on E_n that is homogeneous of degree $k \geqq 1$. If $K(x) = P(x)/|x|^{n+k}$ then*

$$\hat{K}(t) = \left\{i^{-k}\pi^{n/2}\Gamma\left(\frac{k}{2}\right)\Big/\Gamma\left(\frac{n+k}{2}\right)\right\}P(t)/|t|^k = \gamma_{k,0}P(t)/|t|^k.$$

PROOF. We must show that

$$\text{(i)} \quad \gamma_{k,0}\int_{E_n}\{P(t)/|t|^k\}\varphi(t)\,dt = \text{P.V.}\int_{E_n}\{P(t)/|t|^{n+k}\}\hat{\varphi}(t)\,dt$$

for all testing functions φ. Theorem 4.1 asserts that

$$\text{(ii)} \quad \gamma_{k,\alpha}\int_{E_n}\{P(t)/|t|^{k+\alpha}\}\varphi(t)\,dt = \int_{E_n}\{P(t)/|t|^{n+k-\alpha}\}\hat{\varphi}(t)\,dt$$

when $0 < \alpha < n$. As $\alpha \to 0$ the left side of (ii) approaches the left side of (i); therefore, it suffices to show that

$$\lim_{\alpha\to 0}\int_{E_n}\{P(t)/|t|^{n+k-\alpha}\}\hat{\varphi}(t)\,dt = \text{P.V.}\int_{E_n}\{P(t)/|t|^{n+k}\}\hat{\varphi}(t)\,dt.$$

Since the integrals of $P(t)$ over the surfaces of spheres centered at the origin are zero and, as was observed above, $K(t)[\hat{\varphi}(t) - \hat{\varphi}(0)]$ is locally integrable we have, by the dominated convergence theorem,

$$\begin{aligned}\lim_{\alpha\to 0}&\int_{E_n}\{P(t)/|t|^{n+k-\alpha}\}\hat{\varphi}(t)\,dt\\ &= \lim_{\alpha\to 0}\int_{|t|\leqq 1}\{P(t)/|t|^{n+k-\alpha}\}\{\hat{\varphi}(t) - \hat{\varphi}(0)\}\,dt + \int_{|t|>1}K(t)\hat{\varphi}(t)\,dt\\ &= \int_{|t|\leqq 1}K(t)[\hat{\varphi}(t) - \hat{\varphi}(0)]\,dt + \int_{|t|>1}K(t)\hat{\varphi}(t)\,dt.\end{aligned}$$

But it was shown above that the last expression is the desired limit P.V. $\int_{E_n}\{P(t)/|t|^{n+k}\}\hat{\varphi}(t)\,dt$ and the theorem is proved.

If we let $Y^{(k)}(x') = P^{(k)}(x)/|x|^k$, where $x \neq 0$, $x' = x/|x|$ and $P^{(k)}$ is a harmonic polynomial of degree $k \geqq 1$ (i.e., $Y^{(k)}$ is a spherical harmonic of degree k), we put $\Omega(x') = \sum_{k=1}^m Y^{(k)}(x')$. It then follows from Theorem 4.5 that $K(x) = \Omega(x')/|x|^n$ defines a principal value distribution whose Fourier transform is the function

$$\hat{K}(x) = \sum_{k=1}^{m}\gamma_{k,0}Y^{(k)}(x').$$

In view of Corollary 2.3 it is natural to consider, more generally, such kernels K obtained from those functions Ω in $L^2(\Sigma_{n-1})$ that lie in the closed subspace of $L^2(\Sigma_{n-1})$ generated by the spherical harmonics $Y^{(k)}$ for $k = 1, 2, 3, \ldots$. This is precisely the space of those L^2 functions satisfying

$$\int_{\Sigma_{n-1}} \Omega(x')\, dx' = 0. \tag{4.6}$$

If we let $K(x) = \Omega(x')/|x|^n$, a simple extension of the argument that followed (4.4) shows that

$$\lim_{\varepsilon \to 0} \int_{0<\varepsilon \leqq |t|} K(t)\varphi(t)\, dt,$$

where φ is a testing function, defines a principal value distribution that will also be denoted by K. We then have the following generalization of Theorem 4.5:

THEOREM 4.7. *Suppose Ω is a function in $L^2(\Sigma_{n-1})$ satisfying* (4.6). *Then there exist unique spherical harmonics $Y^{(k)}$, of degree k, such that*

$$\Omega = \sum_{k=1}^{\infty} Y^{(k)}, \tag{4.8}$$

the series being convergent in the norm of $L^2(\Sigma_{n-1})$. The function whose values at $x \neq 0$ are $K(x) = \Omega(x')/|x|^n$ defines a principal value distribution whose Fourier transform $\hat{K}$ is a function that is homogeneous of degree zero; thus, $\hat{K}(x) = \Omega_0(x') = \Omega_0(x/|x|)$ for $x \neq 0$. Moreover, $\Omega_0 = \sum_{k=1}^{\infty} Y_0^{(k)}$, where $Y_0^{(k)} = \gamma_{k,0} Y^{(k)}$ and the series is convergent in the norm of $L^2(\Sigma_{n-1})$. Furthermore,

$$\sum_{k=1}^{\infty} k^n \| Y_0^{(k)} \|^2 = \sum_{k=1}^{\infty} k^n \int_{\Sigma_{n-1}} | Y_0^{(k)}(x')|^2\, dx' < \infty. \tag{4.9}$$

Conversely, any function, Ω_0, that is homogeneous of degree 0 and whose restriction to Σ_{n-1} is square integrable, satisfies $\int_{\Sigma_{n-1}} \Omega_0(x')\, dx' = 0$ and has a spherical harmonic development satisfying (4.9) *arises as the Fourier transform of a principal value distribution having the form $K(x) = \Omega(x')/|x|^n$ where Ω belongs to $L^2(\Sigma_{n-1})$ and satisfies* (4.6).

PROOF. The representation (4.8) is a consequence of Corollary 2.3 which, together with the orthogonality relations Corollary 2.4, imply that

$$\sum_{k=1}^{\infty} \| Y^{(k)} \|^2 < \infty.$$

From Theorem 4.5 we see that $\gamma_{k,0} \approx k^{-n/2}$ as k tends to ∞; therefore, inequality (4.9) must hold. In order to see that $\hat{K}$ is obtained from the function Ω_0 in the manner described above we consider the kernel $K^{(m)}(x) = \Omega^{(m)}(x')/|x|^n = \sum_{k=1}^m Y^{(k)}(x')/|x|^n$. From Theorem 4.5 we see that the Fourier transform of $K^{(m)}$ (considered as a principal value distribution) is the function, homogeneous of degree 0, whose restriction to Σ_{n-1} is $\sum_{k=1}^m Y_0^{(k)} = \sum_{k=1}^m \gamma_{k,0} Y^{(k)}$. Thus, if φ is a testing function, it follows from the definition of the Fourier transform of tempered distributions that

$$\int_{E_n} \hat{K}^{(m)}(x)\varphi(x)\,dx = \text{P.V.} \int_{E_n} K^{(m)}(x)\hat{\varphi}(x)\,dx.$$

Since $\sum_{k=1}^m Y_0^{(k)} \to \sum_{k=1}^\infty Y_0^{(k)} = \Omega_0$ in $L^2(\Sigma_{n-1})$ as $m \to \infty$ (this is an immediate consequence of (4.9)), it follows that

$$\lim_{m\to\infty} \int_{E_n} \hat{K}^{(m)}(x)\varphi(x)\,dx = \int_{E_n} \Omega_0(x/|x|)\varphi(x)\,dx.$$

On the other hand, the same argument we gave after (4.4) shows that

$$\text{(4.10)} \quad \text{P.V.} \int_{E_n} K^{(m)}(x)\hat{\varphi}(x)\,dx = \int_{|x|\leqq 1} K^{(m)}(x)[\hat{\varphi}(x) - \hat{\varphi}(0)]\,dx + \int_{|x|>1} K^{(m)}(x)\hat{\varphi}(x)\,dx.$$

But the L^2 convergence $\sum_{k=1}^m Y^{(k)} \to \Omega$, as $m \to \infty$, implies that the right side of (4.10) converges to

$$\int_{|x|\leqq 1} K(x)[\hat{\varphi}(x) - \hat{\varphi}(0)]\,dx + \int_{|x|>1} K(x)\hat{\varphi}(x)\,dx = \text{P.V.} \int_{E_n} K(x)\hat{\varphi}(x)\,dx.$$

Thus,

$$\int_{E_n} \Omega_0(x/|x|)\varphi(x)\,dx = \text{P.V.} \int_{E_n} K(x)\hat{\varphi}(x)\,dx$$

for all testing functions φ and it follows from the definition of the Fourier transform of the distribution K and $\hat{K}$ is the function whose values are $\Omega_0(x/|x|)$ when $x \neq 0$.

In order to show the last part we first observe that if Ω_0 has a spherical harmonic development $\sum_{k=1}^\infty Y_0^{(k)}$ satisfying (4.9) then $\sum_{k=1}^\infty Y^{(k)}$, where $Y^{(k)} = \gamma_{k,0}^{-1} Y_0^{(k)}$, converges in $L^2(\Sigma_{n-1})$ to a function Ω having integral zero over Σ_{n-1}. By applying the first part of our theorem to this function Ω we then obtain the desired converse.

More generally, the argument we gave showing that

$$\text{P.V.} \int_{E_n} K(x)\varphi(x)\,dx = \lim_{\varepsilon \to 0} \int_{|x| \geqq \varepsilon > 0} K(x)\varphi(x)\,dx$$

$$= \int_{|x| \leqq 1} K(x)[\varphi(x) - \varphi(0)]\,dx + \int_{|x| > 1} K(x)\varphi(x)\,dx,$$

when φ is a testing function and $K(x) = \Omega(x')/|x|^n$, depended only on the integrability of Ω on Σ_{n-1} and property (4.6). Thus, K is a principal value distribution in this case as well; the following theorem gives an explicit form for its Fourier transform $\hat{K}$:

THEOREM 4.11. *Suppose* $\Omega \in L^1(\Sigma_{n-1})$ *satisfies* (4.6). *Then the function whose values at* $x \neq 0$ *are* $K(x) = \Omega(x')/|x|^n$ *defines a principal value distribution whose Fourier transform* $\hat{K}$ *is a function that is homogeneous of degree zero; thus,* $\hat{K}(x) = \Omega_0(x') = \Omega_0(x/|x|)$ *for* $x \neq 0$. *Moreover,*

$$\Omega_0(x') = -\int_{\Sigma_{n-1}} \Omega(t')\left[\frac{i\pi}{2}\,\text{sgn}\,(x'\cdot t') + \log|x'\cdot t'|\right] dt'$$

for $x' \in \Sigma_{n-1}$, *where* sgn *denotes the signum function on* $(-\infty, \infty)$: $\text{sgn}\,(s) = 1$ *if* $s > 0$, $\text{sgn}\,(s) = -1$ *if* $s < 0$ *and* $\text{sgn}\,(0) = 0$.

PROOF. Let $\hat{K}_\varepsilon^N$ be the Fourier transform of the $L^1(E_n)$ function whose values are $\Omega(x')/|x|^n$, when $0 < \varepsilon < |x| < N$, and is 0 when $|x|$ is outside this interval.

If we let $x = rx'$ and $t = \rho t'$, with x', t' on Σ_{n-1}, we have

$$\hat{K}_\varepsilon^N(x) = \int_{\varepsilon < |t| < N} e^{-2\pi i x\cdot t} \frac{\Omega(t')}{|t|^n}\,dt$$

$$= \int_{\Sigma_{n-1}} \Omega(t')\left\{\int_\varepsilon^N \frac{e^{-2\pi i r\rho(x'\cdot t')}}{\rho}\,d\rho\right\} dt'.$$

Since Ω satisfies (4.6) this last integral equals

$$\int_{\Sigma_{n-1}} \Omega(t')\left\{\int_\varepsilon^N \frac{\cos 2\pi r\rho(x'\cdot t') - \cos 2\pi r\rho}{\rho}\,d\rho\right\} dt'$$

$$- i\int_{\Sigma_{n-1}} \Omega(t')\left\{\int_\varepsilon^N \frac{\sin 2\pi r\rho(x\cdot' t')}{\rho}\,d\rho\right\} dt'.$$

But, as is well-known,

$$\left|\int_\varepsilon^N \frac{\sin 2\pi r\rho(x'\cdot t')}{\rho}\,d\rho\right| = \left|\int_{2\pi r\varepsilon(x'\cdot t')}^{2\pi rN(x'\cdot t')} \frac{\sin s}{s}\,ds\right|$$

is dominated by a constant independent of ε, N, r and $x' \cdot t'$ and

$$\lim_{\varepsilon \to 0, N \to \infty} \int_{\varepsilon}^{N} \frac{\sin 2\pi r\rho(x' \cdot t')}{\rho} \, d\rho = \frac{\pi}{2} \operatorname{sgn}(x' \cdot t').$$

Moreover,

$$\int_{\varepsilon}^{N} \frac{\cos 2\pi r\rho(x' \cdot t') - \cos 2\pi r\rho}{\rho} \, d\rho = \int_{2\pi r|x' \cdot t'|\varepsilon}^{2\pi r|x' \cdot t'|N} \frac{\cos s}{s} \, ds - \int_{2\pi r\varepsilon}^{2\pi rN} \frac{\cos s}{s} \, ds$$

$$= \int_{2\pi r|x' \cdot t'|\varepsilon}^{2\pi r\varepsilon} \frac{\cos s}{s} \, ds - \int_{2\pi r|x' \cdot t'|N}^{2\pi rN} \frac{\cos s}{s} \, ds.$$

Each of these last two integrals has absolute value dominated by $\log (1/|x' \cdot t'|)$ while an easy calculation shows that

$$\lim_{\varepsilon \to 0} \int_{2\pi r|x' \cdot t'|\varepsilon}^{2\pi r\varepsilon} \frac{\cos s}{s} \, ds = \log \frac{1}{|x' \cdot t'|}$$

and

$$\lim_{N \to \infty} \int_{2\pi r|x' \cdot t'|N}^{2\pi rN} \frac{\cos s}{s} \, ds = 0.$$

We have shown, therefore, that

$$\lim_{\varepsilon \to 0, N \to \infty} \int_{\varepsilon}^{N} \frac{e^{-2\pi i r\rho(x' \cdot t')}}{\rho} \, d\rho = -\log |x' \cdot t'| - i\frac{\pi}{2} \operatorname{sgn}(x' \cdot t')$$

and that this convergence is dominated by a constant (independent of ε, N, r, and $x' \cdot t'$) times $1 + \log (1/|x' \cdot t'|)$.

It is an immediate consequence of Fubini's theorem that the integral $\int_{\Sigma_{n-1}} |\Omega(t')|[1 + \log (1/|x' \cdot t'|)] \, dt'$ is finite for almost every x' in Σ_{n-1} and represents an integrable function on Σ_{n-1}. Thus, by the Lebesgue dominated convergence theorem

$$\lim_{\varepsilon \to 0, N \to \infty} \hat{K}_{\varepsilon}^{N}(x') = -\int_{\Sigma_{n-1}} \Omega(t') \left[\frac{i\pi}{2} \operatorname{sgn}(x' \cdot t') + \log |x' \cdot t'| \right] dt'.$$

On the other hand, by the multiplication formula,

$$\int_{E_n} \hat{K}_{\varepsilon}^{N}(x') \varphi(x) \, dx = \int_{\varepsilon < |x| < N} \frac{\Omega(x')}{|x|^n} \hat{\varphi}(x) \, dx$$

whenever φ is a testing function. Theorem 4.11 now follows easily by passing to the limit as $\varepsilon \to 0$ and $N \to \infty$ and using the definition of the Fourier transform of the principal value distribution K.

COROLLARY 4.12. *If $\Omega \in L^2(\Sigma_{n-1})$ then the function Ω_0 is continuous.*

PROOF. By Theorem 4.7 Ω has a spherical harmonic development $\sum_{k=1}^{\infty} Y^{(k)}$ and Ω_0 has the development $\sum_{k=1}^{\infty} Y_0^{(k)} = \sum_{k=1}^{\infty} \gamma_{k,0} Y^{(k)}$. Let $\Omega^{(m)} = \sum_{k=1}^{m} Y^{(k)}$ and $\Omega_0^{(m)} = \sum_{k=1}^{m} Y_0^{(k)}$. Then $\Omega_0^{(m)}$ is continuous and, by Theorem 4.11,

$$\Omega_0(x') - \Omega_0^{(m)}(x') = \\ - \int_{\Sigma_{n-1}} [\Omega(t') - \Omega^{(m)}(t')] \left[\frac{i\pi}{2} \operatorname{sgn}(x' \cdot t') + \log |x' \cdot t'|\right] dt'.$$

Thus, by Schwarz's inequality,

$$|\Omega_0(x') - \Omega_0^{(m)}(x')| \\ \leqq \left(\int_{\Sigma_{n-1}} |\Omega(t') - \Omega^{(m)}(t')|^2 \, dt'\right)^{1/2} \left(\int_{\Sigma_{n-1}} \left(\frac{\pi^2}{4} + |\log |x' \cdot t'||^2\right) dt'\right)^{1/2}.$$

But the integral

$$\int_{\Sigma_{n-1}} \left(\frac{\pi^2}{4} + |\log |x' \cdot t'||^2\right) dt'$$

is finite and is independent of x', while $\int_{\Sigma_{n-1}} |\Omega(t') - \Omega^{(m)}(t')|^2 \, dt' \to 0$ as $m \to \infty$. Consequently, $\{\Omega_0^{(m)}\}$ converges uniformly to Ω_0. Since $\Omega_0^{(m)}$ is continuous, for $m = 1, 2, 3, \ldots$, it follows that Ω_0 is continuous and the corollary is proved.

The principal value distributions $\Omega(x)/|x|^n$ just presented give an example of the distributions of type (L^2, L^2) discussed in Chapter I, Section 3. As we shall see in Chapter VI, these distributions are also of type (L^p, L^p) for $1 < p < \infty$.

Our next application concerns itself with yet another summability method for integrals. In the first chapter (see (1.12) we introduced the Φ means

$$M_{\varepsilon,\Phi}(h) = \int_{E_n} \Phi(\varepsilon x) h(x) \, dx$$

of the integral $\int_{E_n} h$ whenever Φ belongs to C_0 and satisfies $\Phi(0) = 1$. When $\Phi(x) = e^{-|x|}$ we obtained the Abel means, while the Gauss–

Weierstrass means were obtained when $\Phi(x) = e^{-|x|^2}$. An equally important method of summability, *Bochner–Riesz summability*, is obtained when

$$\Phi_\delta(x) = \Phi(x) = \begin{cases} (1 - |x|^2)^\delta & \text{if } |x| \leqq 1 \\ 0 & \text{if } |x| > 1 \end{cases}$$

for $\delta > 0$. In this case it is customary to put $\varepsilon = 1/R$ and, thus,

$$M_{\varepsilon,\Phi}(h) = M_{(1/R),\Phi}(h) = \int_{|x| \leqq R} \left(1 - \frac{|x|^2}{R^2}\right)^\delta h(x)\,dx.$$

If f belongs to $L^1(E_n)$, $\varphi = \hat{\Phi}$ and $h(x) = \hat{f}(x)e^{2\pi i t\cdot x}$ then (see Theorem 1.16 of Chapter I)

$$S_R^\delta(t;f) = S_R^\delta(t) = \int_{|x| \leqq R} \hat{f}(x)e^{2\pi i x\cdot t}\left(1 - \frac{|x|^2}{R^2}\right)^\delta dx$$

$$= \int_{E_n} f(x)R^n\varphi(R(x - t))\,dx$$

are the Bochner–Riesz means of the integral $\int_{E_n} \hat{f}(x)e^{2\pi i x\cdot t}\,dx$ defining the inverse Fourier transform of f. If φ belongs to $L^1(E_n)$ then, by Theorem 1.18 of Chapter I, S_R^δ converges to f in the norm, while if, in addition, φ satisfies the hypotheses of Theorem 1.25 of Chapter I we have

$$\lim_{R\to\infty} S_R^\delta(t) = f(t) \quad \text{a.e.}$$

We shall show that this form of Fourier inversion does hold when $\delta > (n - 1)/2$. [11]

In order to do this we will calculate the Fourier transform $\varphi = \hat{\Phi}$. The following identity involving Bessel functions is useful for performing this calculation:

LEMMA 4.13. *If* $\mu > -\frac{1}{2}$ *then*

$$J_{\mu+\nu+1}(t) = \frac{t^{\nu+1}}{2^\nu\Gamma(\nu + 1)} \int_0^1 J_\mu(ts)s^{\mu+1}(1 - s^2)^\nu\,ds$$

whenever $\nu > -1$ *and* $t > 0$.

[11] The number $(n - 1)/2$ is called the *critical index* for Bochner–Riesz summability. We shall encounter it again in Chapter VII.

PROOF. If we develop e^{its} into the power series $\sum_{j=0}^{\infty}(its)^j/j!$ it follows easily from the definition of $J_k(t)$, for real $k > -\frac{1}{2}$, that was given in the third section that

$$J_k(t) = \sum_{j=0}^{\infty}(-1)^j \frac{(t/2)^{k+2j}}{j!\,\Gamma(j+k+1)}. \tag{4.14}$$ [12]

Thus, by (4.14),

$$\begin{aligned}\int_0^1 J_\mu(ts)s^{\mu+1}(1-s^2)^\nu\,ds &= \int_0^1\left\{\sum_{j=0}^{\infty}(-1)^j\frac{(ts/2)^{\mu+2j}}{j!\,\Gamma(j+\mu+1)}\right\}s^{\mu+1}(1-s^2)^\nu\,ds\\ &= \sum_{j=0}^{\infty}(-1)^j\frac{(t/2)^{\mu+2j}}{j!\,\Gamma(j+\mu+1)}\frac{1}{2}\int_0^1 r^{\mu+j}(1-r)^\nu\,dr\\ &= \frac{2^\nu\Gamma(\nu+1)}{t^{\nu+1}}\sum_{j=0}^{\infty}(-1)^j\frac{(t/2)^{\mu+\nu+1+2j}}{j!\,\Gamma(\mu+\nu+j+2)}\\ &= \frac{2^\nu\,\Gamma(\nu+1)}{t^{\nu+1}}J_{\mu+\nu+1}(t)\end{aligned}$$

and the lemma is proved.

THEOREM 4.15. *If Φ is the function defined by*

$$\Phi(t) = \begin{cases}(1-|t|^2)^\delta & \text{if } |t| \leqq 1\\ 0 & \text{if } |t| > 1,\end{cases}$$

where $\delta > 0$, then

$$\hat{\Phi}(x) = \pi^{-\delta}\Gamma(\delta+1)|x|^{-(n/2)-\delta}J_{n/2+\delta}(2\pi|x|).$$

PROOF. By Theorem 3.3

$$\hat{\Phi}(x) = 2\pi|x|^{-[(n-2)/2]}\int_0^1(1-|s|^2)^\delta J_{(n-2)/2}(2\pi|x|s)s^{n/2}\,ds.$$

Now applying Lemma 4.13 with $\nu = \delta$ and $\mu = (n-2)/2$ we see that the last expression equals

$$2\pi|x|^{-[(n-2)/2]}(2\pi)^{-\delta-1}2^\delta\Gamma(\delta+1)|x|^{-\delta-1}J_{(n-2)/2+\delta+1}(2\pi|x|),$$

which clearly reduces to the desired expression for $\hat{\Phi}(x)$.

[12] In deriving this formula it is useful to observe that the integral defining J_k is equal to $2\int_0^1(\cos ts)(1-s^2)^{(2k-1)/2}\,ds$. The coefficients of the power series (4.14) will then involve the expressions $\int_0^1 s^{2j}(1-s^2)^{(2k-1)/2}\,ds$ which can be easily evaluated by using the well-known relation $\Gamma(x)\Gamma(y)/\Gamma(x+y) = \int_0^1 u^{x-1}(1-u)^{y-1}\,du$.

It follows from (3.12) and (4.14) that $\varphi = \hat{\Phi}$ belongs to $L^1(E_n)$ provided $\delta > (n-1)/2$; moreover, φ satisfies the additional hypotheses of Theorems 1.18 and 1.25 of the first chapter and we obtain:

COROLLARY. 4.16. *If $f \in L^p(E_n)$, $1 \leqq p < \infty$, and $S_R^\delta = \varphi_{(1/R)} * f$, where $\varphi_{(1/R)}(x) = R^n\varphi(Rx)$ for $R > 0$ and $\delta > (n-1)/2$, then*

(a) $\|S_R^\delta - f\|_p \to 0$ *as* $R \to \infty$;

(b) $\lim_{R\to\infty} S_R^\delta(x) = f(x)$ *for all x in the Lebesgue set of f (in particular, this equality holds for almost every $x \in E_n$).*

5. Further Results

5.1. Theorem 1.1 generalizes easily to a result describing the relation between the Fourier transform and an arbitrary nonsingular linear transformation of E_n onto itself. Let σ be such a transformation and define $\tilde{\sigma}$ to be the inverse of the adjoint of σ (equivalently, $\tilde{\sigma}$ is the adjoint of the inverse of σ). The transformation $\tilde{\sigma}$ is called the *contragredient* of σ. If we define the *action* of σ on the function f, $R_\sigma f$, by letting $(R_\sigma f)(x) = f(\sigma x)$ we then have the following result: If $f \in L^1(E_n)$ then $(R_\sigma f)\hat{} = |\det \sigma|^{-1} \times R_{\tilde{\sigma}}\hat{f}$. That is, $\mathscr{F}R_\sigma = |\det \sigma|^{-1}R_{\tilde{\sigma}}\mathscr{F}$. The proof is no harder than that of Theorem 1.1: since the Jacobian in the change of variables $s = \sigma t$ is $|\det \sigma|^{-1}$, we have

$$(R_\sigma f)\hat{}(t) = \int_{E_n} e^{-2\pi i t\cdot x} f(\sigma x)\,dx = |\det \sigma|^{-1}\int_{E_n} e^{-2\pi i (t\cdot\sigma^{-1}s)} f(s)\,ds$$

$$= |\det \sigma|^{-1}\int_{E_n} e^{-2\pi i(\tilde{\sigma}t\cdot s)} f(s)\,ds = |\det \sigma|^{-1}(R_{\tilde{\sigma}}\hat{f})(t).$$

5.2. That orthogonal transformations and the Fourier transform are closely connected is further indicated by the following result (communicated to us by R. Coifman): Suppose σ is a continuous one-to-one transformation of E_n onto itself such that R_σ commutes with the Fourier transform; then σ is an orthogonal transformation. To see this we first observe that, by assumption, $(R_\sigma \mathscr{F} f)(0) = (\mathscr{F} R_\sigma f)(0)$ for all $f \in L^1(E_n)$. Applying this equality to the function defined by $f(x) = g(x)e^{-2\pi i\sigma^{-1}(x)\cdot y}$ for $x \in E_n$, where $y \in E_n$ is fixed, we obtain

$$0 = \int_{E_n} [e^{-2\pi i x\cdot\sigma(0)} e^{-2\pi i\sigma^{-1}(x)\cdot y} g(x) - e^{-2\pi i x\cdot y} g(\sigma x)]\,dx.$$

But, again by assumption, $\int_{E_n} e^{-2\pi i x\cdot y} g(\sigma x)\,dx = \int_{E_n} e^{-2\pi i x\cdot\sigma y} g(x)\,dx$; thus,

$$0 = \int_{E_n} [e^{-2\pi i(x\cdot h + \sigma^{-1}(x)\cdot y)} - e^{-2\pi i x\cdot\sigma(y)}]g(x)\,dx$$

for all g in $L^1(E_n)$, where $h = \sigma(0)$. Consequently, $e^{-2\pi i(x\cdot h + \sigma^{-1}(x)\cdot y)} = e^{-2\pi i x\cdot\sigma(y)}$ for all x, y in E_n. Thus, $x\cdot h + \sigma^{-1}(x)\cdot y = x\cdot\sigma(y) + k$ for some integer k; however, the continuity of σ implies k is constant. Putting $x = 0$ we then have $\sigma^{-1}(0)\cdot y = k$ for all $y \in E_n$ which implies $k = 0$. Therefore, $h = \sigma(0) = 0$ since $\sigma^{-1}(0) = 0$. We have shown that $\sigma^{-1}(x)\cdot y = x\cdot\sigma(y)$ or, equivalently, $x\cdot y = \sigma(x)\cdot\sigma(y)$ for all $y \in E_n$. Hence, σ is an isometry mapping 0 onto 0. But this clearly implies that σ is an orthogonal transformation.

5.3. Let $\beta(x, y)$, $x, y \in E_n$, be a nondegenerate real symmetric quadratic form (not necessarily positive). Then the mapping $t \to \hat{x}(t) = e^{2\pi i\beta(x,t)}$, where $x \in E_n$ is fixed, is a *character* (by this we mean a continuous map of E_n into the complex numbers of absolute value 1 such that $\hat{x}(s + t) = \hat{x}(s)\hat{x}(t)$ whenever s and t belong to E_n). With multiplication of characters defined by letting $\hat{x}_1\hat{x}_2(t) = \hat{x}_1(t)\hat{x}_2(t)$ the collection of all characters is a group, the *character group* of E_n. The mapping $x \to \hat{x}$ is an isomorphism of E_n into its character group. It is not hard to show that this mapping is onto; thus, we can "identify" E_n with its character group. This identification clearly depends on the bilinear form β.

The definition of character and character group can obviously be extended to any locally compact abelian group G; however, it is not true in general that G and its character group $\hat{G}$ are isomorphic. In abstract harmonic analysis the Fourier transform of a function f in $L^1(G)$ is defined to be the function on $\hat{G}$ whose value at $\hat{x}$ is $\hat{f}(\hat{x}) = \int_G f(t)\overline{\hat{x}(t)}\,d\mu(t)$, where μ is Haar measure. In terms of the identification between E_n and $\hat{E}_n$ described above we obtain a "Fourier transform" $T = T_\beta$ for each nondegenerate real symmetric quadratic form $\beta : (Tf)(x) = \int_{E_n} \overline{\hat{x}(t)} f(t)\,dt = \int_{E_n} e^{-2\pi i\beta(x,t)} f(t)\,dt$. When $\beta(x, t) = x\cdot t$ we obtain the Fourier transform we have been studying. In this chapter we exploited the close connection between the Fourier transform and the orthogonal group. In general, T_β is similarly related to the *orthogonal group associated with* β: the group $0_\beta(n)$ of all linear transformations σ of E_n leaving the quadratic form β invariant (that is, $\beta(\sigma x, \sigma y) = \beta(x, y)$ for all x, y in E_n). If $\sigma \in 0_\beta(n)$ and R_σ denotes the action of σ on functions whose domain is E_n (see 5.1) the argument used to establish Theorem 1.1 also shows that $R_\sigma T_\beta = T_\beta R_\sigma$. Moreover, result 5.2 also extends to this situation: If σ is a continuous one-to-one transformation of E_n onto itself such that R_σ commutes with T_β then $\sigma \in 0_\beta(n)$.

5.4. We have already stressed the significance of the group of dilations. The role of these transformations is further highlighted by considering the Mellin transform, in terms of which some of the results of this chapter can be reinterpreted.

Suppose f is a complex-valued function on the interval $(0, \infty)$ which satisfies the integrability condition

$$\int_0^\infty \frac{|f(x)|}{x}\,dx < \infty.$$

Then its Mellin transform, $\mathscr{M}_f(s) = \mathscr{M}(s)$, is defined by

$$\mathscr{M}(s) = \int_0^\infty f(x)x^s \frac{dx}{x},$$

for $\operatorname{Re}(s) = 0$. If f_1 and f_2 both satisfy this integrability condition, then so does their *multiplicative convolution*, $f_1 \cdot f_2$, defined by

$$(f_1 \cdot f_2)(x) = \int_0^\infty f_1(x/y)f_2(y)\,\frac{dy}{y},$$

and we have $\mathscr{M}_{f_1 \cdot f_2}(s) = \mathscr{M}_{f_1}(s)\mathscr{M}_{f_2}(s)$. The Mellin transform is the multiplicative analogue of the Fourier transform, and many of its elementary properties can be deduced from the Fourier transform on E_1 by making the change ot variables $x \to e^x$, taking E_1 to $(0, \infty)$. In this connection it may be of interest to state the following identities.

(a) $$\mathscr{M}_f(s) = 2^{a+s-\mu-1}\Gamma\left(\frac{a+s}{2}\right)\Big/\Gamma\left(\mu - \frac{a+s}{2} + 1\right),$$

if $f(x) = x^{a-\mu}J_\mu(x)$, with $a > 0$, $\mu > a - \frac{1}{2}$. The case when μ is integral or half-integral is essentially equivalent with Theorem 4.1 (if we assume Theorem 3.10 as known).

(b) Let $f_1(x) = x^{-\mu+1}J_\mu(x)$, and $f_2(x) = x^{-2\mu-2\nu+1}(x^2 - 1)^\nu/2^\nu\Gamma(\nu + 1)$, for $x \geqq 1$, $f_2(x) = 0$, if $0 < x < 1$. Then $f(x) = (f_1 \cdot f_2)(x) = x^{-\mu-\nu}J_{\mu+\nu}(x)$, as long as $\nu > -1$, and $\mu > -\frac{1}{2}$. This identity is equivalent with the statement of Lemma 4.13. An alternative proof for that lemma can be carried out by showing that $\mathscr{M}_f(s) = \mathscr{M}_{f_1}(s)\mathscr{M}_{f_2}(s)$.

For a proof of (a) and further facts concerning the Mellin transform, see Titchmarsh [2].

5.5. In the second chapter we showed that a harmonic function has continuous derivatives of all orders (see Theorem 1.7 in that chapter). It is an immediate consequence of the Poisson integral representation of harmonic functions (see Theorem 1.10 and Corollary 1.11) together with Theorem 2.10 that a harmonic function is real-analytic.

5.6. Lemma 2.11 is equivalent to the statement that the only differential polynomials with constant coefficients that commute with rotations are polynomials in the Laplacian. More precisely, the lemma is equivalent to the following statement, when $n > 1$.

If $P(D)$, $D = (\partial/\partial x_1, \ldots, \partial/\partial x_n)$, is a differential polynomial on E_n with constant coefficients, then $P(D)R_\sigma = R_\sigma P(D)$ for all rotations σ if and only if $P(D) = c_0 I + c_1\Delta + \cdots + c_m\Delta^m$, where $\Delta = \partial^2/\partial x_1^2 + \cdots + \partial^2/\partial x_n^2$ is the Laplace operator.

5.7. We defined the ultraspherical polynomials P_k^λ in terms of the generating function $(1 - 2rt + r^2)^{-\lambda}$. When $\lambda = 0$, however, it is more appropriate to consider the *Tchebichef* polynomials T_k whose generating function is

$$\log(1 - 2rt + r^2) = -2\sum_{k=1}^{\infty} k^{-1}T_k(t)r^k.$$

If this is done it is not necessary to exclude the two-dimensional case, as was done often in the second section. For example, Theorem 2.14 is valid when $n = 2$ if we employ the Tchebichef polynomials.

5.8. An important connection between the group $SO(n)$ of rotations on E_n and spherical harmonics arises in the study of the unitary representations of this group. The mapping $\sigma \to R_{\sigma^{-1}}$, where $R_{\sigma^{-1}}$ is the operator on $L^2(\Sigma_{n-1})$ defined by $(R_{\sigma^{-1}}f)(x) = f(\sigma^{-1}x)$, is a unitary representation of $SO(n)$. When we restrict $R_{\sigma^{-1}}$ to the subspace $\mathscr{H}_k$, $k = 0, 1, 2, \ldots$, we obtain an irreducible representation of $SO(n)$. When the dimension n is 3 all the irreducible representations of $SO(n)$ can be obtained in this way. The group $SO(n - 1)$ can be identified in a natural way with the subgroup $G(n, x')$ of $SO(n)$ consisting of those rotations σ leaving a fixed point $x' \in \Sigma_{n-1}$ fixed. We see from Lemma 2.8 that in each of the spaces $\mathscr{H}_k$ there exists a nonzero vector (viz., $Z_{x'}$) left invariant by all the operators R_σ, where σ ranges over the subgroup $G(n, x')(\sim SO(n - 1))$. Among all the irreducible representations of $SO(n)$, the ones described above are characterized, up to a unitary equivalence, by the existence of such a vector that is invariant under the action of $SO(n - 1)$ (see Weyl [1] and Boerner [1]).

5.9. Corollary 4.12 extends to the case $\Omega \in L^p(\Sigma_{n-1})$, $p > 1$. The proof is not essentially different: instead of Schwarz's inequality one can employ Hölder's inequality and, instead of using the L^2-norm for approximating functions by spherical harmonics, one can avail himself of the L^p norm. A further refinement of this argument shows that the result still holds when $|\Omega| \log^+ |\Omega|$ is integrable (where $\log^+ x = \log x$ if $x \geqq 1$, while $\log^+ x = 0$ if $0 < x < 1$).

5.10. The analogues of the subspaces $\mathfrak{H}_k$ can obviously be defined for $L^p(E_n)$, $1 \leqq p \leqq \infty$. Let us remark that if $f_0(|x|)P(x)$ belongs to $L^p(E_n)$,

$1 \leqq p \leqq 2$, then its Fourier transform $\hat{f}$ is given by the formula (3.10), where the convergence is in the $L^{p'}(E_n)$ norm (see (1.2) of Chapter V). The estimates in Lemma 3.11 for the Bessel functions then have the following consequences: (a) $\hat{f}(x)$ is equivalent to a function continuous when $x \neq 0$, if $1 \leqq p < 2n/(n+1)$; (b) when $k < (1/p - \frac{1}{2})n - \frac{1}{2}$, then $\hat{f}$ is of class $C^{(k)}$ away from the origin; (c) when $p = 1$, the condition can be modified to require $k \leqq (n-1)/2$ (results of this type were communicated to us by N. Varapolous). These facts are simple corollaries of Lemma 3.11, Hölder's inequality and the relation $(t^{-k}J_k(t))' = -t^{-k}J_{k+1}(t)$ (see (3.2)).

Bibliographical Notes

We introduced only those properties of Bessel functions that were needed for our study of the Fourier transform. For a thorough treatment of this subject see Watson [1] or the Bateman Manuscript Project [1], Vol. 2. We also refer the reader to the latter for further details concerning spherical harmonics, Gegenbauer and Tchebichef polynomials. For earlier studies of the action of the Fourier transform on the spaces $\mathfrak{H}_k$ see Hecke [1], Bochner [6], and Herz [1]. For another proof of Theorem 4.1 see Calderón [3]. The Fourier transforms of the kernels $K(x) = \Omega(x')/|x|^n$ were first thoroughly studied by Mihlin [1], and Calderón and Zygmund [5]. The properties of the rotation group can be exploited further to obtain various results related to those in this chapter. See Vilenkin [1], Coifman and G. Weiss [1].

CHAPTER V

Interpolation of Operators

Let T be a linear operator mapping some linear space $\mathscr{A}$ into another linear space $\mathscr{B}$. Suppose A_0, A_1 are Banach subspaces of $\mathscr{A}$ and B_0, B_1 are Banach subspaces of $\mathscr{B}$ such that when T is restricted to A_i it maps this space continuously into B_i, $i = 0, 1$. When this is the case, it can be shown often that there exist infinitely many pairs of "intermediate" Banach spaces (A, B), with $A \subset \mathscr{A}$ and $B \subset \mathscr{B}$, such that T restricted to A is a bounded operator mapping into B. General theorems of this kind are called *interpolation theorems* for linear operators and they have many important applications in Fourier analysis. We have already encountered an example of this situation when $\mathscr{A} = \mathscr{B}$ is the space of tempered distributions, T is the Fourier transform, $A_0 = L^1(E_n)$, $B_0 = L^\infty(E_n)$ and $A_1 = L^2(E_n) = B_1$; it is then true that if T is restricted to $L^p(E_n)$, $1 \leqq p \leqq 2$, it maps this space boundedly into $L^q(E_n)$, where $(1/p) + (1/q) = 1$. This follows from a basic interpolation theorem concerning L^p spaces: the M. Riesz interpolation theorem. We shall study this theorem and some of its applications in the first section of this chapter. There is another type of interpolation theorem in which we do not need assumptions as strong as boundedness at the "end-points" (A_0, B_0) and (A_1, B_1); moreover, we can assume certain conditions weaker than the linearity of T. This theorem, known as the Marcinkiewicz theorem, applies to the maximal function operator $f \to m_f$ that was studied in the second chapter. The second section of this chapter is devoted to applications of the Marcinkiewicz theorem. This theorem is best understood when it is extended to a collection of function spaces that includes the L^p classes. This collection, the $L(p, q)$ spaces, as well as the extended version of the Marcinkiewicz theorem is the subject of the third section. In the fourth section we develop an extension of the M. Riesz theorem in which the operators vary, in a sufficiently smooth manner, with the "intermediate" pairs of Banach spaces.

1. The M. Riesz Convexity Theorem and Interpolation of Operators Defined on L^p Spaces

In the previous chapters we encountered several operators acting on L^p spaces. One class of such operators consists of the convolution operators determined by functions in $L^p(E_n)$; that is, if we fix $f \in L^p(E_n)$ we obtain the operator T that maps $g \in L^1(E_n)$ into the function $f * g$. From Theorem 1.3 of Chapter I we see that T is a bounded operator defined on $L^1(E_n)$

having values in $L^p(E_n)$; furthermore, the operator norm, $\|T\|$,[1] does not exceed $\|f\|_p$. On the other hand, an immediate application of Hölder's inequality shows that T also maps $L^{p'}(E_n)$ boundedly into $L^\infty(E_n)$ when $(1/p) + (1/p') = 1$; as in the last case, the operator norm $\|T\|$ does not exceed $\|f\|_p$. Since T is defined on $L^1(E_n)$ and on $L^{p'}(E_n)$ there is a natural extension of T to all the spaces $L^r(E_n)$ when $1 \leqq r \leqq p'$: If $g \in L^r(E_n)$ then we can find $g_1 \in L^1(E_n)$ and $g_2 \in L^{p'}(E_n)$ such that $g = g_1 + g_2$ and we can define Tg to be $Tg_1 + Tg_2$ (it is clear that Tg does not depend on the particular splitting of g as $g = g_1 + g_2$).[2] It is natural, therefore, to ask whether T maps $L^r(E_n)$ boundedly into some space $L^q(E_n)$. We have already stated without proof (see (4.3) in the first chapter) that this is the case when $(1/q) = (1/p) + (1/r) - 1$, and $\|T\|$ is, again, less than or equal to $\|f\|_p$. That is,

$$\|f * g\|_q \leqq \|f\|_p \|g\|_r \tag{1.1}$$

whenever $f \in L^p(E_n)$ and $g \in L^r(E_n)$. We shall show, in this section, that this inequality (known as *Young's inequality*) follows easily from a general result concerning linear operators acting on L^p-spaces.

This general result can also be applied to the Fourier transform $\mathscr{F}$. We have treated $\mathscr{F}$ mostly as a linear operator defined on $L^1(E_n)$ and $L^2(E_n)$. Theorem 1.1, part (a), of the first chapter asserts that the first space is mapped boundedly into $L^\infty(E_n)$ with operator norm $\leqq 1$, while the Plancherel theorem tells us that the second space is mapped onto itself with operator norm equal to 1. For the same reasons we gave in the previous paragraph, $\mathscr{F}$ has a natural extension to the "intermediate" spaces $L^p(E_n)$, $1 \leqq p \leqq 2$ (see, also, the remarks made toward the end of Section 2 in the first chapter). We shall show that this space is mapped by $\mathscr{F}$ boundedly into $L^{p'}(E_n)$, where $(1/p) + (1/p') = 1$, with operator norm $\leqq 1$. That is, if $f \in L^p(E_n)$, $1 \leqq p \leqq 2$, then $\hat{f} = \mathscr{F}f$ belongs to $L^{p'}(E_n)$ and

$$\|\hat{f}\|_{p'} \leqq \|f\|_p. \tag{1.2}$$

This result is known as the *inequality of Hausdorff–Young*.

In order to state the above mentioned general result concerning linear operators, the *M. Riesz convexity theorem*, we need to introduce some new concepts and notation. Suppose $(M, \mathscr{M}, \mu)$ is a measure space.[3] We let

[1] If T is a bounded transformation from L^r into L^s we shall denote its norm by $\|T\|$. That is, $\|T\| = \inf\{k \geqq 0\colon \|Tf\|_s \leqq k\|f\|_r \text{ for all } f \in L^r\}$.

[2] For example, by letting $g_1(x) = g(x)$ when $|g(x)| \geqq 1$ and $g_1(x) = 0$ when $|g(x)| < 1$, we obtain such a decomposition.

[3] M is a set, $\mathscr{M}$ the σ-algebra of measurable subsets of M and μ is a measure defined on $\mathscr{M}$. We shall consider only σ-finite measure spaces; thus, all the standard results of measure theory, such as the Radon–Nikodym theorem and the F. Riesz representation theorem, are valid.

$L^p(M)$, $1 \leqq p < \infty$, denote the space of all those complex-valued functions f satisfying $\|f\|_p = (\int_M |f|^p \, d\mu)^{1/p} < \infty$. As is the case when $M = E_n$ (see the beginning of Chapter I) we use the same notation, $L^p(M)$, for the space of equivalence classes obtained by considering two such functions to be equivalent when they agree almost everywhere; by defining the norm of the class containing the function f to be $\|f\|_p$ we then obtain a Banach space. The same remarks apply to the space $L^\infty(M)$ of essentially bounded measurable functions if we define $\|f\|_\infty$, for $f \in L^\infty(M)$, to be the essential supremum of f.

If f is a measurable function, a *truncation of* f is any one of the functions g defined by letting $g(x) = f(x)$ if $r_1 < |f(x)| \leqq r_2$ and $g(x) = 0$ otherwise, where r_1 and r_2 are nonnegative. Suppose T is an operator mapping a linear space D of measurable functions on $(M, \mathscr{M}, \mu)$ into measurable functions defined on another measure space $(N, \mathscr{N}, \nu)$. We assume that D contains all characteristic functions of sets of finite measure and has the property that whenever $f \in D$ and g is a truncation of f, then $g \in D$. If there exists a constant $k > 0$ such that

$$\left(\int_N |Tf|^q \, d\nu\right)^{1/q} = \|Tf\|_q \leqq k\|f\|_p = k\left(\int_M |f|^p \, d\mu\right)^{1/p}$$

for all $f \in D \cap L^p(M)$ we say that T is of *type* (p, q). The least k for which this inequality holds is called the (p, q)*-norm* of T.

The M. Riesz convexity theorem can now be formulated in the following way:

THEOREM 1.3. *Suppose a linear operator T is of type (p_i, q_i), with (p_i, q_i)-norm k_i, $i = 0, 1$, then T is of type (p_t, q_t) with (p_t, q_t)-norm $k_t \leqq k_0^{1-t}k_1^t$ provided $1/p_t = (1 - t)/p_0 + t/p_1$ and $1/q_t = (1 - t)/q_0 + t/q_1$, with $0 \leqq t \leqq 1$.*[4]

Before proving this theorem we show how the two inequalities, (1.1) and (1.2), follow immediately from it. We observed that the operator $T: g \to f * g$ is of types $(p_0, q_0) = (1, p)$ and $(p_1, q_1) = (p', \infty)$ (its domain D includes all the spaces $L^r(E_n)$, $1 \leqq r \leqq p'$); moreover, the $(1, p)$-norm and the (p', ∞)-norm do not exceed $\|f\|_p$. Thus, T is of type (p_t, q_t), $0 \leqq t \leqq 1$, where

$$\frac{1}{p_t} = \frac{1-t}{p_0} + \frac{t}{p_1} = 1 - t + \frac{t}{p'} = 1 - \frac{t}{p}$$

and

$$\frac{1}{q_t} = \frac{1-t}{q_0} + \frac{t}{q_1} = \frac{1-t}{p} + \frac{t}{\infty} = \frac{1-t}{p}.$$

[4] In case p_0, p_1, q_0 or q_1 is ∞ we adopt the usual convention: We let $1/\infty = 0$.

Putting $r = p_t$ and $q = q_t$ we then have

$$\frac{1}{q} = \frac{1-t}{p} = \frac{1}{p} + \left(1 - \frac{t}{p}\right) - 1 = \frac{1}{p} + \frac{1}{r} - 1.$$

Moreover, the (r, q)-norm of T does not exceed $\|f\|_p^{1-t}\|f\|_p^t = \|f\|_p$. But this is precisely inequality (1.1). Similarly, since the Fourier transform is of types $(1, \infty)$ and $(2, 2)$, with norms not larger than 1, an application of Theorem 1.3 with $t = 2/p'$ gives us the Hausdorff–Young inequality (1.2).

It is often useful to consider the following geometrical interpretation of the M. Riesz convexity theorem: Let Q be the square in the plane whose vertices are the points $(0, 0)$, $(1, 0)$, $(1, 1)$, and $(0, 1)$. If T is an operator satisfying the hypotheses of Theorem 1.3, we can associate with T the points $x_0 = (1/p_0, 1/q_0)$ and $x_1 = (1/p_1, 1/q_1)$ in order to indicate that this operator is of types (p_0, q_0) and (p_1, q_1). As t ranges from 0 to 1, the point $x_t = (1/p_t, 1/q_t)$ traverses the line segment joining x_0 and x_1; Theorem 1.3 asserts that, for each point (α, β) of this segment, T is of type $(1/\alpha, 1/\beta)$.

The justification for calling Theorem 1.3 a *convexity* theorem is that, if $\varphi(t)$ is the logarithm of the (p_t, q_t)-norm of T, then φ is a convex function. Moreover, the set of all points $(1/p, 1/q)$ such that T is of type (p, q) is a convex set. We shall show that Theorem 1.3 can be derived from a classical result in the theory of functions whose conclusion also involves a logarithmically convex function. This result, a Phragmen–Lindelöf theorem (see (5.2) of Chapter II), is known as *the three lines theorem* and can be formulated in the following way:

LEMMA 1.4. *Suppose F is a bounded continuous complex-valued function on the closed strip $S = \{x + iy = z \in \mathbf{C} : 0 \leqq x \leqq 1\}$ that is analytic in the interior of S. If $|F(iy)| \leqq m_0$ and $|F(1 + iy)| \leqq m_1$ for all y then $|F(x + iy)| \leqq m_0^{1-x}m_1^x$ for all $z = x + iy \in S$. That is, if we define $k_x = \sup\{|F(x + iy)| : -\infty < y < \infty\}$ for $0 \leqq x \leqq 1$, then $\varphi(x) = \log k_x$ is a convex function.*

PROOF. We can obviously assume that m_0 and m_1 are positive. Then by considering the function whose values are $F(z)/m_0^{1-z}m_1^z$, if necessary, we can reduce the problem to the case $m_0 = 1 = m_1$. Thus, we can suppose that $|F(iy)| \leqq 1$ and $|F(1 + iy)| \leqq 1$ for all y. We want to show that $|F(z)| \leqq 1$ for all $z \in S$. If we assume that $\lim_{|y| \to \infty} F(x + iy) = 0$ uniformly for $0 \leqq x \leqq 1$ this result follows easily from the maximum principle: for we can then find $y_0 > 0$ such that $|F(x + iy)| \leqq 1$ for $|y| \geqq y_0$, while $|F(z)| \leqq 1$ on the boundary of the rectangle with vertices iy_0, $1 + iy_0$, $1 - iy_0$, $-iy_0$. In general, we can apply this result to the function $F_n(z) = F(z)e^{(z^2-1)/n}$, $n = 1, 2, \ldots$, and, because F is bounded,

$|F_n(z)| = |F(x + iy)|e^{-y^2/n}e^{(x^2-1)/n} \leqq |F(x + iy)|e^{-y^2/n} \to 0$ as $y \to \infty$, uniformly for $0 \leqq x \leqq 1$, $|F_n(iy)| \leqq 1$ and $|F_n(1 + iy)| \leqq 1$. Therefore, $|F_n(z)| \leqq 1$ and letting n tend to ∞, we obtain the desired inequality.

We now pass to the proof of Theorem 1.3. In order to do this we let $\alpha_j = 1/p_j$, $\beta_j = 1/q_j$, $j = 0, 1$, $\alpha = 1/p_t$ and $\beta = 1/q_t$. Consequently, if $\alpha(z) = (1 - z)\alpha_0 + z\alpha_1$ and $\beta(z) = (1 - z)\beta_0 + z\beta_1$, for $z \in \mathbf{C}$, we have $\alpha(j) = \alpha_j$, $\beta(j) = \beta_j$, $j = 0, 1$, $\alpha(t) = \alpha$ and $\beta(t) = \beta$.

First, we estimate $\|Tf\|_q$ for f a finite linear combination of characteristic functions of subsets of M having finite measure. Such a simple function, f, belongs to the domain of T by definition. Since

$$\|Tf\|_q = \sup \left| \int_N (Tf)g\, d\nu \right|,$$

where the supremum is taken over all simple functions g satisfying $\|g\|_{q'} = 1$, $(1/q) + (1/q') = 1$, it suffices to show that the absolute value of each such integral

$$I = \int_N (Tf)g\, d\nu$$

is less than or equal to $k_0^{1-t}k_1^t\|f\|_p$. Dividing by $\|f\|_p$, if this norm is not zero, we can reduce the problem further to the case $\|f\|_p = 1$.

Suppose, then, that $f = \sum_{j=1}^m a_j\chi_{E_j}$ and $g = \sum_{k=1}^n b_k\chi_{F_k}$ are simple functions satisfying the above conditions. We also assume that $p_t < \infty$ and $q_t > 1$; thus, $\alpha > 0$ and $\beta < 1$. If $a_j = |a_j|e^{i\theta_j}$ and $b_k = |b_k|e^{i\varphi_k}$ we define

$$f_z = \sum_{j=1}^m |a_j|^{\alpha(z)/\alpha}e^{i\theta_j}\chi_{E_j}$$

and

$$g_z = \sum_{k=1}^n |b_k|^{(1-\beta(z))/(1-\beta)}e^{i\varphi_k}\chi_{F_k} \qquad \text{for } z \in \mathbf{C}.$$

One then obtains an entire function by putting $F(z) = \int_N (Tf_z)g_z\, d\nu$, and it follows from the definitions of $\alpha(z)$ and $\beta(z)$ that $F(t) = I$. From the linearity of T we see that

$$F(z) = \sum_{j,k}^{m,n} |a_j|^{\alpha(z)/\alpha}|b_k|^{(1-\beta(z))/(1-\beta)}\gamma_{jk},$$

where $\gamma_{jk} = e^{i(\theta_j + \varphi_k)}\int_N (T\chi_{E_j})\chi_{F_k}\, d\nu$. Since each of these summands is bounded in the strip S, it follows that the entire function F is bounded when restricted to this strip. If we can show that $|F(iy)| \leqq k_0$ and $|F(1 + iy)| \leqq k_1$ for all y, then the desired inequality, $|I| \leqq k_0^{1-t}k_1^t$, will be an immediate consequence of Lemma 1.4.

We can obtain these two estimates by first observing that, since $\alpha(iy) = \alpha_0 + iy(\alpha_1 - \alpha_0)$ and $1 - \beta(iy) = (1 - \beta_0) - iy(\beta_1 - \beta_0)$, we have $|f_{iy}|^{p_0} = |e^{i \arg f}|f|^{iy(\alpha_1 - \alpha_0)/\alpha}|f|^{(p/p_0)}|^{p_0} = |f|^p$ and

$$|g_{iy}|^{q_0'} = |e^{i \arg g}|g|^{-iy(\beta_1 - \beta_0)/(1-\beta)}|g|^{(q'/q_0')}|^{q_0'} = |g|^{q'}.$$

Thus, applying Hölder's inequality and the fact that T is of type (p_0, q_0) with norm k_0,

$$|F_{iy}| \leqq \|Tf_{iy}\|_{q_0}\|g_{iy}\|_{q_0'} \leqq k_0\|f_{iy}\|_{p_0}\|g_{iy}\|_{q_0'}$$

$$= k_0\left(\int_M |f|^p \, d\mu\right)^{1/p_0}\left(\int_N |g|^{q'} \, d\nu\right)^{1/q_0} = k_0\|f\|_p^{(p/p_0)}\|g\|_{q'}^{(q'/q_0')} = k_0.$$

A similar calculation shows that $|F(1 + iy)| \leqq k_1$. We can conclude, therefore, that $\|Tf\|_q \leqq k_0^{1-t}k_1^t\|f\|_p$ for all simple f in $L^p(M)$.

In order to establish this inequality for the general $f \in D \cap L^p(M)$, where D is the domain of T, we shall show that we can find a sequence $\{f_n\}$ of simple functions such that $\|f_n - f\|_p \to 0$ and $(Tf_n)(x) \to (Tf)(x)$ a.e. as $n \to \infty$. If this were the case, an application of Fatou's lemma and the result we just obtained for simple functions would give us the inequality $\|Tf\|_q \leqq \lim_{n\to\infty} \|Tf_n\|_q \leqq \lim_{n\to\infty} \{k_0^{1-t}k_1^t\|f_n\|_p\} = k_0^{1-t}k_1^t\|f\|_p$, and the theorem would be proved. In order to find such a sequence we can assume $f \geqq 0$ (for we may consider, separately, the positive and negative parts of Re $\{f\}$ and $\mathscr{I}\{f\}$). After relabelling, if necessary, we can also assume that $p_0 \leqq p_1$. Let f^0 and f^1 be the truncations of f defined by letting

$$f^0(x) = \begin{cases} f(x) & \text{when } f(x) > 1 \\ 0 & \text{when } f(x) \leqq 1 \end{cases}$$

and $f^1 = f - f_0$. Since $(f^0)^{p_0} \leqq f^p$, $(f^1)^{p_1} \leqq f^p$ and D contains all the truncations of f we have $f^0 \in D \cap L^{p_0}(M)$ and $f^1 \in D \cap L^{p_1}(M)$. If $\{g_m\}$ is an increasing sequence of nonnegative simple functions converging to f it follows from the monotone convergence theorem of Lebesgue that $\lim_{m\to\infty} \|g_m - f\|_p = 0$. For the same reason $\|g_m^0 - f^0\|_{p_0}$ and $\|g_m^1 - f^1\|_{p_1}$ tend to 0 as m tends to ∞, where g_m^0 and g_m^1 are the truncations (at 1) obtained from g_m in the same way as were f^0 and f^1 obtained from f. Since T is of types (p_0,q_0) and (p_1,q_1) we must have $\|Tg_m^0 - Tf^0\|_{q_0} \to 0$ and $\|Tg_m^1 - Tf^1\|_{q_1} \to 0$ as $m \to \infty$. Thus, there exists a subsequence of $\{Tg_m^0\}$ converging almost everywhere to Tf^0. Considering only those indices involved in this subsequence we can then find a subsequence $\{g_{m_n}^1\}$ such that $Tg_{m_n}^1$ converges almost everywhere to Tf^1. If we now let $f_n = g_{m_n}^0 + g_{m_n}^1$ we have a sequence $\{f_n\}$ satisfying the desired properties: $\lim_{n\to\infty} \|f_n - f\|_p = 0$ and $\lim_{n\to\infty} (Tf_n)(x) = (Tf^0)(x) + (Tf^1)(x) = (Tf)(x)$ almost everywhere.

We still need to remove the restrictions $\alpha > 0$ and $\beta < 1$. The excluded cases, however, are simpler than the ones we have considered. If $\alpha = 0$ and $\beta = 1$ then at least one of the pairs (p_0, q_0) and (p_1, q_1) must be $(\infty, 1)$ and there is nothing to prove. If $\alpha > 0$ and $\beta = 1$ then the above proof is still valid if we define g_z to be g for all $z \in \mathbf{C}$. Similarly, if $\alpha = 0$ and $\beta < 1$, putting $f_z = f$ allows us to carry out the argument we gave. This completes the proof of Theorem 1.3.

The M. Riesz convexity theorem is a powerful tool in analysis and, in addition to (1.1) and (1.2), many important inequalities can be derived from it. In some cases, however, one encounters a family of operators $\{T_t\}$, $0 \leqq t \leqq 1$, mapping $L^{p_t}(M, d\mu_t)$ boundedly into $L^{q_t}(N, d\nu_t)$, where both the operators and the measures vary in a "smooth" way with t. In the fourth section we shall derive an extension of Theorem 1.3 that can be applied to such families of operators. In other cases one encounters (not necessarily linear) operators defined on L^{p_t}-spaces, $0 \leqq t \leqq 1$, that are not of types (p_0, q_0) and (p_1, q_1), yet they do map L^{p_t} boundedly into L^{q_t} when $0 < t < 1$. In the next section we shall show that certain weaker conditions at the "end points" (p_i, q_i), $i = 0, 1$, are usually sufficient in order to guarantee (strong) type (p_t, q_t) for $0 < t < 1$.

2. *The Marcinkiewicz Interpolation Theorem*

In the second chapter we introduced the maximal function of Hardy and Littlewood, m_f, of a function $f \in L^p(E_n)$, $1 \leqq p \leqq \infty$. We showed (see Corollary 3.11 of Chapter II) that the operator M mapping f into m_f is of type (p, p) for $1 < p \leqq \infty$. The proof of this result depended on two "end point" results: One is the easily established fact that M is of type (∞, ∞) with norm 1; the other result is Theorem 3.10 which describes the behavior of M restricted to $L^1(E_n)$. When f belongs to $L^1(E_n)$ we showed that the distribution function,[5] λ, of m_f satisfies the inequality

$$\lambda(s) \leqq \frac{c\|f\|_1}{s} \tag{2.1}$$

for all $s > 0$, where the constant c is independent of f. If there existed an integrable function φ on $(0, \infty)$ such that $\lambda(s) \leqq \varphi(s)\|f\|_1$, in view of the equality $\|m_f\|_1 = \int_0^\infty \lambda(s)\, ds$ (see (3.9) of Chapter II), we could conclude that the operator M is of type $(1, 1)$. But (2.1) cannot be improved for the general function in $L^1(E_n)$ (this can be easily seen by calculating m_f when f is the characteristic function of the interval $(0, 1) \subset E_1$).

[5] If $(M, \mathscr{M}, \mu)$ is a measure space and g is a measurable function on M, its *distribution function* λ is defined by letting its value at a nonnegative real number s be $\lambda(s) = \mu\{x \in M; |g(x)| > s\}$. In the third section of Chapter II we introduced this concept when $M = E_n$ and μ is Lebesgue measure.

The argument that was used to show that M is of type (p, p), $1 < p \leqq \infty$, can be extended to a general theorem, the Marcinkiewicz interpolation theorem, concerning operators mapping L^p spaces into L^q spaces. This result extends the M. Riesz convexity theorem in the sense mentioned at the end of Section 1 (that weaker end-point conditions still guarantee boundedness for intermediate values, and that operators that are somewhat more general than linear are involved); on the other hand, it does not apply to all the pairs (p, q) considered in Theorem 1.3, and we do not obtain similar conclusions concerning the convexity of the norms.

In order to state the Marcinkiewicz theorem we need to introduce some new concepts. Suppose T is an operator defined on a linear space D of measurable functions on $(M, \mathscr{M}, \mu)$ having values that are measurable functions defined on another measure space $(N, \mathscr{N}, \nu)$. We say that T is *subadditive* if

$$|[T(f_1 + f_2)](x)| \leqq |[Tf_1](x)| + |[Tf_2](x)| \tag{2.2}$$

for almost every x in N, whenever f_1 and f_2 belong to D. If D contains all finite linear combinations of characteristic functions of sets of finite measure and contains all the truncations of its members, we say that T is of *weak type* (p, q), $1 \leqq p \leqq \infty$ and $1 \leqq q < \infty$, if there exists a constant k, independent of $f \in L^p(M) \cap D$, such that

$$\lambda(s) \leqq \left(\frac{k\|f\|_p}{s}\right)^q, \tag{2.3}$$

where λ is the distribution function of Tf. When $q = \infty$ inequality (2.3) is replaced by the condition $\|Tf\|_q \leqq k\|f\|_p$. The least k for which these inequalities are valid is called the *weak* (p, q) *norm* of T.

It is easy to see that an operator that is of type (p, q) must be of weak type (p, q). When $q = \infty$ this is obvious. Suppose, then, that $q < \infty$, $f \in L^p(M) \cap D$, $s > 0$ and $\lambda(s) = \nu\{x \in N; |(Tf)(x)| > s\} = \nu(E_s)$ is the distribution function of Tf; then

$$s^q\lambda(s) = s^q \int_{E_s} d\nu \leqq \int_{E_s} |Tf|^q \, d\nu \leqq \|Tf\|_q^q \leqq (k\|f\|_p)^q$$

and we obtain (2.3) by dividing by s^q.

The Marcinkiewicz interpolation theorem can now be formulated in the following way:

THEOREM 2.4. *Suppose T is a subadditive operator of weak type (p_j, q_j), where $1 \leqq p_j \leqq q_j \leqq \infty$ for $j = 0, 1$ and $q_0 \neq q_1$, then T is of type (p_t, q_t)*

provided

$$\frac{1}{p_t} = \frac{1-t}{p_0} + \frac{t}{p_1} \qquad \text{and} \qquad \frac{1}{q_t} = \frac{1-t}{q_0} + \frac{t}{q_1},$$

when $0 < t < 1$.

In the next section we shall introduce a class of function spaces (that includes the L^p spaces) which forms a natural setting for an extension of Theorem 2.4. Since we will give a proof of this extension in Section 3 we do not prove the Marcinkiewicz theorem here. Instead, we devote the rest of this section to an important application of this interpolation theorem.

In the third chapter we introduced the Hilbert transform and mentioned some of its properties (see (6.14) in Chapter III). In particular, we stated that it is an operator of type (p, p), $1 < p < \infty$. We shall show that this result follows from the Marcinkiewicz theorem by first establishing that it is of type $(2, 2)$ and then showing that it is of weak type $(1, 1)$. Our previous treatment of this operator was very brief; consequently, we need to develop some of its properties further.

Suppose $f \in L^p(-\infty, \infty)$, $1 \leqq p < \infty$, and is real-valued. Then, by Theorems 2.1 and 3.16 of the second chapter, the Poisson integral of f

$$u(x + iy) = u(x, y) = \frac{1}{\pi} \int_{-\infty}^{\infty} f(t) \frac{y}{(x - t)^2 + y^2}\, dt$$

is a harmonic function in the upper half-plane E_2^+ such that $u(z) = u(x + iy)$ tends to $f(x_0)$ nontangentially for almost all $x_0 \in (-\infty, \infty)$ and

$$\left(\int_{-\infty}^{\infty} |u(x + iy)|^p\, dx \right)^{1/p} \leqq \|f\|_p \tag{2.5}$$

for all $y > 0$.

It is not hard to show that there exists a (unique) harmonic function v, defined on E_2^+, such that $F = u + iv$ is analytic in this region and $\lim_{y \to \infty} v(x + iy) = 0$ for all $x \in (-\infty, \infty)$. This function can be obtained by convolving f with the conjugate Poisson kernel

$$Q(x, y) = \frac{1}{\pi} \frac{x}{x^2 + y^2},$$

$-\infty < x < \infty$, $y > 0$ (see (6.13), Chapter III). That is,

$$v(x, y) = v(x + iy) = \frac{1}{\pi} \int_{-\infty}^{\infty} f(t) \frac{x - t}{(x - t)^2 + y^2}\, dt$$

$$= \frac{1}{\pi} \int_{-\infty}^{\infty} f(x - t) \frac{t}{t^2 + y^2}\, dt.$$

Since

$$P(x, y) + iQ(x, y) = \frac{1}{\pi}\left\{\frac{y}{x^2 + y^2} + i\frac{x}{x^2 + y^2}\right\} = \frac{i}{\pi}\frac{1}{x + iy} = \frac{i}{\pi z}$$

is analytic in E_2^+, it follows immediately that so is $F(z) = u(z) + iv(z)$. The fact that $v(x + iy)$ tends to 0 as $y \to \infty$ is obvious.

LEMMA 2.6. *The function $F(z)$, $z = x + iy \in E_2^+$, has a nontangential limit $f(x_0) + i\tilde{f}(x_0)$ at almost every point $x_0 \in (-\infty, \infty)$.*

PROOF. By considering the negative and positive parts of f separately we can reduce the problem to the case $f \leqq 0$. Thus, $u \leqq 0$ and the function $G = \exp\{u + iv\} = e^F$ has absolute value $e^u \leqq 1$. By Theorem 2.5 of Chapter II, therefore, $G(z)$ has nontangential limits a.e. on the boundary. These limits cannot be 0 on a set of positive measure since u is a Poisson integral. Thus, $v(z)$ must have nontangential limits modulo 2π at almost all boundary points $x_0 \in (-\infty, \infty)$. That is, the limit points of $v(z)$, as z tends nontangentially to x_0, have the form $a + 2k\pi$, where k is an integer. But it is an immediate consequence of the continuity of $v(z)$, $z \in E_2^+$, that, if the values $v(z)$ have two distinct limit points, a and b, as z approaches x_0 nontangentially, then all points between a and b must also be limit points of this set of values. Hence, $v(z)$ can have a nontangential limit modulo 2π only if it has a nontangential limit. This proves the lemma.

If, as in Lemma 2.6, we let $\tilde{f}(x_0)$ denote the nontangential limit of $v(z)$ we obtain an almost everywhere defined function. The mapping $f \to \tilde{f}$ (which we have shown to be defined for all f in $L^p(-\infty, \infty)$, $1 \leqq p < \infty$) is called the *Hilbert transform*. In the third chapter (see (6.13) in Chapter III) we indicated how one can show that the Hilbert transform is of type (2, 2). We first observed that, for each $y > 0$,

$$\hat{Q}_y(t) = \int_{-\infty}^{\infty} e^{-2\pi i t x} Q(x, y)\, dx = (-i \operatorname{sgn} t) e^{-2\pi|yt|}.$$

Thus, since $v(\cdot, y)$ is the convolution of $Q(\cdot, y)$ and f,

$$\hat{v}(t) = \int_{-\infty}^{\infty} e^{-2\pi i t x} v(x, y)\, dx = (-i \operatorname{sgn} t) e^{-2\pi|yt|}\hat{f}(t)$$

for a.e. t. By the Plancherel theorem and Fatou's lemma, therefore,

$$\|\tilde{f}\|_2^2 = \int_{-\infty}^{\infty} |\tilde{f}(x)|^2\, dx \leqq \liminf_{y\to 0} \int_{-\infty}^{\infty} |v(x + iy)|^2\, dx$$

$$= \liminf_{y\to 0} \int_{-\infty}^{\infty} |\hat{v}_y(t)|^2\, dt \leqq \int_{-\infty}^{\infty} |\hat{f}(t)|^2\, dt = \int_{-\infty}^{\infty} |f(t)|^2\, dt = \|f\|_2^2.$$

This is the desired inequality showing that the Hilbert transform is of type (2, 2):

$$\|\tilde{f}\|_2 \leqq \|f\|_2 \tag{2.7}$$

for all $f \in L^2(-\infty, \infty)$.[6]

LEMMA 2.8. *The Hilbert transform is of weak-type* (1, 1).

PROOF. Considering the positive and negative parts of f in $L^1(-\infty, \infty)$ separately, we can reduce the problem to $f \geqq 0$. As before, we put

$$\begin{aligned} F(z) &= F(x + iy) \\ &= u(x, y) + iv(x, y) = \int_{-\infty}^{\infty} f(t)\{P(x - t, y) + iQ(x - t, y)\}\, dt \end{aligned} \tag{2.9}$$

for $z \in E_2^+$. Then, for $s > 0$, $w(x, y) = \log |1 + sF(z)|$ is harmonic in E_2^+ and bounded in each proper sub-half-space $\{x + iy\,; y \geqq y_0 > 0\}$. Thus, by Lemma 2.7 of Chapter II,

$$\begin{aligned} w(x, y) &= \pi \log |1 + sF(x + iy)| \\ &= \int_{-\infty}^{\infty} \frac{(y - \eta)}{(x - \xi)^2 + (y - \eta)^2} \log |1 + sF(\xi + i\eta)|\, d\xi \end{aligned}$$

whenever $0 < \eta < y$. Letting $\eta \to 0$ and applying Lemma 2.6 together with Fatou's lemma we see that the last integral is greater than or equal to

$$\int_{-\infty}^{\infty} \frac{y}{(x - \xi)^2 + y^2} \log \sqrt{(1 + sf(\xi))^2 + (s\tilde{f}(\xi))^2}\, d\xi.$$

Consequently, multiplying by y and using elementary properties of the logarithm function we obtain

$$\begin{aligned} &\frac{1}{\pi} \int_{-\infty}^{\infty} \frac{y^2}{(x - \xi)^2 + y^2} \log \sqrt{(1 + sf(\xi))^2 + (s\tilde{f}(\xi))^2}\, d\xi \\ &\quad \leqq y \log |1 + sF(x + iy)| \leqq y \log (1 + s|F(x + iy)|) \leqq ys|F(x + iy)|. \end{aligned}$$

On the other hand, it follows immediately from the representation (2.9) that $\pi yF(x + iy)$ tends to $\|f\|_1 = \int_{-\infty}^{\infty} f(\xi)\, d\xi$ as y tends to ∞. Thus, letting

[6] A modification of the argument shows that inequality (2.7) can be reversed (thus the Hilbert transform is a unitary operator on $L^2(-\infty, \infty)$). This question will be considered again in a more general context in Chapter VI when we introduce the Riesz transforms.

y tend to ∞ in the first and last terms of the above inequality we have

$$(2.10) \qquad \int_{-\infty}^{\infty} \log \sqrt{(1 + sf(\xi))^2 + (s\tilde{f}(\xi))^2}\, d\xi \leqq s\|f\|_1.$$

If $E_\tau = \{\xi \in (-\infty, \infty) : |\tilde{f}(\xi)| > \tau\}$, $\tau > 0$, and $m(E_\tau)$ denotes the Lebesgue measure of E_τ (2.10) implies

$$(\log s\tau) m(E_\tau) \leqq \int_{E_\tau} \log |s\tilde{f}(\xi)|\, d\xi \leqq \int_{-\infty}^{\infty} \log \sqrt{(1 + sf(\xi))^2 + (s\tilde{f}(\xi))^2}\, d\xi$$
$$\leqq s\|f\|_1.$$

We now choose $s = e/\tau$ and obtain the desired weak type inequality

$$m(E_\tau) \leqq e\|f\|_1/\tau.$$

Theorem 2.4 can now be applied to the Hilbert transform with $p_0 = 1 = q_0$ and $p_1 = 2 = q_1$ and we obtain the fact that there exists a constant A_p, independent of $f \in L^p(-\infty, \infty)$, $1 < p \leqq 2$, such that $\|\tilde{f}\|_p \leqq A_p\|f\|_p$. It is not hard to see that this inequality is also valid when $2 \leqq p < \infty$. First we observe that the Hilbert transform, being a limit of convolution operators, commutes with translations. The inequality

$$(2.11) \qquad \|\tilde{f}\|_p \leqq A_p\|f\|_p,$$

$f \in L^p(-\infty, \infty)$, $1 < p < \infty$, is then a consequence of the just established fact that it is valid when $1 < p \leqq 2$ and Theorem 3.20 of Chapter I.

3. $L(p, q)$ Spaces

An operator of weak type (r, p) is one that maps an L^r space into a class of functions f having distribution functions λ satisfying

$$(3.1) \qquad \sup_{s>0} \Phi(s) = \sup_{s>0} \{s[\lambda(s)]^{1/p}\} = A < \infty.$$

This inequality can be restated by saying that the L^∞ norm of Φ is finite. We have already observed that

$$(3.2) \qquad \|f\|_p = \left(\int_0^\infty s^p\, d\{-\lambda(s)\}\right)^{1/p} = \left(p \int_0^\infty s^{p-1}\lambda(s)\, ds\right)^{1/p}$$

(see (3.6) in Chapter II). Thus, f belongs to L^p if and only if Φ belongs to $L^p(0, \infty)$ with respect to the measure $d\nu(s) = [1/\lambda(s)]\, d\{-\lambda(s)\}$. In fact, we have

$$\|f\|_p = \left(\int_0^\infty [\Phi(s)]^p\, d\nu(s)\right)^{1/p}.$$

Suppose that, instead of the L^∞-norm or the L^p-norm of Φ, we considered an L^q-norm, $1 \leqq q < \infty$; that is, we imposed the condition

$$\left(\int_0^\infty [\Phi(s)]^q \, d\nu(s)\right)^{1/q} < \infty. \tag{3.3}$$

We could then ask whether operators mapping L^r spaces into classes of functions satisfying (3.3) can be interpolated. In this section we shall study function spaces defined by conditions of this type and shall show that they provide a natural setting for the Marcinkiewicz theorem.

The expression occurring in (3.3) does not define a norm. Nevertheless, there are inequalities equivalent to (3.3) that, in most cases, involve the finiteness of a norm. In order to obtain these conditions we introduce the *nonincreasing rearrangement* f^* of a measurable function f defined on the measure space $(M, \mathcal{M}, \mu)$. If λ denotes the distribution function of f, then f^* is defined for $t > 0$ by letting

$$f^*(t) = \inf \{s : \lambda(s) \leqq t\}.$$

It is clear that the function λ is nonincreasing[7]; if it were a continuous and strictly decreasing function onto the positive reals it would have an inverse function having these same properties. It follows immediately from the above definition that, in this case, f^* is this inverse function. The following series of lemmas are devoted to establishing some of the basic properties of the functions λ and f^*. In the sequel we shall assume that $\lim_{s\to\infty} \lambda(s) = 0$ and, hence, $f^*(t) < \infty$ for all $t > 0$.

LEMMA 3.4. (i) *λ and f^* are nonincreasing and continuous on the right,*
(ii) *$\lambda(f^*(t)) \leqq t$ for all $t > 0$;*
(iii) *f and f^* have the same distribution function.*

PROOF. The fact that λ is nonincreasing was observed above. It is also immediate from its definition that f^* must also be nonincreasing. Continuity on the right for λ follows from the continuity from below of the measure μ and the fact that

$$\lim_{s > s_0,\, s\to s_0} \{x \in M : |f(x)| > s\} = \{x \in M : |f(x)| > s_0\}.$$

Inequality (ii) follows immediately from this continuity on the right.

In order to show that f^* is continuous on the right at t_0 we first observe that this is immediate if $f^*(t_0) = 0$ (since, in this case, $f^*(t) = 0$ for $t > t_0$ because f^* is nonnegative and nonincreasing). If $f^*(t_0)$ is positive let α be such that $f^*(t_0) > \alpha > 0$ and $\{\varepsilon_n\}$ a sequence of positive real numbers

[7] Since $\{x \in M : |f(x)| > s_1\} \subset \{x \in M : |f(x)| > s_2\}$ when $s_1 > s_2$ and the measure μ is monotone.

decreasing to 0. From the definition of f^* we must have $\lambda(f^*(t_0) - \alpha) > t_0$. Thus, there exists n_0 such that $\lambda(f^*(t_0) - \alpha) > t_0 + \varepsilon_n$ for $n \geqq n_0$. But this implies $f^*(t_0) - \alpha < f^*(t_0 + \varepsilon_n)$ for $n \geqq n_0$; otherwise, for some $n \geqq n_0, f^*(t_0) - \alpha \geqq f^*(t_0 + \varepsilon_n)$ and, by the nonincreasing property of λ and (ii), we would have the contradiction $\lambda(f^*(t_0) - \alpha) \leqq \lambda(f^*(t_0 + \varepsilon_n)) \leqq t_0 + \varepsilon_n$. Thus, since f^* is nonincreasing $f^*(t_0) - \alpha < f^*(t_0 + \varepsilon_n) \leqq f^*(t_0)$ for all $n \geqq n_0$. This shows that f^* is continuous on the right.

In order to show property (iii) we first observe that it follows from the definition of f^* that $f^*(t) > s$ if and only if $t < \lambda(s)$. Thus, $E_s^* = \{t > 0 : f^*(t) > s\}$ is precisely the interval $(0, \lambda(s))$. Property (iii) now follows from the fact that the value of the distribution function of f^* at s is the Lebesgue measure of the set E_s^*.

LEMMA 3.5. *Suppose $\{f_m\}$ is a sequence of measurable functions such that, for all $x \in M$, $|f_m(x)| \leqq |f_{m+1}(x)|$, $m = 1, 2, 3, \ldots$. If f is a measurable function satisfying $|f(x)| = \lim_{m\to\infty} |f_m(x)|$ for $x \in M$ then*

(i) *for each $s > 0$, $\lambda_m(s)$ increases monotonically to $\lambda(s)$ as m tends to ∞, where λ_m and λ are the distribution functions of f_m and f;*

(ii) *for each $t > 0$, $f_m^*(t)$ increases monotonically to $f^*(t)$.*

PROOF. It is clear that

$$E_s^{(m)} = \{x \in M : |f_m(x)| > s\} \subset E_s = \{x \in M : |f(x)| > s\}$$

and $\bigcup_{m=1}^{\infty} E_s^{(m)} = E_s$. Thus, since the measure μ is monotone and continuous from below we must have $\lambda_m(s) = \mu(E_s^{(m)}) \leqq \mu(E_s) = \lambda(s)$ and $\lim_{m\to\infty} \lambda_m(s) = \lambda(s)$. This proves (i). It now follows from the definition of the nonincreasing rearrangement that $f_m^*(t) \leqq f_{m+1}^*(t) \leqq f^*(t)$ for $m = 1, 2, 3, \ldots$. Let $l = \lim_{m\to\infty} f_m^*(t)$. Since $f_m^*(t) \leqq l$ we must have $\lambda_m(l) \leqq \lambda_m(f_m^*(t)) \leqq t$ (the last inequality being a consequence of part (ii) of Lemma 3.4). Thus, $\lambda(l) = \lim_{m\to\infty} \lambda_m(l) \leqq t$ and this implies $f^*(t) \leqq l$. But, from the inequality $f_m^*(t) \leqq f^*(t)$ we obtain $l \leqq f^*(t)$. It therefore follows that $l = f^*(t)$ and the lemma is proved.

This lemma is particularly useful in proving other properties of the functions λ and f^* since it allows us to reduce these proofs to the case when f is a simple function. The distribution function and the nonincreasing rearrangement of such a function is particularly simple to describe. Suppose $f = \sum_{j=1}^{n} c_j \chi_{E_j}$, where $E_j \in \mathcal{M}, \mu(E_j) > 0$ and $E_j \cap E_k = \emptyset$ if $j \neq k$. By relabelling, if necessary, we can assume $c_1 > c_2 > \cdots > c_n >$

$c_{n+1} = 0$. Let $d_j = \mu(E_1) + \cdots + \mu(E_j)$, $1 \leqq j \leqq n$, and define d_0 to be 0. Then, the distribution function λ of f has the form

$$\lambda(s) = \begin{cases} d_j & \text{if } c_{j+1} \leqq s < c_j,\ 1 \leqq j \leqq n \\ 0 & \text{if } c_1 \leqq s. \end{cases} \tag{3.6}$$

It follows that

$$f^*(t) = \begin{cases} c_j & \text{if } d_{j-1} \leqq t < d_j,\ 1 \leqq j \leqq n \\ 0 & \text{if } d_n \leqq t. \end{cases} \tag{3.7}$$

By inspection, therefore, we see that for $p > 0$

$$\sup_{s>0} s[\lambda(s)]^{1/p} = \sup_{1 \leqq j \leqq n} d_j^{1/p} c_j = \sup_{t>0} t^{1/p} f^*(t).$$

From this we can easily show that the supremum in (3.1) can be expressed in terms of the nonincreasing rearrangement of the function in question:

LEMMA 3.8. *If f is measurable, λ is its distribution function and $p > 0$ then* (3.1) *holds if and only if*

$$\sup_{t>0} t^{1/p} f^*(t) = A < \infty.$$

PROOF. We have just seen that Lemma 3.8 is true when f is simple. If f is only measurable we can find a sequence $\{f_m\}$ of nonnegative simple functions that increase monotonically to $|f|$ at all the points of M; the lemma then follows immediately from Lemma 3.5.

In view of inequality (3.3) and the comments made before it, it is natural to consider the space $L(p, q)$ of all those measurable functions f satisfying

$$\|f\|^*_{pq} = \left(\frac{q}{p} \int_0^\infty [t^{1/p} f^*(t)]^q \frac{dt}{t}\right)^{1/q} < \infty,$$

when $1 \leqq p < \infty$, $1 \leqq q < \infty$, and

$$\|f\|^*_{pq} = \sup_{t>0} t^{1/p} f^*(t) < \infty$$

when $1 \leqq p \leqq \infty$ and $q = \infty$.[8]

When f is simple we see immediately from (3.7) that

$$\|f\|_p = \left(\int_M |f|^p \, d\mu\right)^{1/p} = \left(\int_0^\infty [f^*(t)]^p \, dt\right)^{1/p} = \|f^*\|_p, \quad 1 \leqq p < \infty, \tag{3.9}$$

[8] These definitions make sense also when $0 < p < 1$ and $0 < q < 1$; however, since we shall be interested in converting $L(p, q)$ into a Banach space, we do not consider these cases. Moreover, in the above definition we adopt the usual convention that $1/\infty = 0$. Observe that we have defined $L(\infty, q)$ only when $q = \infty$.

and

(3.9′) $$\|f\|_\infty = \|f^*\|_\infty.$$

From Lemma 3.5 we obtain the fact that these equalities hold for all $f \in L^p(M)$, $1 \leqq p \leqq \infty$. Since $\|f\|^*_{pp} = (\int_0^\infty [t^{1/p}f^*(t)]^p\,(dt/t))^{1/p} = (\int_0^\infty [f^*(t)]^p\,dt)^{1/p} = \|f^*\|_p$ it follows that $L(p,p) = L^p(M)$ and $\|\ \|^*_{pp}$ is a norm. In general, however, $\|\ \|^*_{pq}$ is not a norm since the Minkowski inequality may fail. Despite this, we will show that $\|\ \|^*_{pq}$ can be used in order to introduce a topology on $L(p,q)$ which, in most cases, can be shown to be that of a Banach space.

There are several reasons for introducing the constant q/p in the definition of $\|\ \|^*_{pq}$. One of the consequences of this definition is the fact that $\|\chi_E\|^*_{pq}$ is independent of q whenever E is a measurable subset of M of finite measure and χ_E its characteristic function. In fact, we have

(3.10) $$\|\chi_E\|^*_{pq} = \{\mu(E)\}^{1/p}$$

for $1 \leqq q \leqq \infty$. This equality is a reflection of the fact that each $L(p,q)$ space, for p fixed, is a "type of L^p space." The following theorem gives us a precise formulation of this assertion. It also shows that, in some sense, $L(p,1)$ is the "smallest" normed linear space for which (3.10) holds, while $L(p,\infty)$ is the "largest" such space.

THEOREM 3.11. *If $f \in L(p,q_1)$ and $q_1 \leqq q_2$ then*

(i) $$\|f\|^*_{pq_2} \leqq \|f\|^*_{pq_1};$$

consequently,

(ii) $$L(p,q_1) \subset L(p,q_2).$$ [9]

[9] In order to obtain this inclusion relation it suffices to show that there exists a constant B, independent of $f \in L(p,q_1)$, such that $\|f\|^*_{pq_2} \leqq B\|f\|_{pq_1}$. Short proofs establishing this inequality can be found. For example:

$$\begin{aligned}\left(\int_0^\infty [t^{1/p}f^*(t)]^{q_2}\frac{dt}{t}\right)^{q_1/q_2} &\leqq \left(\sum_{k=-\infty}^{\infty} [f^*(2^{k-1})]^{q_2}\int_{2^{k-1}}^{2^k} t^{(q_2/p)-1}\,dt\right)^{q_1/q_2}\\ &\leqq B'\sum_{k=-\infty}^{\infty}[f^*(2^{k-1})]^{q_1}2^{kq_1/p_2}\\ &\leqq B''\sum_{k=-\infty}^{\infty}\int_{2^{k-2}}^{2^{k-1}}[t^{1/p}f^*(t)]^{q_1}\frac{dt}{t}\\ &\leqq B'''\{\|f\|^*_{pq_1}\}^{q_1}.\end{aligned}$$

Inequality (i), however, is more subtle and requires a somewhat more difficult proof. Equality (3.10) shows that (i) is best possible.

Suppose $\|\ \|$ *is an order preserving norm (we mean by this that* $\|g\| \leqq \|f\|$ *whenever* $|g(x)| \leqq |f(x)|$ *a.e.) defined on the simple functions on* M*; then*

(iii) $\|\chi_E\| \leqq \{\mu(E)\}^{1/p}$ *for all* $E \in \mathcal{M}$ *implies* $\|f\| \leqq \|f\|^*_{p1}$ *for all simple functions* f*;*

(iv) $\{\mu(E)\}^{1/p} \leqq \|\chi_E\|$ *for all* $E \in \mathcal{M}$ *implies* $\|f\|^*_{p\infty} \leqq \|f\|$ *for all simple functions* f*.*

PROOF. Inequality (i) can be shown in a particularly simple way when $q_2 = \infty$. Since f^* is nonincreasing,

$$t^{1/p}f^*(t) = f^*(t)\left\{(q_1/p)\int_0^t u^{(q_1/p)-1}\,du\right\}^{1/q_1}$$

$$\leqq \left\{(q_1/p)\int_0^t [u^{1/p}f^*(u)]^{q_1}\frac{du}{u}\right\}^{1/q_1} \leqq \|f\|^*_{pq_1}.$$

By taking the supremum of $t^{1/p}f^*(t)$, over all $t > 0$, we obtain the desired inequality.

When $q_2 < \infty$, it follows from Lemma 3.5 that it suffices to establish the lemma for f simple. We assume, therefore, that f^* has the form (3.7), in which case

$$\|f\|^*_{pq} = \left\{\sum_{j=1}^n c_j^q(d_j^{(q/p)} - d_{j-1}^{(q/p)})\right\}^{1/q}.$$

Setting $a_j = c_j^{q_2}$, $b_j = d_j^{q_2/p}$ and $\theta = q_1/q_2$, therefore, our inequality is equivalent to

$$\text{(a)} \qquad \sum_{j=1}^n a_j(b_j - b_{j-1}) \leqq \left\{\sum_{j=1}^n a_j^\theta(b_j^\theta - b_{j-1}^\theta)\right\}^{1/\theta},$$

where $a_1 > a_2 > \cdots > a_n > 0$, $0 = b_0 < b_1 < \cdots < b_n$ and $0 < \theta \leqq 1$. We shall prove (a) by induction. When $n = 1$ this inequality reduces to $a_1b_1 \leqq \{a_1^\theta b_1^\theta\}^{1/\theta}$, which is obviously true.

Assume that (a) is true for $n = N$. We let

$$\varphi(x) = \left\{\sum_{j=1}^N a_j^\theta(b_j^\theta - b_{j-1}^\theta) + x^\theta(b_{N+1}^\theta - b_N^\theta)\right\}^{1/\theta} = (A + x^\theta B)^{1/\theta}$$

and

$$l(x) = \sum_{j=1}^N a_j(b_j - b_{j-1}) + x(b_{N+1} - b_N) = A_1 + xB_1$$

for $0 \leqq x \leqq a_N$.

We want to show that $\varphi(a_{N+1}) \geqq l(a_{N+1})$ whenever $0 < a_{N+1} < a_N$. By the induction assumption $\varphi(0) \geqq l(0)$ and $\varphi(a_N) \geqq l(a_N)$ (this last inequality reduces to (a) with b_N replaced by b_{N+1}). On the other hand, since the derivative $\varphi'(x) = B(Ax^{-\theta} + B)^{(1/\theta)-1}$ of φ is decreasing on the positive real axis, the function φ must be concave. Therefore, the fact that it dominates the linear function l at the end points 0 and a_N implies that $\varphi(x) \geqq l(x)$ for $0 \leqq x \leqq a_N$. This proves part (i). Part (ii) is then an immediate consequence.

Since $\| \ \|$ is order preserving it suffices to show (iii) when $f = \sum_{j=1}^N c_j\chi_{E_j}$ is a nonnegative simple function. We can assume that $c_1 > c_2 > \cdots > c_n > c_{n+1} = 0$ and the sets $E_1, \ldots, E_n$ are pairwise disjoint. If $k = 1, 2, \ldots, n$ define $f_k = b_k\chi_{F_k}$, where $F_k = \bigcup_{j=1}^k E_j$ and $b_k = c_k - c_{k+1}$. Then, clearly,

$$\text{(b)} \qquad f^*(t) = \sum_{k=1}^n f_k^*(t), \qquad \text{for } t > 0,$$

and

$$\begin{aligned} \|f\| &\leqq \sum_{k=1}^n \|f_k\| = \sum_{k=1}^n b_k\|\chi_{F_k}\| \\ &\leqq \sum_{k=1}^n b_k\{\mu(F_k)\}^{1/p} = \sum_{k=1}^n \frac{1}{p}\int_0^\infty t^{1/p}f_k^*(t)\,\frac{dt}{t} \\ &= \frac{1}{p}\int_0^\infty t^{1/p}\sum_{k=1}^n f_k^*(t)\,\frac{dt}{t} = \frac{1}{p}\int_0^\infty t^{1/p}f^*(t)\,\frac{dt}{t} = \|f\|_{p1}^*. \end{aligned}$$

This proves (iii).

In order to show (iv) we observe that, in terms of the notation developed before equality (3.6), we have

$$\|f\|_{p\infty}^* = \sup_{1 \leqq j \leqq n} d_j^{1/p}c_j$$

(this fact follows from the definition of $\| \ \|_{p\infty}$ and the equalities displayed immediately after (3.7)). Suppose that the supremum on the right is assumed at $j = k$; thus,

$$\|f\|_{p\infty}^* = d_k^{1/p}c_k.$$

Let $g = c_k\chi_{F_k}$. Then $0 \leqq g \leqq f$ and, thus,

$$\|f\|_{p\infty}^* = d_k^{1/p}c_k = c_k\{\mu(F_k)\}^{1/p} \leqq c_k\|\chi_{F_k}\| = \|g\| \leqq \|f\|$$

and the theorem is proved.

It is useful to associate with each space $L(p, q)$ the point $(1/p, 1/q)$ of the square $Q = \{(x_1, x_2) \in E_2 : 0 \leqq x_1, x_2 \leqq 1\}$. As a consequence of Theorem

3.11, part (ii), we see that if $x_1 = 1/p$ is fixed and $x_2 = 1/q$ is allowed to vary from 1 to 0 we obtain the points associated with an increasing family of function spaces. When we reach the diagonal $\{(x_1, x_2) \in Q : x_1 = x_2\}$ of the square Q, the associated space is L^p (see (3.9)). In view of the definition of a weak type operator that was made in Section 2, it is natural to refer to $L(p, \infty)$, the largest of the spaces encountered in this manner, as *weak* L^p. The justification for this terminology is that the inequality (2.3) that was used in the definition of an operator of weak type (r, p) is, by Lemma 3.8, equivalent to

$$\|Tf\|^*_{p\infty} \leqq k\|f\|_r.$$

Since $\|f\|_r = \|f\|^*_{rr}$ and $r \geqq 1$, it follows from inequality (i) of Theorem 3.11 that $\|f\|_r \leqq \|f\|^*_{r1}$. Thus, an operator T of weak type (r, p) satisfies the weaker inequality

$$\|Tf\|^*_{p\infty} \leqq k\|f\|^*_{r1} \tag{3.12}$$

for all f belonging to the domain of T and to $L(r, 1)$. The interpolation theorem that will be proved in this section asserts, in part, that end point conditions like (3.12), even though weaker than weak type conditions, are sufficient in order to guarantee the conclusion of the Marcinkiewicz theorem (2.4). Closely connected with part (iii) of Theorem 3.11 is the fact that the hypotheses of the Marcinkiewicz theorem can be reduced still further. In fact, we have the following result, showing that the conclusion of Theorem 2.4 depends only on knowing that the weak type inequalities at the end points hold for characteristic functions of sets of finite measure.

THEOREM 3.13. *Suppose T is a linear operator which maps the finite linear combinations of characteristic functions χ_E of sets $E \subset M$ of finite measure into a vector space B that is endowed with an order preserving norm $\|\ \|$. If*

$$\|T\chi_E\| \leqq C\|\chi_E\|^*_{r1} = C\{\mu(E)\}^{1/r},$$

where C is independent of E, then there exists a constant A such that

$$\|Tf\| \leqq A\|f\|^*_{r1}$$

for all f in the domain of T.

PROOF. If $f \geqq 0$ belongs to the domain of T we can represent it as a sum $f = \sum_{k=1}^n f_k$, where the functions f_k are the multiples of characteristic functions defined in the proof of Theorem 3.11 so that equality (b) is satisfied. Then

$$\|Tf\| = \left\|\sum_{k=1}^n Tf_k\right\| \leqq \sum_{k=1}^n \|Tf_k\| \leqq \sum_{k=1}^n Cb_k\|\chi_{F_k}\|^*_{r1}.$$

But, as was shown in the proof of Theorem 3.11, part (iii), the last sum equals $C\|f\|_{r1}^{*}$. If $f = f_1 + if_2$ is complex valued we obtain the theorem with $A = 4C$ by applying the result that has been just established to the positive and negative parts of f_1 and f_2.[10]

The interpolation theorem we have been discussing is a consequence of the following classical estimate, known as *Hardy's inequality.*

LEMMA 3.14. *If* $q \geqq 1$, $r > 0$ *and* g *is a nonnegative function defined on* $(0, \infty)$ *then*

(i) $\left(\int_0^\infty \left[\int_0^t g(u)\,du\right]^q t^{-r-1}\,dt\right)^{1/q} \leqq (q/r)\left(\int_0^\infty [ug(u)]^q u^{-r-1}\,du\right)^{1/q}$;

(ii) $\left(\int_0^\infty \left[\int_t^\infty g(u)\,du\right]^q t^{r-1}\,dt\right)^{1/q} \leqq (q/r)\left(\int_0^\infty [ug(u)]^q u^{r-1}\,du\right)^{1/q}$.

PROOF. We first show that (i) implies (ii). If we apply (i) to the function $g_1(u) = u^{-2}g(u^{-1})$ we obtain that

$$\int_0^\infty \left[\int_0^t g_1(u)\,du\right]^q t^{-r-1}\,dt = \int_0^\infty \left[\int_{1/t}^\infty g(v)\,dv\right]^q t^{-r-1}\,dt$$
$$= \int_0^\infty \left[\int_s^\infty g(v)\,dv\right]^q s^{r-1}\,ds$$

is less than or equal to $(q/r)^q$ times

$$\int_0^\infty [ug_1(u)]^q u^{-r-1}\,du = \int_0^\infty [vg(v)]^q v^{r-1}\,dv.$$

Thus, it suffices to establish inequality (i).

If we apply Jensen's inequality, with $\varphi(x) = |x|^q$ and $d\mu(u) = u^{(r/q)-1}\,du$ (see Section 4, Example (2), in the second chapter) we obtain

$$\left[\int_0^t g(u)\,du\right]^q = \left[\int_0^t g(u)u^{1-(r/q)}u^{(r/q)-1}\,du\right]^q$$
$$\leqq \left(\frac{q}{r}\right)^{q-1} t^{r(1-1/q)} \int_0^t [g(u)]^q u^{q-r-1+r/q}\,du.$$

[10] Part (iii) of Theorem 3.11 is the special case of Theorem 3.13 obtained when T is the identity operator. It is clear that the proof of Theorem 3.11 is valid for *sublinear* operators T; that is, operators that are subadditive and positive homogeneous ($|Taf| = |a|\,|Tf|$ for all scalars a).

Thus,

$$\int_0^\infty \left[\int_0^t g(u)\,du\right]^q t^{-r-1}\,dt \leqq \left(\frac{q}{r}\right)^{q-1} \int_0^\infty t^{-1-r/q}\left[\int_0^t [g(u)]^q u^{q-r-1+r/q}\,du\right] dt$$

$$= \left(\frac{q}{r}\right)^{q-1} \int_0^\infty [g(u)u]^q u^{-r-1+r/q}\left(\int_u^\infty t^{-1-r/q}\,dt\right) du$$

$$= \left(\frac{q}{r}\right)^{q} \int_0^\infty [g(u)u]^q u^{-r-1}\,du$$

and Hardy's inequality is proved.

We can now state and prove the principal result of this section. We say that a subadditive operator T is of *restricted weak type* (r, p) if its domain D contains all the truncations of its members as well as all finite linear combinations of characteristic functions of sets of finite measure and satisfies inequality (3.12) whenever $f \in D \cap L(r, 1)$. In view of Theorem 3.13 such an operator may be considered to be one that satisfies the weak type (r, p) conditions when it is restricted to characteristic functions of sets of finite measure.

THEOREM 3.15. *Suppose T is a subadditive operator of restricted weak types (r_j, p_j), $j = 0, 1$, with $r_0 < r_1$ and $p_0 \neq p_1$, then there exists a constant $B = B_\theta$ such that*

$$\|Tf\|^*_{pq} \leqq B\|f\|^*_{rq}$$

for all f belonging to the domain of T and to $L(r, q)$, where $1 \leqq q \leqq \infty$,

$$\frac{1}{p} = \frac{(1-\theta)}{p_0} + \frac{\theta}{p_1}, \qquad \frac{1}{r} = \frac{1-\theta}{r_0} + \frac{\theta}{r_1} \qquad \textit{and} \qquad 0 < \theta < 1.$$

PROOF. For $f \in L(r, q) \cap D$ we define

$$f^t(x) = \begin{cases} f(x) & \text{if } |f(x)| > f^*(t^\gamma) \\ 0 & \text{if } |f(x)| \leqq f^*(t^\gamma) \end{cases}$$

and $f_t(x) = f(x) - f^t(x)$, where

$$\gamma = \frac{(1/p_0) - (1/p)}{(1/r_0) - (1/r)} = \frac{(1/p) - (1/p_1)}{(1/r) - (1/r_1)}.$$

We then have the two inequalities

(a)
$$(f^t)^*(s) \leqq \begin{cases} f^*(s) & \text{if } 0 < s < t^\gamma \\ 0 & \text{if } t^\gamma \leqq s \end{cases} \qquad \text{and}$$
$$f_t^*(s) \leqq \begin{cases} f^*(t^\gamma) & \text{if } 0 < s < t^\gamma \\ f^*(s) & \text{if } t^\gamma \leqq s. \end{cases}$$

When f is simple (a) can be derived readily from formulas analogous to (3.7); the general case then follows easily from this one. Since the operator T is subadditive $|[Tf](y)| = |[T(f^t + f_t)](y)| \leqq |[Tf^t](y)| + |[Tf_t](y)|$ for almost every $y \in N$. Thus, if $s > 0$,

$$\{y \in N : |[Tf](y)| > (Tf^t)^*(s) + (Tf_t)^*(s)\}$$
$$\subset \{y \in N : |[Tf^t](y)| > (Tf^t)^*(s)\} \cup \{y \in N : |[Tf_t](y)| > (Tf_t)^*(s)\}.$$

If λ, λ^t, and λ_t are the distribution functions of Tf, Tf^t, and Tf_t, respectively, it follows from this inclusion relation and Lemma 3.4, part (ii),

$$\lambda((Tf^t)^*(s) + (Tf_t)^*(s)) \leqq \lambda^t((Tf^t)^*(s)) + \lambda_t((Tf_t)^*(s)) \leqq s + s = 2s.$$

By the definition of $(Tf)^*$, therefore, we have

(b)
$$(Tf^t)^*(s) + (Tf_t)^*(s) \geqq (Tf)^*(2s) \qquad \text{for all } s > 0.$$

Suppose $r_1 < \infty$ and $q < \infty$. Then, from (b) with $s = t$, a change of variables and Minkowski's inequality, we obtain

$$\|Tf\|_{pq}^* = (q/p)^{1/q}\left\{\int_0^\infty [t^{1/p}(Tf)^*(t)]^q \frac{dt}{t}\right\}^{1/q}$$
$$= (q/p)^{1/q}2^{1/p}\left\{\int_0^\infty [t^{1/p}(Tf)^*(2t)]^q \frac{dt}{t}\right\}^{1/q}$$
$$\leqq (q/p)^{1/q}2^{1/p}\left\{\left(\int_0^\infty [t^{1/p}(Tf^t)^*(t)]^q \frac{dt}{t}\right)^{1/q} + \left(\int_0^\infty [t^{1/p}(Tf_t)^*(t)]^q \frac{dt}{t}\right)^{1/q}\right\}.$$

Since T is of restricted weak types (r_0, p_0) and (r_1, p_1) we have $t^{1/p_0}(Tf^t)^*(t) \leqq k_0\|f^t\|_{r_01}^*$ and $t^{1/p_1}(Tf_t)^*(t) \leqq k_1\|f_t\|_{r_11}^*$ for all $t > 0$. Thus, the above sum within brackets is not larger than

$$k_0\left(\int_0^\infty [t^{(1/p)-(1/p_0)}\|f^t\|_{r_01}^*]^q \frac{dt}{t}\right)^{1/q} + k_1\left(\int_0^\infty [t^{(1/p)-(1/p_1)}\|f_t\|_{r_11}^*]^q \frac{dt}{t}\right)^{1/q}.$$

But, it follows from the first part of (a), a change of variables and Lemma 3.14, part (i), that

$$\left(\int_0^\infty [t^{(1/p)-(1/p_0)}\|f^t\|_{r_01}^*]^q \frac{dt}{t}\right)^{1/q}$$
$$\leq \left(\int_0^\infty \left[t^{(1/p)-(1/p_0)}\left(\frac{1}{r_0}\int_0^{t^\gamma} f^*(s)s^{(1/r_0)-1}\,ds\right)\right]^q \frac{dt}{t}\right)^{1/q}$$
$$= |\gamma|^{-(1/q)}(1/r_0)\left(\int_0^\infty \left[u^{(1/r)-(1/r_0)}\int_0^u f^*(s)s^{(1/r_0)-1}\,ds\right]^q \frac{du}{u}\right)^{1/q}$$
$$\leq \frac{|\gamma|^{-(1/q)}}{1-(r_0/r)}\left(\int_0^\infty [u^{1/r_0}f^*(u)]^q u^{(q/r)-(q/r_0)}\frac{du}{u}\right)^{1/q}$$
$$= \frac{1}{1-(r_0/r)}\left(\frac{r}{|\gamma|q}\right)^{1/q}\|f\|_{rq}^*.$$

A similar argument, involving the second parts of (a) and Lemma 3.14. gives us

$$\left(\int_0^\infty [t^{(1/p)-(1/p_1)}\|f_t\|_{r_11}^*]^q \frac{dt}{t}\right)^{1/q}$$
$$\leq \left(\int_0^\infty \left[t^{(1/p)-(1/p_1)}\left(\frac{1}{r_1}\int_0^{t^\gamma} f^*(t^\gamma)s^{(1/r_1)-1}\,ds + \frac{1}{r_1}\int_{t^\gamma}^\infty f^*(s)s^{(1/r_1)-1}\,ds\right)\right]^q \frac{dt}{t}\right)^{1/q}$$
$$\leq \left(\int_0^\infty [t^{(1/p)-(1/p_1)}f^*(t^\gamma)t^{\gamma/r_1}]^q \frac{dt}{t}\right)^{1/q}$$
$$+ \frac{1}{r_1}\left(\int_0^\infty \left[t^{(1/p)-(1/p_1)}\int_{t^\gamma}^\infty f^*(s)s^{(1/r_1)-1}\,ds\right]^q \frac{dt}{t}\right)^{1/q}$$
$$= |\gamma|^{-(1/q)}\left\{\left(\int_0^\infty [u^{1/r}f^*(u)]^q \frac{du}{u}\right)^{1/q}\right.$$
$$\left. + \frac{1}{r_1}\left(\int_0^\infty \left[u^{(1/r)-(1/r_1)}\int_u^\infty f^*(s)s^{(1/r_1)-1}\,ds\right]^q \frac{du}{u}\right)^{1/q}\right\}$$
$$\leq \left(\frac{r}{|\gamma|q}\right)^{1/q}\|f\|_{rq}^* + \frac{|\gamma|^{-(1/q)}}{(r_1/r)-1}\left(\int_0^\infty [f^*(u)u^{1/r_1}]^q u^{(q/r)-(q/r_1)}\frac{du}{u}\right)^{1/q}$$
$$= \left(\frac{r}{|\gamma|q}\right)^{1/q}\left(\frac{r_1}{r_1-r}\right)\|f\|_{rq}^*.$$

We have shown, therefore, that

$$\|Tf\|^*_{pq} \leqq B\|f\|^*_{rq}$$

with

$$B = \left(\frac{r}{|\gamma|p}\right)^{1/q} 2^{1/p}\left(\frac{rk_0}{r - r_0} + \frac{r_1k_1}{r_1 - r}\right).$$

When $r_1 < \infty$ and $q = \infty$ we estimate $t^{1/p}(Tf)^*(t)$, for $t > 0$, using inequalities (a) and (b), just as was done in the above proof. We obtain positive constants c_1, c_2, and c_3 (in fact, $c_1 = 2^{1/p}k_0$, and $c_2 = c_3 = 2^{1/p}k_1$) such that

$$t^{1/p}(Tf^*)(t) \leqq c_1 t^{(1/p)-(1/p_0)} \int_0^{t^\gamma} f^*(s)s^{(1/r_0)-1}\,ds$$

$$+ c_2 t^{(1/p)-(1/p_1)} \int_0^{t^\gamma} f^*(t^\gamma)s^{(1/r_1)-1}\,ds + c_3 t^{1/p-1/p_1} \int_{t^\gamma}^{\infty} f^*(s)s^{(1/r_1)-1}\,ds.$$

If we now use the fact that $f^*(s)s^{1/r} \leqq \|f\|^*_{r\infty}$ we have

$$t^{1/p}(Tf^*)(t) \leqq \left\{\frac{c_1}{(1/r_0) - (1/r)} + c_2 r_1 + \frac{c_3}{(1/r) - (1/r_1)}\right\} \|f\|^*_{r\infty} = B\|f\|^*_{r\infty},$$

and, thus, $\|Tf\|^*_{p\infty} \leqq B\|f\|^*_{r\infty}$.

In the remaining case, $r_1 = \infty$ and $q = \infty$, the proof follows the same lines as that of the above cases, except for the necessity of using the estimate $\|f_t\|^*_{\infty\infty} \leqq f^*(t^\gamma)$.

Theorem 3.15 obviously contains the Marcinkiewicz interpolation theorem as a special case; moreover, it gives us sharper results since it involves the spaces $L(p, q)$. For example, if we apply it to the Fourier transform we obtain the following strengthening of the inequality of Hausdorff–Young (see (1.2)):

COROLLARY 3.16. *If* $f \in L^p(E_n)$, $1 < p \leqq 2$, *then* $\hat{f}$ *belongs to* $L(p', p)$ *and there exists a constant* $B = B_p$ *such that*

$$\|\hat{f}\|^*_{p'p} \leqq B\|f\|_p,$$

where $(1/p) + (1/p') = 1$.

The remainder of this section will be devoted to showing that most of the spaces $L(p, q)$ have a norm that is closely connected to $\|\ \|^*_{pq}$; moreover, with this norm, $L(p, q)$ is a Banach space. The following lemmas will be needed to establish these facts.

For notational convenience we restrict our attention to *nonatomic measure spaces*. This means that $\mathscr{M}$ does not contain *atoms* (sets of positive measure all of whose measurable subsets have measure zero). The applications we shall make of the material developed in this section will involve only the nonatomic measure spaces $(E_n, \mathscr{B}, m)$, where $\mathscr{B}$ consists of the Borel subsets of E_n and m is Borel–Lebesgue measure. The only place where this restriction is used is in the proof of part (iii) of the following lemma.

LEMMA 3.17. *If* $E \in \mathscr{M}$ *and* $F_s = \{x \in M : |f(x)| > s \geqq 0\}$ *(so that* $\lambda(s) = \mu(F_s)$*) then*

(i) $\int_E |f|\, d\mu \leqq \int_0^{\mu(E)} f^*(t)\, dt;$

(ii) $\int_{F_s} |f|\, d\mu = \int_0^{\lambda(s)} f^*(t)\, dt;$

(iii) *If* $\mu(M) \geqq t > 0$ *then there exists a set* $E_t \in \mathscr{M}$ *such that* $\mu(E_t) = t$ *and*

$$\int_{E_t} |f|\, d\mu = \int_0^t f^*(u)\, du.$$

PROOF. It is clear that if $|f_1| \leqq |f|$ then $f_1^* \leqq f^*$. Thus, if χ_E is the characteristic function of E and $f_1 = f\chi_E$, then

$$\text{(a)} \qquad \int_0^{\mu(E)} f_1^*(t)\, dt \leqq \int_0^{\mu(E)} f^*(t)\, dt.$$

On the other hand, the distribution function of f_1 must be bounded by $\mu(E)$ and, consequently, $f_1^*(t) = 0$ for $t > \mu(E)$. Using (3.9) with $p = 1$, therefore,

$$\text{(b)} \qquad \int_E |f|\, d\mu = \int_M |f_1|\, d\mu = \int_0^\infty f_1^*(t)\, dt = \int_0^{\mu(E)} f_1^*(t)\, dt.$$

Inequality (a) and inequality (b) imply (i).

In order to show (ii) we first observe that if we put

$$g(t) = \begin{cases} f^*(t) & \text{for } 0 < t < \lambda(s) \\ 0 & \text{for } \lambda(s) \leqq t, \end{cases}$$

then g and $h = f\chi_{F_s}$ have the same distribution function (when f is simple this follows immediately from Lemma 3.4, part (iii), and equalities (3.6) and (3.7). The general case follows from this and Lemma 3.5). Thus,

$$\int_{F_s} |f|\, d\mu = \int_M |h|\, d\mu = \int_0^\infty g(u)\, du = \int_0^{\lambda(s)} f^*(u)\, du.$$

Finally, if $0 < t \leqq \mu(M)$, let $G = \{x \in M : |f(x)| > f^*(t)\}$ and $H = \{x \in M : |f(x)| \geqq f^*(t)\}$. We claim that

$$\mu(G) \leqq t \leqq \mu(H).$$

The first inequality is a consequence of part (ii) of Lemma 3.4 and the fact that $\mu(G) = \lambda(f^*(t))$. The second inequality is particularly easy to establish with the help of (3.6) and (3.7) when f is simple; the general case then follows from Lemma 3.5. Since we are assuming that the measure space $(M, \mathscr{M}, \mu)$ is not atomic we can find a set E_t in $\mathscr{M}$ such that $G \subset E_t \subset H$ and $\mu(E_t) = t$. Let $h = f\chi_{E_t}$. Then $h^* = f^*$ when these functions are restricted to $(0, t)$, while $h(u) = 0$ for $u \geqq t$. Therefore,

$$\int_{E_t} |f|\, d\mu = \int_M |h|\, d\mu = \int_0^\infty h^*(u)\, du = \int_0^t f^*(u)\, du.$$

An immediate corollary of parts (i) and (iii) of this lemma is

$$\sup_{E \in \mathscr{M},\, \mu(E) \leqq t} \int_E |f|\, d\mu = \int_0^t f^*(u)\, du. \tag{3.18}$$

If f is not zero almost everywhere and $t > 0$, then the expression on the left in (3.18) is positive. It follows, therefore, that the mapping $f \to t^{-1} \int_0^t f^*(u)\, du$ is a norm defined on each of the spaces $L^p(M)$, $1 \leqq p \leqq \infty$ (the fact that it is finite-valued is a consequence of Hölder's inequality:

$$\int_E |f|\, d\mu = \int_M \chi_E |f|\, d\mu \leqq \|f\|_p \|\chi_E\|_{p'} \leqq t^{1/p'} \|f\|_p,$$

where $(1/p) + (1/p') = 1$). This norm is closely related to the Hardy–Littlewood maximal function introduced in Chapter II, Section 3. In order to see this relation we make a few simple observations concerning the maximal function.

The maximal function introduced in the second chapter was defined to be the supremum over all integral averages of $|f|$ taken over spheres with center at x. In the one-dimensional case it is not hard to show that its basic properties, developed in the third section of that chapter, are shared by the more general "nonsymmetric" maximal function defined by letting its value at x be

$$\sup \frac{1}{h + k} \int_{x-h}^{x+k} |f(u)|\, du, \tag{3.19}$$

where the supremum is taken over all nonnegative h and k satisfying

$h + k > 0$. This is a consequence of the inequalities

$$\frac{1}{h+k}\int_{x-h}^{x+k}|f(u)|\,du \leqq \frac{1}{h+k}\int_{x-(h+k)}^{x+h+k}|f(u)|\,du$$
$$\leqq 2\sup_{r>0}\frac{1}{2r}\int_{|x-u|\leqq r}|f(u)|\,du$$

which show that this nonsymmetric maximal function is no larger than twice the one we introduced earlier. In order to avoid introducing new notation, we let $m = m_f$ denote the function defined by (3.19).

Suppose g is a nonnegative function defined on the real line such that $g(u) = 0$ for $u < 0$ and g is nonincreasing when restricted to $[0, \infty)$. If $t > 0$ it then follows from the fact that $g(u) = 0$ for $u < 0$ that $m(t) = m_g(t)$ is the supremum over all integral averages

$$\frac{1}{h+k}\int_{t-h}^{t+k} g(u)\,du$$

such that $t - h \geqq 0$. Since g is nonincreasing, however, a simple argument shows that each such average is not larger than

$$\frac{1}{t}\int_0^t g(u)\,du.$$

If we apply these arguments to the function g whose values at $u \geqq 0$ are $f^*(u)$ we obtain the fact that

$$m_g(t) = m_{f^*}(t) = m(t) = \frac{1}{t}\int_0^t f^*(u)\,du$$

for all $t > 0$.

It is clear from the nonincreasing property of f^* that $m(t) \geqq f^*(t)$ for all $t > 0$. Thus, if we define

$$\|f\|_{pq} = \left(\frac{q}{p}\int_0^\infty [t^{1/p}m(t)]^q\,\frac{dt}{t}\right)^{1/q}$$

when $1 \leqq p < \infty$ and $1 \leqq q < \infty$, and

$$\|f\|_{pq} = \sup_{t>0} t^{1/p}m(t)$$

when $1 \leqq p \leqq \infty$ and $q = \infty$, we must have

$$\|f\|^*_{pq} \leqq \|f\|_{pq}. \tag{3.20}$$

One of the advantages of using $\| \|_{pq}$ instead of $\| \|_{pq}^*$ is that the former is a norm. This follows from the fact, observed immediately following (3.18), that the mapping $t \to m(t) = m_{f^*}(t)$ is a norm for each $t > 0$. From (3.9), with $p = 1$, we see that $\|f\|_{1\infty} = \|f\|_1$; on the other hand, a simple calculation shows that a function f satisfying $\|f\|_{1q} < \infty$, for $1 \leqq q < \infty$, must be zero almost everywhere. When $1 < p \leqq \infty$, however, $\| \|_{pq}$ is "equivalent" to $\| \|_{pq}^*$. More precisely, we have the following result[11]:

THEOREM 3.21. *If $f \in L(p, q)$, $1 < p \leqq \infty$, then*

$$\|f\|_{pq}^* \leqq \|f\|_{pq} \leqq \frac{p}{p-1} \|f\|_{pq}^*.$$

PROOF. The first inequality is just inequality (3.20) and has already been proved. When $1 < p < \infty$ and $1 \leqq q < \infty$ the second inequality follows from Lemma 3.14, part (i):

$$\|f\|_{pq} = \left(\frac{q}{p} \int_0^\infty \left[\int_0^t f^*(u)\, du\right]^q t^{-q(1-1/p)-1}\, dt\right)^{1/q}$$

$$\leqq \frac{p}{p-1} \left(\frac{q}{p} \int_0^\infty [uf^*(u)]^q u^{-q(1-1/p)-1}\, du\right)^{1/q} = \frac{p}{p-1} \|f\|_{pq}^*.$$

For the remaining cases, $1 < p \leqq \infty$ and $q = \infty$, we have

$$t^{1/p} m(t) = t^{(1/p)-1} \int_0^t f^*(u)\, du = t^{(1/p)-1} \int_0^t u^{-1/p} u^{1/p} f^*(u)\, du$$

$$\leqq \|f\|_{p\infty}^* t^{(1/p)-1} \int_0^t u^{-1/p}\, du = \frac{p}{p-1} \|f\|_{p\infty}^*.$$

THEOREM 3.22. *$L(p, q)$, $1 < p \leqq \infty$, $1 \leqq q \leqq \infty$, with the norm $\| \|_{pq}$ is a Banach space.*

PROOF. We need to show that $L(p, q)$ is complete. When $p = \infty$ we only defined $L(p, q)$ for $q = \infty$; since $L(\infty, \infty) = L^\infty$ the theorem, in this case, follows from the completeness of L^∞. Suppose, therefore, that $1 < p < \infty$ and that $\{f_n\}$ is a Cauchy sequence in $L(p, q)$. By Theorem 3.21, part (i) of Theorem 3.11, and Lemma 3.8 we must then have

$$\sup_{s>0} s^p \mu(\{x \in M : |f_n(x) - f_m(x)| > s\}) \to 0 \qquad \text{as } n, m \to \infty.$$

[11] It is important to use both $\| \|_{pq}^*$ and $\| \|_{pq}$ since each has its advantages. The advantage that $\| \|_{pq}$ is a norm is offset by the fact that the former is often easier to handle; moreover, as was just observed, $\| \|_{1q}$ cannot be used to define a satisfactory extension of L^1.

In particular, the sequence $\{f_n\}$ is fundamental in measure; therefore, there exists a subsequence $\{f_{n_k}\}$ and a function f such that $f_{n_k}(x) \to f(x)$ a.e. Given $\delta > 0$ there exists $j_0 = j_0(\delta)$ such that $\|f_n - f_j\|_{pq} < \delta$ for $j, n \geqq j_0$. Let $g_k = f_{n_k} - f_j$ and $g = f - f_j$; it follows from Fatou's lemma that

$$\int_E |g| \, d\mu \leqq \liminf_{k \to \infty} \int |g_k| \, d\mu$$

for any measurable set E. From (3.18) and another application of Fatou's lemma, therefore,

$$\|f - f_j\|_{pq} = \|g\|_{pq} \leqq \delta$$

for $j \geqq j_0$. This proves the theorem. It can be shown that $L(p, q)$ cannot be normed when $p = 1$, and $1 < q \leqq \infty$. For this, see 5.12 below.

4. Interpolation of Analytic Families of Operators

In this section we will extend the M. Riesz convexity theorem to the case where the operators being interpolated vary with the indices p and q in a sufficiently smooth way. We suppose that to each z in the strip $S = \{z \in \mathbf{C} : 0 \leqq \operatorname{Re} z \leqq 1\}$ there is assigned a linear operator T_z on the space of simple functions in $L^1(M)$ into measurable functions on N in such a way that $(T_z f)g$ is integrable on N whenever f is a simple function in $L^1(M)$ and g is a simple function in $L^1(N)$. We shall say that the family $\{T_z\}$ is admissible if the mapping

$$z \to \int_N (T_z f) g \, d\nu$$

is analytic in the interior of S, continuous on S and there exists a constant $a < \pi$ such that

$$e^{-a|y|} \log \left| \int_N (T_z f) g \, d\nu \right|,$$

$z = x + iy$, is uniformly bounded above in the strip S.

The extension of Theorem 1.3 is, then, the following one:

THEOREM 4.1. *Suppose* $\{T_z\}$, $z \in S$, *is an admissible family of linear operators satisfying*

$$\|T_{iy} f\|_{q_0} \leqq M_0(y) \|f\|_{p_0} \qquad \text{and} \qquad \|T_{1+iy} f\|_{q_1} \leqq M_1(y) \|f\|_{p_1}$$

for all simple functions f *in* $L^1(M, m, \mu)$, *where* $1 \leqq p_j, q_j \leqq \infty$, $M_j(y)$, $j = 0, 1$, *are independent of* f *and satisfy*

$$\sup_{-\infty < y < \infty} e^{-b|y|} \log M_j(y) < \infty$$

for some $b < \pi$. *Then, if* $0 \leqq t \leqq 1$, *there exists a constant* M_t *such that*

$$\|T_t f\|_{q_t} \leqq M_t \|f\|_{p_t}$$

for all such simple functions f *provided* $1/p_t = (1 - t)/p_0 + t/p_1$ *and* $1/q_t = (1 - t)/q_0 + t/q_1$.

The proof of this theorem is very similar to that of the M. Riesz convexity theorem once we have established the following extension of the three lines theorem (Lemma 1.4):

LEMMA 4.2. *Suppose F is a continuous function on S that is analytic in the interior of S satisfying*

$$\sup_{0 \leqq x \leqq 1,\, -\infty < y < \infty} e^{-a|y|} \log |F(x + iy)| < \infty \tag{4.3}$$

for some $a < \pi$. *Then*

$$\log |F(x)| \leqq \tfrac{1}{2} \sin \pi x \int_{-\infty}^{\infty} \left\{ \frac{\log |F(iy)|}{\cosh \pi y - \cos \pi x} + \frac{\log |F(1 + iy)|}{\cosh \pi y + \cos \pi x} \right\} dy \tag{4.4}$$

whenever $0 < x < 1$.

PROOF. For $\zeta \in D = \{z \in \mathbf{C} : |z| \leqq 1\}$, $\zeta \neq 1, -1$, let

$$h(\zeta) = (1/\pi i) \log i(1 + \zeta)/(1 - \zeta).$$

The function h is easily seen to be the composition of the conformal mapping $\zeta \to w = i(1 + \zeta)/(1 - \zeta)$ of the closed unit disk D, minus the points 1 and -1, onto the upper half plane $\{w \in \mathbf{C} : w \neq 0, \operatorname{Im} w \geqq 0\}$ and of the conformal mapping $w \to z = (\log w)/\pi i$ of this half-plane onto S. Thus, h maps $D - \{1, -1\}$ conformally onto S. Moreover, for $z \in S$

$$\zeta = h^{-1}(z) = \frac{e^{\pi i z} - i}{e^{\pi i z} + i}.$$

If we define $G(\zeta) = F(h(\zeta))$ for $\zeta \in D - \{1, -1\}$ we obtain an analytic function on the open unit disk that is continuous on the closed unit disk with the exception of the points 1 and -1. If $0 \leqq \rho < R < 1$, an application of the Poisson–Jensen formula gives us, for $\zeta = \rho e^{i\theta}$,

$$\log |G(\zeta)| \leqq \frac{1}{2\pi} \int_{-\pi}^{\pi} \frac{R^2 - \rho^2}{R^2 - 2R\rho \cos(\theta - \varphi) + \rho^2} \log |G(Re^{i\varphi})| \, d\varphi. \tag{4.5}$$

Condition (4.3), expressed in terms of G and $\eta = h^{-1}(x + iy)$, implies

$$\log |G(\eta)| \leqq A\{|1 + \eta|^{-(a/\pi)} + |1 - \eta|^{-(a/\pi)}\}$$

for some constant A, independent of $\eta \in D$. Since $a/\pi < 1$ this inequality, with $\eta = Re^{i\varphi}$, allows us to apply the dominated convergence theorem of

Lebesgue to the integral in (4.5) and we see, by letting R tend to 1, that

$$\text{(4.6)} \qquad \log |G(\zeta)| \leqq \frac{1}{2\pi} \int_{-\pi}^{\pi} \frac{1 - \rho^2}{1 - 2\rho \cos(\theta - \varphi) + \rho^2} \log |G(e^{i\varphi})| \, d\varphi$$

for $\zeta = \rho e^{i\theta}$, $\rho < 1$. We now claim that the desired inequality (4.4) can be obtained from (4.6) by a change of variables. First we note that the condition $0 < x = h(\rho e^{i\theta}) < 1$ can be easily translated to a condition on ρ and θ: since, in this case, we must have

$$\rho e^{i\theta} = h^{-1}(x) = \frac{e^{\pi i x} - i}{e^{\pi i x} + i} = -i \frac{\cos \pi x}{1 + \sin \pi x} = \left\{\frac{\cos \pi x}{1 + \sin \pi x}\right\} e^{-i(\pi/2)},$$

it follows that $\rho = (\cos \pi x)/(1 + \sin \pi x)$ and $\theta = -(\pi/2)$, when $0 < x \leqq \frac{1}{2}$, while $\rho = -(\cos \pi x)/(1 + \sin \pi x)$ and $\theta = \pi/2$, when $\frac{1}{2} \leqq x < 1$. In the first case we have

$$\begin{aligned} \frac{1 - \rho^2}{1 - 2\rho \cos(\theta - \varphi) + \rho^2} &= \frac{1 - \rho^2}{1 - 2\rho \cos(\varphi + \pi/2) + \rho^2} \\ &= \frac{1 - \rho^2}{1 + 2\rho \sin \varphi + \rho^2} \\ &= \frac{\sin \pi x}{1 + \cos \pi x \sin \varphi}; \end{aligned}$$

moreover, this equality is also valid when $\frac{1}{2} \leqq x < 1$. As we can see from the relation $e^{i\varphi} = h^{-1}(iy) = (e^{-\pi y} - i)/(e^{-\pi y} + i)$, as φ ranges from $-\pi$ to 0, y ranges from $+\infty$ to $-\infty$. Furthermore, $\sin \varphi = -1/(\cosh \pi y)$ and $d\varphi = -[\pi/(\cosh \pi y)] \, dy$. We therefore obtain

$$\begin{aligned} \frac{1}{2\pi} \int_{-\pi}^{0} & \frac{1 - \rho^2}{1 - 2\rho \cos(\theta - \varphi) + \rho^2} \log |G(e^{i\varphi})| \, d\varphi \\ &= \frac{1}{2} \int_{-\infty}^{\infty} \frac{\sin \pi x}{\cosh \pi y - \cos \pi x} \log |F(iy)| \, dy. \end{aligned}$$

Similarly, as φ ranges from 0 to π, $h(e^{i\varphi})$ describes the points $1 + iy$ with $-\infty < y < \infty$ and we obtain

$$\begin{aligned} \frac{1}{2\pi} \int_{0}^{\pi} & \frac{1 - \rho^2}{1 - 2\rho \cos(\theta - \varphi) + \rho^2} \log |G(e^{i\varphi})| \, d\varphi \\ &= \frac{1}{2} \int_{-\infty}^{\infty} \frac{\sin \pi x}{\cosh \pi y + \cos \pi x} \log |F(1 + iy)| \, dy. \end{aligned}$$

This proves the lemma.

Before showing how (4.2) can be used in order to prove Theorem 4.1 we make some observations. First, we show that Lemma 4.2 is, indeed, an extension of the three lines theorem. We have

$$\frac{1}{2}\int_{-\infty}^{\infty}\frac{\sin \pi x}{\cosh \pi y + \cos \pi x}\,dy$$

$$= \frac{1}{2}\int_{-\infty}^{\infty}\frac{\tan(\pi/2)x}{(1+\tan^2(\pi/2)x\tanh^2(\pi/2)y)\cosh^2(\pi/2)y}\,dy$$

$$= \frac{1}{\pi}\int_{-1}^{1}\frac{\tan(\pi/2)x}{1+s^2\tan^2(\pi/2)x}\,ds = \frac{2}{\pi}\int_{0}^{1}\frac{\tan(\pi/2)x}{1+s^2\tan^2(\pi/2)x}\,ds$$

$$= \frac{2}{\pi}\int_{0}^{\tan(\pi/2)x}\frac{du}{1+u^2} = \frac{2}{\pi}\tan^{-1}(\tan(\pi/2)x) = x.$$

Since $\sin \pi(1-x) = \sin \pi x$ and $\cos \pi(1-x) = -\cos \pi x$, we also have

$$\frac{1}{2}\int_{-\infty}^{\infty}\frac{\sin \pi x}{\cosh \pi y - \cos \pi x}\,dy = 1 - x.$$

Consequently, if condition (4.3) were replaced by the conditions $|F(iy)| \leqq m_0$, $|F(1+iy)| \leqq m_1$ and the fact that $|F|$ is uniformly bounded in S, then the equalities we have just established and (4.4) imply $|F(x+iy)| \leqq m_0^{1-x}m_1^x$. That is, the three lines theorem (Lemma 1.4) is a special case of Lemma 4.2.

It is clear from the proof of (4.2) that the integral on the right side of inequality (4.4) is the "conformal image" of a corresponding Poisson integral on the unit circle. In fact, a small change in the argument used to establish (4.2) gives us the following solution to the Dirichlet problem for the strip: If f is defined on the vertical lines $x = 0$ and $x = 1$ in such a way that $e^{-a|y|}f(y)$ and $e^{-a|y|}f(1+iy)$ are bounded for some $a < \pi$, then

$$u(x, y) = \tfrac{1}{2}\sin \pi x \int_{-\infty}^{\infty}\left\{\frac{f(iy)}{\cosh \pi y - \cos \pi x} + \frac{f(1+iy)}{\cosh \pi y + \cos \pi x}\right\}dy$$

defines a harmonic function in the interior of S having (nontangential) boundary values f. The conditions $e^{-a|y|}f(y) = O(1)$ and $e^{-a|y|}f(1+iy) = O(1)$ imply the existence of the above integral (this also explains the role played by condition (4.3)).

We now pass to the proof of Theorem 4.1. Let $f \in L^1(M)$ and $g \in L^1(N)$

be simple functions satisfying

$$\|f\|_p = 1 = \|g\|_{q'},$$

where $p = p_t, q = q_t$ and $1/q + 1/q' = 1$. Suppose that $f = \sum_{j=1}^m a_j\chi_{E_j}$ and $g = \sum_{k=1}^n b_k\chi_{F_k}$. Using the same notation that was introduced in the proof of the M. Riesz convexity theorem (Theorem 1.3), we construct the function

$$F(z) = \int_N (T_zf_z)g_z\,dv = \sum_{j,k}^{m,n} |a_j|^{\alpha(z)/\alpha}|b_k|^{(1-\beta(z))/(1-\beta)}\gamma_{jk}(z),$$

where $\gamma_{jk}(z) = e^{i(\theta_j+\varphi_k)}\int_N (T_z\chi_{E_j})\chi_{F_k}\,dv$. It follows from the assumption that $\{T_z\}$ is an admissible family that F satisfies the hypotheses of Lemma 4.2. Moreover, since

$$|f_{iy}|^{p_0} = |f|^p = |f_{1+iy}|^{p_1} \qquad |g_{iy}|^{q_0'} = |g|^{q'} = |g_{1+iy}|^{q_1'}$$

an application of Hölder's inequality gives us $|F(iy)| \leqq M_0(y)$ and $|F(1 + iy)| \leqq M_1(y)$. Thus, by Lemma 4.2,

$$\left|\int_N (T_tf)g\,dv\right| = |F(t)|$$

$$\leqq \exp\left[\tfrac{1}{2}\sin\pi t\int_{-\infty}^{\infty}\left\{\frac{\log M_0(y)}{\cosh\pi y - \cos\pi t} + \frac{\log M_1(y)}{\cosh\pi y + \cos\pi t}\right\}dy\right] = M_t.$$

Since $\|T_tf\|_{p_t} = \sup_{\|g\|_{q_t'}=1} |\int_N (T_tf)g\,dv|$, this proves the theorem.

5. *Further Results*

5.1. The theory of interpolation of linear operators has been extended in many ways to general Banach spaces. The setting for these extensions can be described in the following manner: Let V be a topological vector space and A^0, A^1 two Banach spaces, with norms $\|\ \|_0$ and $\|\ \|_1$, that are continuously embedded in V. We can convert $A^0 \cap A^1$ and $A^0 + A^1 = \{a = a_0 + a_1 : a_0 \in A^0 \text{ and } a_1 \in A^1\}$ into Banach spaces by introducing the norm $\|a\|_{A^0\cap A^1} = \max\{\|a\|_0, \|a\|_1\}$ in the former space and $\|a\|_{A^0+A^1} = \inf\{\|a_0\|_0 + \|a_1\|_1 : a_0 \in A^0, a_1 \in A^1 \ni a_0 + a_1 = a\}$ in the latter. An *intermediate space* between A^0 and A^1 is any Banach space A satisfying $A^0 \cap A^1 \subset A \subset A^0 + A^1$. For example, when $A^0 = L^{p_0}$, $A^1 = L^{p_1}$, $1 \leqq p_0, p_1 \leqq \infty$, $1/p_t = (1-t)/p_0 + t/p_1$, $0 \leqq t \leqq 1$, then L^{p_t} is an intermediate space between L^{p_0} and L^{p_1}. Suppose an intermediate space A is invariant under each linear transformation T of $A^0 + A^1$ into itself having the property that its restriction to $A^j, j = 0, 1$, is a bounded operator of A^j into itself. If, moreover, the restriction of any such T to A is also a bounded operator, we say that A is a *space of linear interpolation* between A^0 and A^1. It follows from the M. Riesz convexity theorem that L^{p_t} is a space of linear interpolation between L^{p_0} and L^{p_1}.

We can then formulate the general problems in the abstract theory of interpolation of linear operators in the following way:

(i) Given two Banach spaces A^0 and A^1, how can "all" the spaces of linear interpolation between A^0 and A^1 be characterized?

(ii) How can one construct spaces of linear interpolation between two given Banach spaces A^0 and A^1?

(iii) Suppose A is a space of linear interpolation between A^0 and A^1. If B^0 and B^1 are two other Banach spaces, does there exist a space of linear interpolation B (between B^0 and B^1) such that each linear transformation of $A^0 + A^1$ into $B^0 + B^1$ whose restriction to A^j maps this space continuously into B^j, $j = 0, 1$, has the property that it maps A continuously into B?

(iv) Suppose A is a space of linear interpolation that has been constructed by some specific method from A^0 and A^1 and the "same" method has been employed to obtain B from two other Banach spaces B^0 and B^1. Is it true that each linear transformation from $A^0 + A^1$ into $B^0 + B^1$ that maps A^j continuously into B^j, $j = 0, 1$, also maps A continuously into B?

An answer to the first problem has been obtained by Gagliardo (see [1]). Moreover, Gagliardo's method enabled him to obtain an affirmative answer to problem (iii). These results are very general and are not as closely related to the material developed in this chapter as are two constructive theories of interpolation (that is, answers to problems (ii) and (iv)) developed by Lions [1], Calderón [2], Lions and Peetre [1]. The next few paragraphs will be devoted to a description of these methods of interpolation.

5.2. We begin by describing the *complex method of interpolation* due to Calderón. Let B be a Banach space and D a domain in the complex plane. A mapping $z \to b(z)$, from D into B, is said to be *analytic* if $z \to l[b(z)]$ is an analytic complex-valued function on D for each continuous linear functional l on B. We introduce an auxiliary space $\mathscr{F}(A^0, A^1)$ consisting of all functions f defined on the strip $S = \{z \in \mathbf{C} : 0 \leqq \operatorname{Re} z \leqq 1\}$ with values in $A^0 + A^1$ satisfying:

(1) f is analytic in the interior $S^0 = \{z \in \mathbf{C} : 0 < \operatorname{Re} z < 1\}$ of S;

(2) f is continuous and $\|f\|_{A^0 + A^1}$ is bounded on S;

(3) $f(it) \in A^0$ for $-\infty < t < \infty$, $t \to f(it)$ is continuous as a function from $(-\infty, \infty)$ into A^0 and $t \to \|f(it)\|_0$ is bounded;

(4) $f(1 + it) \in A^1$ for $-\infty < t < \infty$, $t \to f(1 + it)$ is continuous as a function from $(-\infty, \infty)$ into A^1 and $t \to \|f(1 + it)\|_1$ is bounded.

$\mathscr{F}(A^0, A^1)$ is a Banach space if we introduce the norm $\|f\|_{\mathscr{F}} = \|f\|_{\mathscr{F}(A^0,A^1)} = \max\{\sup_t \|f(it)\|_0, \sup_t \|f(1+it)\|_1\}$. Suppose $0 \leqq t \leqq 1$ and $N_t = \{f \in \mathscr{F} : f(t) = 0\}$ we then define A_t to be the space $\mathscr{F}/N_t$. Clearly, we can then identify A_t with the linear space of all elements $f(t) \in A^0 + A^1$ with $f \in \mathscr{F}$. We obtain a Banach space if we introduce on A_t the usual quotient space norm $\|a\|_{A_t} = \inf\{\|f\|_{\mathscr{F}} : f \in \mathscr{F}, f(t) = a\}$ whenever $a \in A_t$. We shall assume that $\{0\} \neq A^0 \cap A^1$.

It is clear that A_t is an intermediate space between A^0 and A^1. It can be shown, moreover, that A_t is a space of linear interpolation between A^0 and A^1. In fact, we obtain the following solution to problem (iv) announced above:

If B^0 and B^1 are two Banach spaces and T is a linear transformation from $A^0 + A^1$ into $B^0 + B^1$ that maps A^j into $B^j, j = 0, 1$, in such a way that $\|Ta\|_j \leqq M_j\|a\|_j$ for all $a \in A^j$ then T maps A_t into B_t and $\|Ta\|_{B_t} \leqq M_0^{(1-t)}M_1^t\|a\|_{A_t}$ for all $a \in A_t$.

Calderón gives another (but related) construction of spaces of linear interpolation between A^0 and A^1 that leads to another solution to problem (iv). When we apply either of these results, however, we must identify the spaces of linear interpolation that have been obtained. In order to better understand the significance of this new problem we present some basic examples of interpolation of linear operators (for the most part, these were known before the formulation of the general theory).

5.3. Suppose $(M, \mathscr{M}, \mu)$ is a measure space and $A^j = L^{p_j}(M)$, $1 \leqq p_j \leqq \infty$ for $j = 0, 1$. We can embed A^0 and A^1 in, say, the space V of locally integrable functions. Then, there exists a norm preserving isomorphism between A_t and $L^{p_t}(M)$, where $1/p_t = (1-t)/p_0 + t/p_1$, $0 \leqq t \leqq 1$, provided $p_t < \infty$. In case $p_t = \infty$, p_t must be either p_0 or p_1; let us suppose $p_t = p_0$ and $p_1 < \infty$. We do not then obtain, in general, a norm preserving isomorphism between A_0 and $A^0 = L^\infty(M)$. It can be shown, however, that A_0 can be identified with the closure in the L^∞ norm of the set of functions $f \in L^\infty(M)$ for which $\mu(\{x \in M : |f(x)| > \alpha\}) < \infty$ for all $\alpha > 0$. If $B^0 = L^{q_0}(N)$ and $B^1 = L^{q_1}(N)$, where $(N, \mathscr{N}, \nu)$ is another measure space, the theorem stated in paragraph 5.2 is a reformulation of the M. Riesz convexity theorem (1.3).

5.4. Suppose $(M, \mathscr{M}, \mu)$ is a measure space and $\mu_j, j = 0, 1$, is a measure on $\mathscr{M}$ that is absolutely continuous with respect to μ. If $A^j = L^{p_j}(M, \mu_j)$ then there exists a norm-preserving isomorphism between A_t and $L^{p_t}(M, \mu_t)$ where $0 < 1/p_t = (1-t)/p_0 + t/p_1$ and $d\mu_t = \alpha_0^{1-t}\alpha_1^t\, d\mu$, with $0 \leqq t \leqq 1$, (α_j being the Radon–Nikodym derivative of $\mu_j, j = 0, 1$, with respect to μ). When $p_t = \infty$ we have to make the modifications described in 5.3. See Stein and G.Weiss [5].

5.5. For $p \geqq 1$ let H^p be the space of all functions $F(z)$ that are analytic in the upper half-plane $\text{Im } z > 0$ and satisfy

$$\|F\|_p = \sup_{y>0}\left(\int_{-\infty}^{\infty} |F(x+iy)|^p\,dx\right)^{1/p} < \infty.$$

This is the H^p space associated with the tube whose base is the positive real line (see Chapter III). If $A^j = H^{p_j}$, $1 \leqq p_j < \infty$, then A_t, $0 \leqq t \leqq 1$, is *equivalent* to H^{p_t} (that is, there exists an invertible linear map of A_t onto H^{p_t}). (See Salem and Zygmund [1], Calderón and Zygmund [2] and G. Weiss [1].) The identification of the spaces of linear interpolation between two H^p spaces associated with higher dimensional tubes is an open problem. This also applies to the H^p spaces of systems of conjugate harmonic functions that are discussed in the next chapter.

5.6. Suppose $0 < \alpha < 2$, we then consider the space $\lambda_\alpha = \lambda_\alpha(E_n)$ consisting of all continuous bounded functions on E_n satisfying $\sup_x |f(x+t) + f(x-t) - 2f(x)| = o(|t|^\alpha)$ as $|t| \to 0$. This space is a Banach space with the norm $\|f\|^{(\alpha)} = \|f\|_\infty + \sup_{x,t}\{|t|^{-\alpha}|f(x+t) + f(x-t) - 2f(x)|\}$. For general $\alpha > 0$ we define λ_α to be the space of all C^k functions, where k is the largest integer strictly smaller than α, satisfying $D^\beta f \in \lambda_{\alpha-k}$, for $0 \leqq |\beta| = \beta_1 + \cdots + \beta_n < \alpha$ (we are using the notation introduced in (1.9) of Chapter I). Then λ_α is a Banach space with the norm $\|f\|^{(\alpha)} = \sum_{|\beta|<\alpha} \|D^\beta f\|^{(\alpha-k)}$. If $A^j = \lambda_{\alpha_j}$, $j = 0, 1$, then A_t is equivalent to λ_α, where $\alpha = (1-t)\alpha_0 + t\alpha_1$. (See Taibleson [1] and Calderón [2].)

5.7. The following example is usually studied in the context of E_n but is simpler to describe in the setting of multiple Fourier series. For any $\alpha \geqq 0$ we define the space $L^p_\alpha(T_n)$ to be the set of those $f \in L^p(T_n)$ with $f(x) \sim \sum_m a_m e^{2\pi i m\cdot x}$ so that $a_0 + \sum_{m\neq 0} |m|^\alpha a_m e^{2\pi i m\cdot x}$ is the Fourier series of a function $f^{(\alpha)} \in L^p(T_n)$ (we are using the notation that is introduced at the beginning of Chapter VII). This space is a Banach space with the norm $\|f\|_p^{(\alpha)} = \|f^{(\alpha)}\|_p$. Suppose $A^j = L^{p_j}_{\alpha_j}$, $1 < p_j < \infty, j = 0, 1$, then A_t is equivalent to L^p_α, where $1/p = (1-t)/p_0 + t/p_1$ and $\alpha = (1-t)\alpha_0 + t\alpha_1$. The proof of this fact makes use of Theorem (4.1) and the fact that $f \to \sum_m |m|^{-i\gamma} a_m e^{2\pi i m\cdot x}$ is a bounded operator on $L^p(T^n)$ for $1 < p < \infty$ when γ is real.

5.8. The classical interpolation theorem of M. Riesz, as formulated in 5.3 or in Section 1 of this chapter, can be extended to the case of Banach-space-valued functions on a measure space. For a detailed statement of this result, see Calderón [2]. Special cases, where the values of the functions are themselves L^q spaces, were obtained earlier by Boas and Bochner [1] and by Benedeck and Panzone [1]. Another extension of the classical interpolation theorem of M. Riesz can be formulated for *Banach Lattices* (these being Banach function spaces such that $|f| \leqq |g|$ implies $\|f\| \leqq \|g\|$. For a precise statement see Calderón [2]). This extension includes the M. Riesz theorem, the example described in 5.4 and an interpolation theorem involving $L(p, q)$ spaces that were defined in the third section.

5.9. The constructive theory of interpolation mentioned in 5.1 which was introduced by Gagliardo, Lions and Peetre can be described in the following way. In terms of the notation introduced in 5.1 we define a norm on $A^0 + A^1$ for each $t > 0$ by letting $K_a(t) = \inf\{\|a_0\|_0 + t\|a_1\|_1 : a_0 + a_1 = a, a_0 \in A^0, a_1 \in A^1\}$. It follows immediately from this definition that K_a is a concave and nonnegative function on the positive real numbers. Suppose now that Φ is a nonnegative (possibly infinite-valued) function defined on the set of all nonnegative Lebesgue measurable functions on $(0, \infty)$ satisfying:

(i) $\Phi(h) = 0$ if and only if $h(t) = 0$ a.e.;
(ii) if $\Phi(h) < \infty$ then $h(t) < \infty$ a.e.;
(iii) $\Phi(ah) = a\Phi(h)$ when $a > 0$;
(iv) if $h(t) \leqq \sum_{j=1}^{\infty} h_j(t)$ a.e. then $\Phi(h) \leqq \sum_{j=1}^{\infty} \Phi(h_j)$.

Such a function Φ is called a *function norm* and the set $A(\Phi) = \{a \in A^0 + A^1 : \Phi(K_a) < \infty\}$ is a subspace of $A^0 + A^1$. The mapping $a \to \Phi(K_a)$ is a norm on $A(\Phi)$ which shall be denoted by $\|a\|_\Phi$. When Φ satisfies certain general conditions the space $A(\Phi)$ is a space of linear interpolation between A^0 and A^1; moreover, we then have an interpolation theorem that gives us a solution to problem (iv) that was posed in 5.1: *Suppose T is a linear transformation from $A^0 + A^1$ into $B^0 + B^1$ that maps A^j into B^j, $j = 0, 1$, in such a way that $\|Ta\|_j \leqq M_j\|a\|_j$ for all $a \in A^j$; then T maps $A(\Phi)$ into $B(\Phi)$ and there exists a constant $M = M(\Phi, M_0, M_1)$ such that $\|Ta\|_\Phi \leqq M\|a\|_\Phi$ for all $a \in A(\Phi)$.* The proof of this theorem is particularly simple if Φ satisfies the following condition: There exists a nonnegative function ψ on $(0, \infty)$ such that $\Phi(h^\lambda) \leqq \psi(\lambda)\Phi(h)$, where $\lambda > 0$ and $h^\lambda(t) = h(\lambda t)$. When this is the case, we have $\|Ta\|_\Phi = \Phi(K_{Ta}) \leqq M_0\Phi(K_a^{M_1/M_0}) \leqq M_0\psi(M_1/M_0)\Phi(K_a) = M_0\psi(M_1/M_0)\|a\|_\Phi$ (the first inequality being a consequence of the hypotheses on T and the second following from our new assumption on Φ). Thus, the interpolation theorem is obtained with $M \leqq M_0\psi(M_1/M_0)$. If, for example, $\psi(\lambda) = \lambda^s$ for $0 \leqq s \leqq 1$ we then have the familiar inequality $M \leqq M_0^{1-s}M_1^s$.

5.10. When $A^0 = L^1$, $A^1 = L^\infty$ and $f \in L^1 + L^\infty$ it is not hard to establish that $K_f(t) = \int_0^t f^*(u)\,du$. The following observations show that it suffices to consider only the case $f \geqq 0$ and, in this case, $K_f(t) = \inf\{\|f_0\|_{L^1} + t\|f_1\|_{L^\infty} : f = f_0 + f_1, f_j \geqq 0$ and $f_j \in A^j, j = 0, 1\}$. Suppose

$f = f_0 + f_1$ then $|f| = (\operatorname{sgn} \bar{f})f = (\operatorname{sgn} \bar{f})f_0 + (\operatorname{sgn} \bar{f})f_1 = g_0 + g_1$; similarly, if $|f| = g_0 + g_1$ then $f = (\operatorname{sgn} f)g_0 + (\operatorname{sgn} f)g_1 = f_0 + f_1$. In either case we have $\|f_0\|_{L^1} + t\|f_1\|_{L^\infty} = \|g_0\|_{L^1} + t\|g_1\|_{L^\infty}$ and, thus, $K_f = K_{|f|}$. When f is real-valued and $f = f_0 + f_1$ then $f = \operatorname{Re} f_0 + \operatorname{Re} f_1$; thus, in estimating K_f it suffices to consider only real-valued f_0 and f_1. Finally, if $f \geqq 0$ and $f = f_0 + f_1$ let $f_0' = \inf\{f_0^+, f\}$ and $f_1' = f - f_0'$. Then, $f = f_0' + f_1'$, $0 \leqq f_0' \leqq f_0^+ \leqq |f_0|$ and $0 \leqq f_1' = f - f_0' \leqq |f - f_0| = |f_1|$. Consequently, $\|f_0'\|_{L^1} + t\|f_1'\|_{L^\infty} \leqq \|f_0\|_{L^1} + t\|f_1\|_{L^\infty}$. We can, therefore, assume that $f = f_0 + f_1$ with all three functions being nonnegative. If $s = \|f_1\|_{L^\infty}$ define $f^{(s)} = f \wedge s = \inf\{f, s\}$ and $g = f - f^{(s)}$. By changing f_1 on a set of measure 0, if necessary, we can assume $0 \leqq f_1(x) \leqq s$ for all x. Since $f_1 \leqq f$ it follows that $f_1 \leqq f^{(s)}$ and, consequently, $g = f - f^{(s)} \leqq f - f_1 = f_0$. This shows that $\|g\|_{L^1} + t\|f^{(s)}\|_{L^\infty} = \|g\|_{L^1} + st \leqq \|f_0\|_{L^1} + st \leqq \|f_0\|_{L^1} + t\|f_1\|_{L^\infty}$. Therefore, for $f \geqq 0$ we have $K_f(t) = \inf_{s \geqq 0}\{\|f - f^{(s)}\|_{L^1} + ts : f^{(s)} = f \wedge s\}$. If λ denotes the distribution function of f and $s_0 = \inf\{s : \lambda(s) < t\}$ we will now show that $K_f(t) = \|f - f^{(s_0)}\|_{L^1} + ts_0$. First suppose $s_1 > s_0$; then $f^{(s_1)}(x) - f^{(s_0)}(x)$ equals 0 when $f(x) \leqq s_0$, equals $f(x) - s_0$ if $s_0 < f(x) \leqq s_1$, and takes on the value $s_1 - s_0$ if $s_1 < f(x)$. Thus, since $\lambda(s_0) \leqq t$,

$$\begin{aligned}
&\|f - f^{(s_1)}\|_{L^1} + ts_1 - (\|f - f^{(s_0)}\|_{L^1} + ts_0) \\
&\qquad = t(s_1 - s_0) - \int (f^{(s_1)} - f^{(s_0)})\, d\mu \\
&\qquad \geqq \lambda(s_0)(s_1 - s_0) - \int_{f(x) > s_0} (f^{(s_1)} - f^{(s_0)})\, d\mu \\
&\qquad \geqq \lambda(s_0)(s_1 - s_0) - \lambda(s_0)(s_1 - s_0) = 0.
\end{aligned}$$

Now suppose $s_1 < s_0$ and $s_1 \leqq s < s_0$. For s in this interval $\lambda(s) \geqq t$; therefore,

$$\begin{aligned}
&\|f - f^{(s_1)}\|_{L^1} + ts_1 - (\|f - f^{(s)}\|_{L^1} + st) \\
&\qquad \geqq \int_{f(x) > s_1} (f^{(s)} - f^{(s_1)})\, d\mu - \lambda(s)(s - s_1) \\
&\qquad \geqq \int_{f(x) > s} (s - s_1)\, d\mu - \lambda(s)(s - s_1) \\
&\qquad = 0.
\end{aligned}$$

By continuity the inequality $\|f - f^{(s_1)}\|_{L^1} + ts_1 \geqq \|f - f^{(s)}\|_{L^1} + ts$ must hold for $s = s_0$ and we have shown that $K_f(t) = \|f - f^{(s_0)}\|_{L^1} + ts_0$. It follows from the definition of f^* that $f^*(u) = s_0$ when $\lambda(s_0) \leqq u < t$. This, together with equality (ii) of Lemma 3.17, gives us

$$K_f(t) = \int (f - f^{(s_0)})\, d\mu + ts_0 = \int_{f(x) > s_0} f\, d\mu - s_0\lambda(s_0) + ts_0$$

$$= \int_0^{\lambda(s_0)} f^*(u)\, du + \int_{\lambda(s_0)}^{t} f^*(u)\, du = \int_0^t f^*(u)\, du.$$

If we now choose the function norm

$$\Phi(h) = \left((q/p) \int_0^\infty (t^{(1/p)-1}h(t))^q(dt/t)\right)^{1/q}$$

we see that the space $A(\Phi)$ defined in 5.9 is $L(p, q)$, with $1 < p \leqq \infty$, and the norm is identical with $\| \ \|_{pq}$.

5.11. In the first section of this chapter we gave a geometrical interpretation of the M. Riesz convexity theorem in terms of the points of the unit square $Q = \{(\alpha, \beta) : 0 \leqq \alpha, \beta \leqq 1\}$. A similar interpretation can be given for the Marcinkiewicz theorem (2.4); however, in this case only the points of the "lower triangle" $\{(\alpha, \beta) \in Q : \alpha \geqq \beta\}$ are involved. One can show by example that this theorem is not valid in the "upper triangle" (see Hunt [1]). In his paper [2] M. Riesz showed that his convexity theorem is valid for the entire unit square Q only if the L^p spaces under consideration consist of complex-valued functions. In case only real-valued functions are considered, M. Riesz gave examples showing that the convexity result fails in the upper triangle and, yet, is valid for the lower triangle. It follows immediately from the complex case, however, that, given the hypotheses of Theorem 1.3, there exists a constant k_t (not necessarily less than or equal to $k_0^{1-t}k_1^t$) such that the operator T is of type (p_t, q_t) with (p_t, q_t)—norm $\leqq k_t$. In this connection, it is of interest to remark that virtually all operators that arise in a natural way in Fourier analysis are of type (or weak type) (p, q) for some indices p and q satisfying $p \leqq q$; thus, interpolation results corresponding to points of the "lower triangle" suffice for our purpose.

5.12. When $p = 1$ and $1 < q \leqq \infty$, then there is no norm equivalent with $\| \ \|_{p,q}^*$, and thus these spaces $L(p, q)$ cannot be normed. Let us show this in the case $q = \infty$. On E_1 we consider the sequence of functions $\{f_k\}$ defined by $f_k(x) = 1/|x - k|$. Then each $f_k \in L(1, \infty)$, and $\|f_k\|_{1,\infty}^*$ is independent of k. Let $f = \sum_{k=1}^N f_k$. If $\| \ \|$ were a norm equivalent with $\| \ \|_{1,\infty}^*$ then we would have $\|f\| \leqq C_1N$, for some constant C_1. However, $f(x) \geqq C_2 \log N$ for $0 \leqq x \leqq N$, and thus $\|f\|_{1,\infty}^* \geqq C_2N \log N$, giving a contradiction when $N \to \infty$. The proof for $1 < q < \infty$ is similar.

Bibliographical Notes

The Marcinkiewicz theorem was first announced in [2], where a sketch of the proof is given for the "main diagonal" $\{(\alpha, \beta) \in Q : \alpha = \beta\}$. Its proof for the "lower triangle," as well as many important applications was first published by Zygmund [2]. The proof of the M. Riesz convexity theorem presented in this chapter was discovered by Thorin [1] and, independently, by Tamarkin and Zygmund [1]. Interpolation of analytic families of operators is due to Stein [2]. A related result, which also makes use of Lemma 4.2 can be found in Hirschman [1]. An extension of this type of interpolation to H^p spaces is found in Stein and G. Weiss [3]. $L(p, q)$ spaces were introduced by Lorentz [2]. Calderón was the first to use the function m_{f^*} and to point out how it induces a norm on $L(p, q)$. For the general theory of these spaces see Hunt [1] and Oklander [1]. Our treatment of the results leading to the interpolation Theorem 3.15 is adapted from Hunt [1]. For related results see Stein and G. Weiss [2], Krein and Semenov [1], Calderón [2] and Lions and Peetre [1]. Two expository articles describing various general methods of interpolation and how they are related to each other are Krein and Petunin [1] and Magenes [1]. Weak type inequalities for the Hilbert transform go back to Kolmogorov [1], Besicovitch [1] and Titchmarsh [1].

CHAPTER VI

Singular Integrals and Systems of Conjugate Harmonic Functions

We have encountered the Hilbert Transform in the third and fifth chapter. In both instances its properties were developed in the context of the theory of functions of a single complex variable. In the first section we develop a part of the real-variable approach to the study of the Hilbert transform. In the second section we show how these results can be used to define and develop basic properties of a class of singular integral operators, the ones with an odd kernel. The third section is devoted to the introduction of a larger class of singular integral operators. Finally, in the fourth section we show that many such operators are related to systems of partial differential equations that extend the Cauchy–Riemann equations. These systems enable us to consider yet another approach to complex methods in Fourier analysis of several variables. In particular, we introduce a theory of H^p-spaces that, in a sense, parallels the one developed in the third chapter.

1. The Hilbert Transform

In the second section of the fifth chapter we introduced the Hilbert transform $\tilde{f}$ of a function $f \in L^p(-\infty, \infty)$, $1 \leqq p < \infty$. We defined it as the limit, as $y > 0$ tends to 0, of $v(x, y) = (1/\pi) \int_{-\infty}^{\infty} f(x - t)[t/(t^2 + y^2)]\, dt$. Using properties of analytic functions we showed that this limit exists for almost all x. In view of this fact, it is natural to wonder if $\tilde{f}$ could not be defined directly by the equality $\tilde{f}(x) = (1/\pi) \int_{-\infty}^{\infty} [f(x - t)/t]\, dt$. Unfortunately, this integral is not defined even for very smooth $f \in L^p(-\infty, \infty)$. If, however, we consider this integral in the principal value sense (see (4.4) in Chapter IV) we do obtain

$$\tilde{f}(x) = \lim_{\varepsilon \to 0} \frac{1}{\pi} \int_{0 < \varepsilon \leqq |t|} \frac{f(x - t)}{t}\, dt \tag{1.1}$$

for almost every real number x. This is an immediate consequence of Lemma 2.6 of Chapter V and the following result:

LEMMA 1.2. *Suppose* $f \in L^p(-\infty, \infty)$, $1 \leqq p < \infty$, *then*

$$\lim_{\varepsilon \to 0} \left\{ \int_{-\infty}^{\infty} f(x-t) \frac{t}{t^2 + \varepsilon^2}\, dt - \int_{0<\varepsilon\leqq|t|} \frac{f(x-t)}{t}\, dt \right\} = 0$$

at each point x *in the Lebesgue set of* f.

PROOF. Let

$$\varphi(t) = \begin{cases} \dfrac{t}{t^2+1} - \dfrac{1}{t} & \text{if } |t| \geqq 1 \\ \dfrac{t}{t^2+1} & \text{if } |t| < 1 \end{cases}$$

and, for $\varepsilon > 0$, $\varphi_\varepsilon(t) = \varepsilon^{-1}\varphi(t/\varepsilon)$. Then

$$\int_{-\infty}^{\infty} f(x-t) \frac{t}{t^2+\varepsilon^2}\, dt - \int_{0<\varepsilon\leqq|t|} \frac{f(x-t)}{t}\, dt = \int_{-\infty}^{\infty} f(x-t)\varphi_\varepsilon(t)\, dt.$$

Since

$$\psi(x) = \sup_{|t|\geqq|x|} |\varphi(t)| = \begin{cases} \dfrac{1}{|x|(1+x^2)} & \text{if } |x| \geqq 1 \\ \dfrac{1}{2} & \text{if } |x| < 1, \end{cases} \tag{1.3}$$

it follows that $\psi \in L^1(-\infty, \infty)$. Thus, by Theorem 1.25 of Chapter I and the fact that $\int_{-\infty}^{\infty} \varphi(t)\, dt = 0$ (since φ is odd), $\lim_{\varepsilon\to 0} \int_{-\infty}^{\infty} f(x-t)\varphi_\varepsilon(t)\, dt = f(x)\int_{-\infty}^{\infty}\varphi(t)\, dt = 0$ and the lemma is established.

In the previous chapter we showed that the Hilbert transform is of type (p, p) for $1 < p < \infty$. We shall show that the *maximal Hilbert transform* which assigns to $f \in L^p(-\infty, \infty)$ the function whose values are

$$\sup_{\varepsilon>0} \frac{1}{\pi} \left| \int_{|t|>\varepsilon} f(x-t) \frac{dt}{t} \right|$$

is also of type (p, p). More precisely, we shall establish the following result:

THEOREM 1.4. *If* $f \in L^p(-\infty, \infty)$, $1 < p < \infty$, *then*

$$(Mf)(x) = \sup_{\varepsilon>0} \frac{1}{\pi} \left| \int_{|t|>\varepsilon} f(x-t)\frac{dt}{t} \right| \leqq \frac{1}{\pi} \|\psi\|_1 m_f(x) + m_{\tilde{f}}(x),$$

where ψ *is the function defined in* (1.3) *and* m_f *and* $m_{\tilde{f}}$ *are the Hardy–Littlewood maximal functions of* f *and* $\tilde{f}$. *In particular, there exists a constant* B_p, *independent of* $f \in L^p(-\infty, \infty)$ *such that*

$$\|M_f\|_p \leqq B_p \|f\|_p.$$

PROOF. We have

$$\int_{0<\varepsilon\leq|t|} \frac{f(x-t)}{t}\,dt = \int_{-\infty}^{\infty} f(x-t)\frac{t}{t^2+\varepsilon^2}\,dt - \int_{-\infty}^{\infty} f(x-t)\varphi_\varepsilon(t)\,dt.$$

Since ψ is radial and nonincreasing, it follows from (3.9) of Chapter II that

$$\sup_{\varepsilon>0}\left|\int_{-\infty}^{\infty} f(x-t)\varphi_\varepsilon(t)\,dt\right| \leq \|\psi\|_1 m_f(x).$$

Thus, the theorem is proved once we show that

$$\sup_{\varepsilon>0}\frac{1}{\pi}\left|\int_{-\infty}^{\infty} f(x-t)\frac{t}{t^2+\varepsilon^2}\,dt\right| \leq m_{\tilde f}(x).$$

But this inequality is a consequence of Theorem 3.10 of Chapter II and the following equality:

LEMMA 1.5. *If $f\in L^p(-\infty,\infty)$, $1<p<\infty$, and $y>0$ then*

$$\frac{1}{\pi}\int_{-\infty}^{\infty} f(x-t)\frac{t}{t^2+y^2}\,dt = \frac{1}{\pi}\int_{-\infty}^{\infty} \tilde f(x-t)\frac{y}{t^2+y^2}\,dt.$$

PROOF. The left side of this equality is the convolution of f with the conjugate Poisson kernel $Q_y(t) = (1/\pi)[t/(t^2+y^2)]$ introduced in (6.13) of Chapter III, while the right side is the convolution of $\tilde f$ with the Poisson kernel $P_y(t) = (1/\pi)[y/(t^2+y^2)]$. Since for each $y>0$, Q_y and P_y belong to $L^{p'}(-\infty,\infty)$, $(1/p)+(1/p')=1$, it follows from Hölder's inequality and (2.11) of Chapter V that it suffices to establish the lemma for f belonging to a dense subset of L^p. Suppose, therefore, that f belongs to the class $\mathscr{S}$ of testing functions introduced in the first chapter. Then, in particular, f belongs to $L^2(-\infty,\infty)$ and, thus, by (2.11) of Chapter V $\tilde f$ also belongs to $L^2(-\infty,\infty)$ and its Fourier transform $(\tilde f)^\wedge$ is defined. It is not hard to show that

$$(\tilde f)^\wedge(x) = (-i\,\mathrm{sgn}\,x)\hat f(x). \tag{1.6}$$

In fact, we showed in (6.13) of the third chapter that

$$\hat Q_y(x) = (-i\,\mathrm{sgn}\,x)e^{-2\pi|yx|}.$$

Thus, $(Q_y * f)^\wedge(x) = \hat Q_y(x)\hat f(x) = (-i\,\mathrm{sgn}\,x)e^{-2\pi|yx|}\hat f(x)$ and, by Plancherel's theorem it follows that, as y tends to 0, $Q_y * f$ tends (in the L^2-norm) to a function g whose Fourier transform is $(-i\,\mathrm{sgn}\,x)\hat f(x)$. On the other hand, we know from Lemma 2.6 of Chapter V that $(Q_y * f)(x)\to \tilde f(x)$ a.e. as $y\to 0$. Thus, $\tilde f$ and g must be equal almost everywhere and (1.6) follows. In order to establish (1.5) it suffices to show that the Fourier transforms of both sides are equal. From (1.6) and Theorem 1.14 of Chapter I we have $(\tilde{f} * P_y)^\wedge(x) = (\tilde{f})^\wedge(x)e^{-2\pi|yx|} = (-i \operatorname{sgn} x)\hat{f}(x)e^{-2\pi|yx|}$. But we observed above that this last expression is equal to $(f * Q_y)^\wedge(x)$. This proves the lemma and, consequently, Theorem 1.4.

Our aim is to extend these results to n dimensions. In order to do so we must first find an appropriate generalization of the Hilbert transform. One way of introducing such a generalization is by introducing the kernel

$$K(t) = \frac{\Omega(t)}{|t|^n}, \qquad t \neq 0, \tag{1.7}$$

where Ω is an odd function that is homogeneous of degree zero (i.e., $\Omega(at) = \Omega(t)$ for all positive real numbers a). When $n = 1$ and $\Omega(t) = (\operatorname{sgn} t)/\pi$ the convolution of f with K (in the principal value sense) yields the Hilbert transform of f. Thus, in view of the results presented in this section, it is natural to inquire whether the operator

$$(Hf)(x) = \lim \int_{\delta \geqq |t| \geqq \varepsilon > 0} f(x - t)K(t)\, dt, \qquad \varepsilon \to 0,\ \delta \to \infty \tag{1.8}$$

is well-defined for $f \in L^p(E_n)$, $1 < p < \infty$, and maps this space boundedly into itself.[1] In the next section we will show, in particular, that this is indeed the case if Ω is integrable when restricted to the surface of the unit sphere Σ_{n-1} (i.e., if $\int_{\Sigma_{n-1}} |\Omega(t')|\, dt' < \infty$). Such an operator is known as a *singular integral operator with odd kernel.*

Another more general kernel is obtained if we merely assume that the restriction of Ω to the surface of the unit sphere is integrable and

$$\int_{\Sigma_{n-1}} \Omega(t')\, dt' = 0.$$

This condition is clearly satisfied when Ω is an odd function. In the third section we shall discuss operators defined by (1.8) when the kernel K is of the form (1.7) with Ω satisfying this more general condition. The study of these operators, however, is more complicated than that of the singular

[1] In the one-dimensional case the integral $\int_{0<\varepsilon \leqq |t|} [f(x - t)/t]\, dt$ is well defined since, for $f \in L^p(-\infty, \infty)$, $1 < p < \infty$, the integrand is the product of a translate of f (which must also belong to L^p) and the L^q function, $(1/p) + (1/q) = 1$, whose values are $1/t$ when $|t| \geqq \varepsilon$ and is 0 otherwise. It is not immediately clear, however, that, if instead of (1.8), we considered the integrals $\int_{|t| \geqq \varepsilon > 0} f(x - t)K(t)\, dt$ we would then have a well-defined object. It can be shown that this integral has meaning, but we choose not to do this here (see (5.5) for details about this matter and related problems concerning the integrals considered in the next section). Observe that under the conditions imposed below, the integral (1.8) converges absolutely for almost every x.

integral operators with odd kernels. In order to show that such an operator is of type (p, p), $1 < p < \infty$, for example, we have to assume more than integrability of Ω over Σ_{n-1}.

2. Singular Integral Operators with Odd Kernels

We suppose that K is the kernel defined by (1.7) and Ω is an odd function, homogeneous of degree 0 and integrable over Σ_{n-1}. We first observe that for each $\varepsilon > 0$ and $f \in L^p(E_n)$, $1 \leqq p < \infty$,

$$(2.1) \qquad \int_{\delta \geqq |t| \geqq \varepsilon} f(x-t)\frac{\Omega(t)}{|t|^n}\,dt = \frac{1}{2}\int_{\Sigma_{n-1}} \Omega(t')\left\{\int_{\delta \geqq |r| \geqq \varepsilon} \frac{f(x-rt')}{r}\,dr\right\} dt'.$$

To show this equality we express t in spherical coordinates $t = rt'$ with $|t| = r$ and $t' \in \Sigma_{n-1}$ and, thus, we have

$$\begin{aligned}\int_{\delta \geqq |t| \geqq \varepsilon} f(x-t)\frac{\Omega(t)}{|t|^n}\,dt &= \int_\varepsilon^\delta \left\{\int_{\Sigma_{n-1}} f(x-rt')\Omega(t')\,dt'\right\}\frac{dr}{r}\\ &= \int_{\Sigma_{n-1}} \Omega(t')\left\{\int_\varepsilon^\delta f(x-rt')\frac{dr}{r}\right\} dt'.\end{aligned}$$

On the other hand, since Ω is odd the last expression is equal to

$$\begin{aligned}\int_{\Sigma_{n-1}} \Omega(-t')\left\{-\int_\varepsilon^\delta f(x-rt')\frac{dr}{r}\right\} dt' &= \int_{\Sigma_{n-1}} \Omega(-t')\left\{\int_{-\delta}^{-\varepsilon} f(x+rt')\frac{dr}{r}\right\} dt'\\ &= \int_{\Sigma_{n-1}} \Omega(t')\left\{\int_{-\delta}^{-\varepsilon} f(x-rt')\frac{dr}{r}\right\} dt'.\end{aligned}$$

Thus,

$$\begin{aligned}2&\int_{\delta \geqq |t| \geqq \varepsilon} f(x-t)\frac{\Omega(t)}{|t|^n}\,dt\\ &= \int_{\delta \geqq |t| \geqq \varepsilon} f(x-t)\frac{\Omega(t)}{|t|^n}\,dt + \int_{\delta \geqq |t| \geqq \varepsilon} f(x-t)\frac{\Omega(t)}{|t|^n}\,dt\\ &= \int_{\Sigma_{n-1}} \Omega(t')\left\{\int_{-\delta}^{-\varepsilon} f(x-rt')\frac{dr}{r}\right\} dt' + \int_{\Sigma_{n-1}} \Omega(t')\left\{\int_\varepsilon^\delta f(x-rt')\frac{dr}{r}\right\} dt'\\ &= \int_{\Sigma_{n-1}} \Omega(t')\left\{\int_{\delta \geqq |r| \geqq \varepsilon} f(x-rt')\frac{dr}{r}\right\} dt'\end{aligned}$$

and (2.1) is established.

Let $\mathbf{1} = (1, 0, \ldots, 0) \in E_n$ be the first vector in the standard basis of E_n and, for $x = (x_1, x_2, \ldots, x_n) \in E_n$, put

$$(H^{(1)}_{\varepsilon,\delta} f)(x) = \int_{\delta \geqq |s| \geqq \varepsilon} f(x - s\mathbf{1}) \frac{ds}{s} = \int_{\delta \geqq |s| \geqq \varepsilon} f(x_1 - s, x_2, \ldots, x_n) \frac{ds}{s}.$$

If σ is a member of the rotation group $SO(n)$ acting on E_n we let R_σ be the operator on functions on E_n defined by

$$(R_\sigma f)(x) = f(\sigma x).$$

LEMMA 2.2. *If $f \in L^p(E_n)$, $1 \leqq p < \infty$, then*

$$\int_{\delta \geqq |t| \geqq \varepsilon} f(x - t) \frac{\Omega(t')}{|t|^n}\, dt = \frac{1}{2} \int_{SO(n)} (R_{\sigma^{-1}} H^{(1)}_{\varepsilon,\delta} R_\sigma f)(x) \Omega(\sigma\mathbf{1})\, d\sigma,$$

where $d\sigma$ is the element of Haar measure on the group $SO(n)$ so normalized that $\int_{SO(n)} d\sigma = |\Sigma_{n-1}|$ is the Lebesgue measure of Σ_{n-1}.

PROOF. By (4.14) of Chapter I we have

$$\int_{\Sigma_{n-1}} \left\{ \int_{\delta \geqq |r| \geqq \varepsilon} \frac{f(x - rt')}{r}\, dr \right\} \Omega(t')\, dt'$$
$$= \int_{SO(n)} \left\{ \int_{\delta \geqq |r| \geqq \varepsilon} \frac{f(x - r\sigma\mathbf{1})}{r}\, dr \right\} \Omega(\sigma\mathbf{1})\, d\sigma$$
$$= \int_{SO(n)} \left\{ \int_{\delta \geqq |r| \geqq \varepsilon} (R_\sigma f)(\sigma^{-1}x - r\mathbf{1}) \frac{dr}{r} \right\} \Omega(\sigma\mathbf{1})\, d\sigma$$
$$= \int_{SO(n)} (R_{\sigma^{-1}} H^{(1)}_{\varepsilon,\delta} R_\sigma f)(x) \Omega(\sigma\mathbf{1})\, d\sigma.$$

The lemma can now be seen to be an immediate consequence of equality (2.1).

An immediate consequence of this lemma is the fact that, for $f \in L^p(E_n)$, $1 \leqq p < \infty$,

$$(2.3) \qquad 2 \cdot \sup_{\substack{\varepsilon > 0 \\ \delta > 0}} \left| \int_{\delta \geqq |t| \geqq \varepsilon} f(x - t) \frac{\Omega(t')}{|t|^n}\, dt \right|$$
$$\leqq \int_{SO(n)} \sup_{\substack{\varepsilon > 0 \\ \delta > 0}} |(R_{\sigma^{-1}} H^{(1)}_{\varepsilon,\delta} R_\sigma f)(x) \Omega(\sigma\mathbf{1})|\, d\sigma.$$

Let $M^{(1)}_g$, for $g \in L^p(E_n)$, $1 < p \leqq \infty$, be defined by

$$M^{(1)}_g(x) = \sup_{r > 0} (2r)^{-1} \int_{|t| \leqq r} |g(x_1 - t, x_2, \ldots, x_n)|\, dt.$$

As was observed in the second chapter, it follows from Theorem 3.7 of that chapter that $M_g^{(1)} \in L^p(E_n)$ and

$$\|M_g^{(1)}\|_p \leqq b\|g\|_p,$$

where $b = b(p, n)$ depends on p and n, but not on g. But, by Theorem 1.4 we have[2]

$$\sup_{\substack{\varepsilon>0\\ \delta>0}} |(R_{\sigma^{-1}} H_{\varepsilon,\delta}^{(1)} R_\sigma f)(x)| \leqq 2\|\psi\|_1 M_{R_\sigma f}^{(1)}(\sigma^{-1}x) + 2\pi M_{H^{(1)} R_\sigma f}^{(1)}(\sigma^{-1}x).$$

Consequently, it follows from Theorem 1.4 and the fact that R_σ is an isometry when restricted to $L^p(E_n)$, that for $1 < p < \infty$

$$\text{(2.4)} \quad \left(\int_{E_n} \left\{\sup_{\substack{\varepsilon>0\\ \delta>0}} |(R_{\sigma^{-1}} H_{\varepsilon,\delta}^{(1)} R_\sigma f)(x)|\right\}^p dx\right)^{1/p}$$
$$\leqq 2(b\|\psi\|_1 + \pi B_p)\|f\|_p = C_p\|f\|_p.$$

Thus, an application of Minkowski's integral inequality, (2.3) and (2.4) give us

$$\text{(2.5)} \quad \left(\int_{E_n} \left\{\sup_{\substack{\varepsilon>0\\ \delta>0}} \left| \int_{\delta \geqq |t| \geqq \varepsilon} f(x - t) \frac{\Omega(t')}{|t|^n} dt \right|\right\}^p dx\right)^{1/p}$$
$$\leqq C_p\left(\int_{\Sigma_{n-1}} |\Omega(t')| \, dt'\right)\|f\|_p = A_p\|f\|_p.$$

We can therefore apply Theorem 3.12 of Chapter II and obtain the following result on the existence and boundedness of singular integral operators with odd kernels:

THEOREM 2.6. *If $f \in L^p(E_n)$, $1 < p < \infty$, then*

$$(Hf)(x) = \lim_{\delta^{-1},\varepsilon \to 0} \int_{\delta \geqq |t| \geqq \varepsilon > 0} f(x - t)K(t) \, dt$$

exists for almost every $x \in E_n$. Furthermore, there exists a constant $c = c(p, n)$ depending on p and the dimension n, but not on f, such that

$$\|Hf\|_p \leqq c\|f\|_p.$$

An important class of singular integral operators with odd kernels is the one consisting of the *M. Riesz transforms*. In n dimensions these are the n

[2] Since $|\int_{\delta \geqq |t| \geqq \varepsilon} f(x - t)(dt/t)| \leqq |\int_{|t| \geqq \varepsilon} f(x - t)(dt/t)| + |\int_{|t| \geqq \delta} f(x - t)(dt/t)|$ it follows from Theorem 1.4 that $\sup_{\varepsilon>0,\delta>0} |\int_{\delta \geqq |t| \geqq \varepsilon} f(x - t)(dt/t)| \leqq 2\|\psi\|_1 m_f(x) + \pi 2 m_{\tilde{f}}(x)$.

singular integral operators $R_1, R_2, \ldots, R_n$ defined by the kernels

$$K_j(x) = c_n(x'_j/|x|^n) = c_n(x_j/|x|^{n+1}),$$

for

$$x = (x_1, x_2, \ldots, x_n) \in E_n, j = 1, 2, \ldots, n \text{ and } c_n = \frac{\Gamma[(n+1)/2]}{\pi^{(n+1)/2}}$$

(when $n = 1$ we obtain only one transform, the Hilbert transform). By Theorem 4.5 of Chapter IV each such kernel induces a tempered distribution whose Fourier transform is the function whose value at $t = (t_1, t_2, \ldots, t_n)$ is

$$\hat{K}_j(t) = -i\frac{t_j}{|t|}.$$

Thus, if φ is a testing function, $(K_j * \varphi)\hat{}(t) = -i(t_j/|t|)\hat{\varphi}(t)$ (see Theorem 3.18 of Chapter I). It follows from Theorem 2.6, therefore, that

$$(R_j f)\hat{}(t) = -i\frac{t_j}{|t|}\hat{f}(t) \tag{2.7}$$

a.e. for any $f \in L^2(E_n)$. Moreover, since $R_j f$ must then also belong to $L^2(E_n)$, we have

$$(R_j^2 f)\hat{}(t) = -\frac{t_j^2}{|t|^2}\hat{f}(t).$$

Hence,

$$\left(\sum_{j=1}^{n} R_j^2 f\right)\hat{}(t) = -\hat{f}(t) \tag{2.8}$$

for all $f \in L^2(E_n)$. It follows, therefore, that

$$\sum_{j=1}^{n} R_j^2 = -I, \tag{2.9}$$

where I is the identity operator (on $L^p(E_n)$, $1 < p < \infty$).

The identity (2.7) and Plancherel's theorem also implies the following "unitary" character of the Riesz transforms

$$\sum_{j=1}^{n} \|R_j f\|_2^2 = \|f\|_2^2. \tag{2.10}$$

3. *Singular Integral Operators with Even Kernels*

We shall now study the singular integral operators associated with kernels of the form (1.7) with Ω satisfying $\int_{\Sigma_{n-1}} \Omega(t')\,dt' = 0$, in addition to being homogeneous of degree 0 and integrable over Σ_{n-1}. Let us express

Ω as the sum

$$\Omega(t') = \frac{\Omega(t') - \Omega(-t')}{2} + \frac{\Omega(t') + \Omega(-t')}{2} = \Omega^{(1)}(t') + \Omega^{(2)}(t'),$$

where $\Omega^{(1)}$ and $\Omega^{(2)}$ are the odd and even parts of Ω. Theorem 2.6 then applies to the operator associated with the kernel $\Omega^{(1)}(t)/|t|^n$ and we can conclude, in particular, that this operator maps $L^p(E_n)$ boundedly into itself. We mentioned at the end of the first section of this chapter, however, that we need to assume more about Ω in order to reach the same conclusion concerning the operator associated with the kernel $K(t) = \Omega(t)/|t|^n$. We will show that such an operator maps $L^p(E_n)$ continuously into itself, $1 < p < \infty$, provided $\|\Omega\| = (\int_{\Sigma_{n-1}} |\Omega(t')|^2 \, dt')^{1/2} < \infty$. More precisely, we will prove the following theorem:

THEOREM 3.1. *Suppose K is a kernel of the form $K(t) = \Omega(t)/|t|^n$, where $\|\Omega\| = (\int_{\Sigma_{n-1}} |\Omega(t')|^2 \, dt')^{1/2} < \infty$, $\int_{\Sigma_{n-1}} \Omega(t') \, dt' = 0$ and Ω is homogeneous of degree 0. Then $\lim_{\varepsilon \to 0} \int_{|t| \geqq \varepsilon > 0} \varphi(x - t)K(t) \, dt = (K * \varphi)(x)$ exists for each $x \in E_n$ and testing function $\varphi \in \mathscr{S}$. Moreover, if $1 < p < \infty$ there exists a constant $c = c(p, n)$, depending only on p and the dimension n, such that*

$$\|K * \varphi\|_p \leqq c \|\varphi\|_p$$

for all testing functions φ.

Since the condition $\|\Omega\| < \infty$ implies that this same property must hold for the even part $\Omega^{(2)}$ and, since we already know the above theorem to be valid for (more general) odd kernels, we can assume Ω to be even. Briefly, the idea of the proof of this theorem is then the following. If T is the singular integral operator associated with K we compose it with the Riesz transform R_j, $1 \leqq j \leqq n$, and show that this composition $R_j T$ is a singular integral operator with odd kernel of the type studied in the last section. Thus, there exists a constant a_p such that $\|R_j T\varphi\|_p \leqq a_p \|\varphi\|_p$ for all $\varphi \in \mathscr{S}$. Hence, by Theorem 2.6 $\|R_j(R_j T\varphi)\|_p \leqq b_p \|R_j T\varphi\|_p \leqq b_p a_p \|\varphi\|_p$. From (2.9), however, we have $\|T\varphi\|_p = \|(\sum_{j=1}^n R_j^2) T\varphi\|_p \leqq \sum_{j=1}^n \|R_j(R_j T\varphi)\|_p \leqq n b_p a_p \|\varphi\|_p = c\|\varphi\|_p$.

The rest of this section will be devoted to making this argument precise. A number of results needed for this proof have already been developed in the fourth section of Chapter IV. The existence of the limit

$$\lim_{\varepsilon \to 0} \int_{|t| \geqq \varepsilon} \varphi(x - t)K(t) \, dt$$

was shown in an argument preceding Theorem 4.7. By Theorem 4.7 of that chapter, the Fourier transform of K, regarded as a principal value distribution, is a function that is homogeneous of degree zero. Thus, there exists Ω_0, defined on Σ_{n-1}, such that $\hat{K}(x) = \Omega_0(x') = \Omega_0(x/|x|)$, $x \neq 0$. Moreover, if $\Omega = \sum_{k=1}^{\infty} Y^{(k)}$ is the development in spherical harmonics of Ω on Σ_{n-1}($Y^{(k)}$ is a spherical harmonic of degree k and the series is convergent in the norm of $L^2(\Sigma_{n-1})$), then Ω_0 has the development $\Omega_0 = \sum_{k=1}^{\infty} Y_0^{(k)}$, where $Y_0^{(k)} = \gamma_k Y^{(k)}$ with

$$\gamma_k = i^{-k}\pi^{n/2}\frac{\Gamma(k/2)}{\Gamma[(n+k)/2]} = O(k^{-(n/2)}). \tag{3.2}$$

By Corollary 4.12 of Chapter IV the function Ω_0 is continuous on Σ_{n-1} and, hence, bounded.[3] Consequently, the function $\hat{K}$ is also bounded. By Theorem 3.18 of Chapter I, therefore, the operator T assigning to each $\varphi \in \mathscr{S}$ the function $T\varphi = K * \varphi$ maps into $L^2(E_n)$, satisfies $\|T\varphi\|_2 = \|K * \varphi\|_2 \leqq \|\hat{K}\|_\infty \|\varphi\|_2$ and $(T\varphi)\hat{}(t) = \hat{K}(t)\hat{\varphi}(t) = \Omega_0(t/|t|)\hat{\varphi}(t)$ for $t \neq 0$ in E_n.

Since $T\varphi$ belongs to $L^2(E_n)$, the Riesz transform $R_jT\varphi$ of $T\varphi$ is well-defined (see Theorem 2.6) and belongs to $L^2(E_n)$. Using (2.7), therefore, we have

$$(R_jT\varphi)\hat{}(t) = -i\frac{t_j}{|t|}(T\varphi)\hat{}(t) = -i\frac{t_j}{|t|}\Omega_0\left(\frac{t}{|t|}\right)\hat{\varphi}(t). \tag{3.3}$$

Let $\omega_0(t') = -it_j'\Omega_0(t')$ for $t' = (t_1', t_2', \ldots, t_n') \in \Sigma_{n-1}$ and consider the function that assigns to $t \neq 0$ in E_n the value $\omega_0(t/|t|)$. We now show that this function is the Fourier transform of a principal value distribution defined by an odd kernel of the type discussed in Section 2. In order to do this we shall make use of the following result:

LEMMA 3.4. *Suppose $Y^{(k)}$, $k \geqq 1$, is a spherical harmonic of degree k and j is an integer satisfying $1 \leqq j \leqq n$. Then there exist spherical harmonics $W^{(k-1)}$ and $W^{(k+1)}$ of degrees $k-1$ and $k+1$, such that*

$$t_j Y^{(k)}(t) = W^{(k-1)}(t) + W^{(k+1)}(t)$$

for all $t = (t_1, t_2, \ldots, t_n) \in \Sigma_{n-1}$.

PROOF. By Corollaries 2.3 and 2.4 of Chapter IV it suffices to show that $\int_{\Sigma_{n-1}} t_j Y^{(k)}(t) W^{(l)}(t)\, dt = 0$ whenever $W^{(l)}$ is a spherical harmonic of degree $l \neq k-1, k+1$. Let $P^{(k)}$ and $Q^{(l)}$ be the corresponding solid spherical harmonics and put $u(x) = x_j P^{(k)}(x)$ and $v(x) = Q^{(l)}(x)$. Then $\Delta v \equiv 0$ and

$$(\Delta u)(x) = x_j(\Delta P^{(k)})(x) + 2\frac{\partial P^{(k)}}{\partial x_j}(x) = 2\frac{\partial P^{(k)}}{\partial x_j}(x).$$

[3] The continuity of Ω_0 was shown to be a consequence of the representation obtained in Theorem 4.11 of the fourth chapter.

Thus, by Green's theorem:

$$(3.5)\quad -2 \int_{|x|\leq 1} Q^{(l)}(x) \frac{\partial P^{(k)}}{\partial x_j}(x)\, dx = \int_{|x|\leq 1} [u(x)(\Delta v)(x) - v(x)(\Delta u)(x)]\, dx = \int_{\Sigma_{n-1}} [u(x') \frac{\partial v}{\partial n}(x') - v(x') \frac{\partial u}{\partial n}(x')]\, dx'.$$

But $\partial P^{(k)}/\partial x_j$ is a solid harmonic of degree $k - 1$; thus, if $l \neq k - 1$ the first integral in (3.5) must be 0. If we write $x = rt$ with $r = |x|$, then $Q^{(l)}(x) = r^l W^{(l)}(t)$ and $(\partial Q^{(l)}/\partial n)(x) = (\partial/\partial r)(r^l W^{(l)}(t)) = lr^{l-1}W^{(l)}(t) = lW^{(l)}(t)$ is the derivative of $Q^{(l)}$ in the direction of the outward directed normal to Σ_{n-1} at t. Similarly, the fact that $u(x) = r^{k+1}t_j Y^{(k)}(t)$ implies that $(\partial u/\partial n)(t) = (k + 1)t_j Y^{(k)}(t)$. Consequently, if $l \neq k - 1$,

$$0 = l \int_{\Sigma_{n-1}} t_j Y^{(k)}(t) W^{(l)}(t)\, dt - (k + 1) \int_{\Sigma_{n-1}} t_j Y^{(k)}(t) W^{(l)}(t)\, dt.$$

We can conclude, therefore, that if, in addition, $l \neq k + 1$ then

$$\int_{\Sigma_{n-1}} t_j Y^{(k)}(t) W^{(l)}(t)\, dt = 0$$

and the lemma is proved.

Since Ω was assumed to belong to $L^2(\Sigma_{n-1})$ the spherical harmonics $\{Y^{(k)}\}$ representing Ω must satisfy $\sum_{k=1}^{\infty} \|Y^{(k)}\|^2 = \|\Omega\|^2 < \infty$. Because of (3.2), therefore,

$$(3.6)\qquad \sum_{k=1}^{\infty} k^n \|Y_0^{(k)}\|^2 = \sum_{k=1}^{\infty} k^n |\gamma_k|^2 \|Y^{(k)}\|^2 < \infty.$$

Let $\sum_{k=1}^{\infty} y_0^{(k)}$ be the development in spherical harmonics of the function ω_0 restricted to Σ_{n-1}. (Since ω_0 is the product of an odd function with an even function, it must be odd. Thus, $\int_{\Sigma_{n-1}} \omega_0 = 0$ and the development $\sum_{k=1}^{\infty} y_0^{(k)}$ does not involve a harmonic of degree 0). It follows immediately from Lemma 3.4 and inequality (3.6) that

$$\sum_{k=1}^{\infty} k^n \|y_0^{(k)}\|^2 < \infty.$$

By Theorem 4.7 of Chapter IV ω_0 is the Fourier transform of a principal value distribution having the form $J(x) = \omega(x')/|x|^n$ where $\omega \in L^2(\Sigma_{n-1})$

and satisfies $\int_{\Sigma_{n-1}} \omega(x')\,dx' = 0$. Actually, this last equality is a consequence of the fact that ω is odd (since the Fourier transform of J is an odd function). Thus, by Theorem 2.6, if $1 < p < \infty$, there exists a constant A_p, independent of $\varphi \in \mathcal{S}$ such that

$$\|J * \varphi\|_p \leqq A_p\|\varphi\|_p.$$

Moreover, since $\varphi \in L^2(E_n)$ using Theorem 3.18 of Chapter I we see that $(J * \varphi)\hat{}(t) = -i\,(t_j/|t|)\Omega_0(t/|t|)\hat{\varphi}(t)$. Since $J * \varphi \in L^p(E_n)$, $1 < p < \infty$, we can apply Theorem 2.6 again and obtain the fact that $[R_j * (J * \varphi)](x) = \lim_{\varepsilon \to 0} \int_{|t| \geqq \varepsilon} (J * \varphi)(x - t)K(t)\,dt$ exists for almost every $x \in E_n$ and

$$\|R_j * (J * \varphi)\|_p \leqq B_p\|\varphi\|_p, \tag{3.7}$$

where B_p is independent of φ. Since, in particular, $J * \varphi \in L^2(E_n)$, approximating $J * \varphi$ by functions in $\mathcal{S}$ and using Theorem 3.18 of Chapter I another time we obtain

$$[R_j * (J * \varphi)]\hat{}(t) = -\frac{t_j^2}{|t|^2}\,\Omega_0\left(\frac{t}{|t|}\right)\hat{\varphi}(t).$$

Hence,

$$\sum_{j=1}^{n} [R_j * (J * \varphi)]\hat{}(t) = -\sum_{j=1}^{n} \frac{t_j^2}{|t|^2}\,\Omega_0\left(\frac{t}{|t|}\right)\hat{\varphi}(t)$$
$$= -\Omega_0\left(\frac{t}{|t|}\right)\hat{\varphi}(t) = -(K * \varphi)\hat{}(t).$$

This shows that $\sum_{j=1}^{n} R_j * (J * \varphi) = -K * \varphi$ and, by (3.7),

$$\|K * \varphi\|_p \leqq \sum_{j=1}^{n} \|R_j * (J * \varphi)\|_p \leqq nB_p\|\varphi\|_p.$$

Theorem 3.1 is thus proved with $c = c(p, n) = nB_p$.

4. $\mathbf{H}^p$ *Spaces of Conjugate Harmonic Functions*

Toward the end of the first section of the third chapter we introduced spaces of certain vector valued functions

$$F(x, y) = (u_1(x, y), \ldots, u_{n+1}(x, y))$$

defined at all points $(x, y) \in E_{n+1}^{+}$, where u_j, $j = 1, 2, \ldots, n + 1$, satisfied the system of partial differential equations (1.6) of Chapter III and the inequalities

$$\int_{E_n} |u_j(x, y)|^p\,dx \leqq A < \infty \qquad \text{for all } y > 0. \tag{4.1}$$

When $p > 1$ the existence of boundary values $\lim_{y \to 0} u_j(x, y) = u_j(\dot{x}, 0)$

(both almost everywhere and in the norm) is an immediate consequence of Theorems 2.5 and 3.16 of Chapter II, and this existence depends only on condition (4.1). In fact, the latter theorem assures us that nontangential boundary values exist almost everywhere. Moreover, part (b) of Theorem 2.5 can be used to show this is the case also when $p = 1$. If we only require, however, that the harmonic functions satisfy condition (4.1) and not the system of equations (1.6), the convergence in the norm may fail. For example, if one of the harmonic functions $u(x, y) = u_j(x, y)$ is the Poisson–Stieltjes integral of the Dirac-δ measure μ (i.e., $\mu(A) = 0$ if $0 \notin A \subset E_n$ and $\mu(A), = 1$ if $0 \in A$) then $u(x, y)$ is the Poisson kernel

$$P(x, y) = c_n \frac{y}{(|x|^2 + y^2)^{(n+1)/2}}$$

and, therefore, $u(x, y)$ tends to 0 as (x, y) tends nontangentially to $(x_0, 0) = x_0 \in E_n$ as long as $x_0 \neq 0$. On the other hand,

$$\lim_{y \to 0} \int_{E_n} |u(x, y) - u(x, 0)|\, dx = \lim_{y \to 0} \int_{E_n} P(x, y)\, dx = 1.$$

When $n = 1$, this situation illustrates an important difference in the boundary behavior of harmonic and analytic functions. If u_1 and u_2 are the real and imaginary parts of an analytic function in E_2^+ satisfying (4.1), then the boundary values in question exist whenever $p > 0$ (this is just the case $n = 1$ of Theorem 5.1 of Chapter III). The fact that the functions u_1 and u_2 are related by the Cauchy–Riemann equations allows one to obtain the existence of boundary values for the wider range of indices p. Equations (1.6) of Chapter III represent perhaps the most direct generalization of the Cauchy–Riemann equations. The basic property enjoyed by these systems of harmonic functions is the fact that there are indices $p < 1$ for which $(|u_1|^2 + |u_2|^2 + \cdots + |u_{n+1}|^2)^{p/2}$ is subharmonic (see Theorem 4.14 below). This extends the fact that $\log |F|$ is subharmonic when F is an analytic function of a complex variable (see Example (3) in Section 4 of Chapter II). The following result shows how this property is used in order to obtain the desired boundary values.

THEOREM 4.2. *Suppose $u_1, u_2, \ldots, u_k$ are real-valued harmonic functions defined in E_{n+1}^+ such that there exists a positive $p_0 < 1$ for which*

$$s = (u_1^2 + u_2^2 + \cdots + u_k^2)^{p_0/2}$$

is subharmonic. Let $F = (u_1, u_2, \ldots, u_k)$ and suppose, further, that inequality (4.1) *is satisfied for all $y > 0$ for some $p > p_0$; then the limits*

$$u_j(x, 0) = \lim_{(w,y) \to (x,0)} u_j(w, y), \tag{4.3}$$

where (w, y) *approaches* $(x, 0)$ *nontangentially, exist for almost every* $x \in E_n$. *Moreover,*

$$(4.4)\quad \lim_{y\to 0} \int_{E_n} |F(x, y) - F(x, 0)|^p \, dx$$

$$= \lim_{y\to 0} \int_{E_n} \left(\sum_{j=1}^{k} (u_j(x, y) - u_j(x, 0))^2\right)^{p/2} dx = 0.$$

PROOF. Let $q = p/p_0 > 1$. Then since (4.1) is satisfied,

$$\int_{E_n} [s(x, y)]^q \, dx = \int_{E_n} |F(x, y)|^p \, dx \leqq A < \infty$$

for all $y > 0$. Since, moreover, s is subharmonic we know from Theorem 4.6 of Chapter II that s has a least-harmonic majorant m which is the Poisson integral of a function $f \in L^q(E_n)$. By Theorem 3.16 of Chapter II, therefore, m has nontangential limits at almost every point of E_n. Thus, since $s \leqq m$, the function s must be nontangentially bounded at almost every point of E_n and the same must be true for $|u_j| \leqq s^{1/p_0}, j = 1, 2, 3, \ldots, k$. We can therefore apply Theorem 3.19 of Chapter II and obtain the existence of the nontangential limits (4.3). The convergence in the norm (4.4) is then a consequence of the existence of these limits and the dominated convergence theorem of Lebesgue since both m and f are dominated by the Hardy–Littlewood maximal function of f which belongs to $L^q(E_n)$ (see Theorems 3.10 and 3.7 of Chapter II). This proves the theorem.

When $p = p_0$ and, thus, $q = 1$ we still have a least-harmonic majorant m of s; however, we can only conclude that m is a Poisson–Stieltjes integral of a finite Borel measure on E_n (see the last part of Theorem 4.6 in Chapter II). Consequently, we cannot apply the dominated convergence theorem as we did above in order to obtain the convergence (4.4) from the existence of the nontangential limits (4.3). On the other hand, these non-tangential limits do exist almost everywhere because the Poisson–Stieltjes integral m has such limits[4] and, thus, s must be nontangentially bounded.

We see from this theorem, therefore, that we do obtain a meaningful extension of the classical H^p space theory whenever the k components of F are so related that there exists a $p_0 < 1$ for which $s = |F|^{p_0}$ is subharmonic. When $n = 2 = k$ we have already observed that if these components satisfy the Cauchy–Riemann equations then $|F|^p$ is subharmonic

[4] If m is the Poisson–Stieltjes integral of the finite Borel measure μ and we introduce the maximal function $M_\mu(x) = \sup_{S_x} (1/|S_x|) \int_{S_x} d\mu(t)$, then it is easily checked that the arguments used in Section 3 of Chapter II can be used to establish the almost everywhere convergence of m if we modify them appropriately by introducing M_μ.

for all $p > 0$ (in fact $\log |F|$ is subharmonic). It is natural, therefore, to consider the problem of determining which systems of linear first order partial differential equations with constant coefficients have the property that their solutions $F = (u_1, u_2, \ldots, u_k)$ are harmonic and have such a subharmonicity property. Now, any such system of partial differential equations can be written in the form

$$\sum_{j=1}^{n} A_j \frac{\partial F}{\partial x_j} = 0, \tag{4.5}$$

where A_j is an $l \times k$ constant matrix and $\partial F/\partial x_j$ is the (column) vector having components $\partial u_i/\partial x_j$, $i = 1, 2, \ldots, k$.[5]

We shall call such a system of equations a *Generalized Cauchy Riemann* (GCR) *System* if each solution $F = (u_1, u_2, \ldots, u_k)$ has harmonic components u_i, $i = 1, 2, \ldots, k$. Such a solution will be called a system of *conjugate harmonic functions*. We shall show that when $n = 2 = k = l$ a linear change reduces such a system to the ordinary Cauchy–Riemann equations. Before doing this, however, we make some general observations.

We say that the system (4.5) is *elliptic* provided

$$\sum_{j=1}^{n} \lambda_j A_j v = 0 \tag{4.6}$$

for a k-dimensional (column) vector v and an n-tuple $\lambda = (\lambda_1, \lambda_2, \ldots, \lambda_n)$ only if either v or λ is zero. We claim that every GCR system is elliptic. If this were not the case we could find nonzero v and λ satisfying (4.6). But then $F(x) = \{\exp \sum_{j=1}^{n} \lambda_j x_j\} v$ is a nonharmonic solution of (4.5); hence, (4.5) could not be a GCR system.

If the system under consideration is elliptic and $\lambda \neq 0$ then the mapping $v \to (\sum_{j=1}^{n} \lambda_j A_j) v$ from k-dimensional Euclidean space into l dimensional Euclidean space must clearly be one to one (otherwise a nonzero vector v would be mapped into the zero vector, contradicting ellipticity). It follows, therefore, that $l \geqq k$. By choosing λ so that $\lambda_j = 1$ and $\lambda_i = 0$, if $i \neq j$, we see that each of the matrices A_j, $j = 1, 2, \ldots, n$, represents a one to one mapping. In particular, if $k = l$ the matrices must be invertible. If, in addition, $n = 2 = k$ the system (4.5) is equivalent to a system of the form

$$\frac{\partial F}{\partial x_1} + A \frac{\partial F}{\partial x_2} = 0, \tag{4.7}$$

[5] When we first introduced a system F satisfying Eqs. (1.6) we assumed that it was a function of the $n + 1$ variables $(x_1, x_2, \ldots, x_n, y)$, where $y > 0$ and $x = (x_1, \ldots, x_n)$ was thought of as a point in E_n. The variable y plays a special role as is evidenced by inequality (4.1) which is clearly needed for extending the notion of H^p spaces. On the other hand, the property that $|F|^p$ is subharmonic for some $p > 0$ does not involve such a distinguished variable. It is for this reason that we introduce these systems for functions in a domain in E_n and make no distinctions between the variables.

where A is a nonsingular matrix (since we can multiply both sides of (4.5), in this case, by A_1^{-1}). Let us now suppose that (4.7) is a GCR system. Then it follows from the harmonicity of F that

$$(A + A^{-1})\frac{\partial^2 F}{\partial x_1 \partial x_2} = 0 \tag{4.8}$$

for any solution F. But, given any two-dimensional vector b, there always exists a solution of (4.7) such that $\partial^2 F/\partial x_1 \partial x_2 \equiv b$. Indeed, $F(x_1, x_2) = 2x_1x_2b - (x_1^2Ab + x_2^2A^{-1}b)$ is such a solution. Thus, since (4.8) must be satisfied we can conclude that $A + A^{-1} = 0$; or equivalently $A^2 = -I$. But this equation implies that A must be of the form

$$\begin{pmatrix} a & b \\ c & -a \end{pmatrix}$$

with $a^2 + bc = -1$. This last equation implies that $bc < 0$; thus, the matrix

$$Q = \begin{pmatrix} 1 & a \\ 0 & c \end{pmatrix}$$

has an inverse and satisfies

$$Q^{-1}AQ = \begin{pmatrix} 0 & -1 \\ 1 & 0 \end{pmatrix}.$$

Hence, if we define G by letting

$$\begin{pmatrix} v_1 \\ v_2 \end{pmatrix} = G = Q^{-1}F = Q^{-1}\begin{pmatrix} u_1 \\ u_2 \end{pmatrix},$$

equation (4.7) is equivalent to

$$\frac{\partial G}{\partial x_1} + \begin{pmatrix} 0 & -1 \\ 1 & 0 \end{pmatrix}\frac{\partial G}{\partial x_2} = Q^{-1}\left[\frac{\partial F}{\partial x_1} + A\frac{\partial F}{\partial x_2}\right] = 0.$$

But these are the classical Cauchy–Riemann equations for v_1 and v_2.

We now introduce a function space associated with any fixed system of generalized Cauchy–Riemann equations. If $p > 0$ and $F = (u_1, u_2, \ldots, u_k)$ is a solution of this system in the domain E_{n+1}^+, we say that F belongs to $\mathbf{H}^p(E_{n+1}^+)$ if there exists $A < \infty$ such that

$$\int_{E_n} |F(x, y)|^p \, dx = \int_{E_n} (|u_1(x, y)|^2 + \cdots + |u_k(x, y)|^2)^{p/2} \, dx \leqq A$$

for all $y > 0$. In view of the above comments, the spaces $\mathbf{H}^p(E_{n+1}^+)$ clearly extend the classical H^p spaces of analytic functions in the upper half-plane. The following result, together with Theorem 4.2, shows that this definition allows us to develop a meaningful theory of these spaces for indices $p < 1$.

THEOREM 4.9. *Suppose F is a solution of the Generalized Cauchy–Riemann system*

$$\sum_{j=1}^{n} A_j \frac{\partial F}{\partial x_j} = 0,$$

then $|F|^p$ is subharmonic for $p \geqq 2 - (1/\alpha)$, where α is a positive number less than 1 depending only on the matrices $A_1, \ldots, A_n$.

PROOF. By Theorem 4.4 of Chapter II it suffices to show that $s = |F|^p$ has a nonnegative Laplacian on the set of all x in the domain of F such that $s(x) > 0$. Since F has harmonic components we have, on this set,

$$\Delta s = p|F|^{p-2}\left\{(p-2)|F|^{-2} \sum_{j=1}^{n} (F \cdot F_j)^2 + \sum_{j=1}^{n} |F_j|^2\right\},$$

where $F_j = \partial F/\partial x_j$ and $F \cdot F_j = \sum_{i=1}^{k} u_i(\partial u_i/\partial x_j)$. Thus, the condition that $\Delta s \geqq 0$ is equivalent to the inequality

$$\sum_{j=1}^{n} (F \cdot F_j)^2 \leqq \frac{1}{2-p} |F|^2 \sum_{j=1}^{n} |F_j|^2. \tag{4.10}$$ [6]

We now claim that there exists a positive $\alpha < 1$, depending only on $A_1, \ldots, A_n$, such that

$$\max_{|v|=1} \sum_{j=1}^{n} (u^{(j)} \cdot v)^2 \leqq \alpha \sum_{j=1}^{n} |u^{(j)}|^2 \tag{4.11}$$

whenever $u^{(1)}, \ldots, u^{(n)}$ are n k-tuples satisfying

$$\sum_{j=1}^{n} A_j u^{(j)} = 0. \tag{4.12}$$

This would certainly imply (4.10). In order to establish this claim we can clearly assume that $\sum_{j=1}^{n} |u^{(j)}|^2 = 1$. If there does not exist $\alpha < 1$ for which (4.11) is valid, by compactness there exists v and $u^{(1)}, \ldots, u^{(n)}$ satisfying (4.12), as well as $|v|^2 = 1 = \sum_{j=1}^{n} |u^{(j)}|^2$, for which $\sum_{j=1}^{n} (u^{(j)} \cdot v)^2 = 1$. By the Cauchy–Schwarz inequality, however,

$$\sum_{j=1}^{n} (u^{(j)} \cdot v)^2 \leqq \sum_{j=1}^{n} |u^{(j)}|^2 |v|^2 = \sum_{j=1}^{n} |u^{(j)}|^2 = 1.$$

Hence, since $(u^{(j)} \cdot v)^2 \leqq |u^{(j)}|^2 |v|^2$, we must have $(u^{(j)} \cdot v)^2 = |u^{(j)}|^2 |v|^2$ for

[6] We are assuming, in deriving (4.10), that $p < 2$; but, if $p \geqq 2$ the fact that $\Delta s \geqq 0$ is obvious.

$j = 1, 2, \ldots, n$. Thus, by the equality case of the Cauchy–Schwarz inequality, there exists $\lambda = (\lambda_1, \lambda_2, \ldots, \lambda_n)$ such that $u^{(j)} = \lambda_j v$. But this would contradict the ellipticity condition (4.6) since we would then have

$$0 = \sum_{j=1}^{n} A_j u^{(j)} = \sum_{j=1}^{n} \lambda_j A_j v$$

with $|\lambda|^2 = (\lambda_1^2 + \cdots + \lambda_n^2) = 1 = |v|^2$. Thus, such a number α does exist and the theorem is proved.

The smallest number α for which Theorem 4.9 holds clearly depends on the GCR system that F satisfies. In many important cases this number can be calculated. For example, if we consider the *M. Riesz system* of partial differential equations

$$\begin{aligned} &\frac{\partial u_1}{\partial x_1} + \frac{\partial u_2}{\partial x_2} + \cdots + \frac{\partial u_n}{\partial x_n} = 0 \\ &\frac{\partial u_i}{\partial x_j} = \frac{\partial u_j}{\partial x_i}, \qquad i, j = 1, 2, \ldots, n, \end{aligned} \tag{4.13}$$

which was introduced in Chapter III (see (1.6)), $\alpha = (n - 1)/n$. More precisely, we show the following result:

THEOREM 4.14. *If $F = (u_1, u_2, \ldots, u_n)$ is a solution of the M. Riesz system* (4.13) *then $|F|^p$ is subharmonic for $p \geqq (n - 2)/(n - 1)$. Moreover, there exists such a solution F for which $|F|^p$ is not subharmonic if* $0 < p < (n - 2)/(n - 1)$.

PROOF. From the proof of Theorem 4.9 we see that it suffices to find the smallest α satisfying (4.11) whenever $u^{(j)} = (u_{j1}, u_{j2}, \ldots, u_{jn})$, $j = 1, 2, \ldots, n$, are n-tuples satisfying

$$\sum_{j=1}^{n} u_{jj} = 0 \qquad \text{and} \qquad u_{ij} = u_{ji}, \qquad i, j = 1, 2, \ldots, n.$$

If we let $\mathscr{M}$ denote the $n \times n$ matrix with coefficients u_{ij} these last conditions are equivalent to the properties that $\mathscr{M}$ is symmetric and has trace 0. Inequality (4.11) can then be written in the form

$$\max_{|v|=1} |\mathscr{M}v|^2 \leqq \alpha |||\mathscr{M}|||^2, \tag{4.15}$$

where $|||\mathscr{M}||| = (\sum_{i,j=1}^{n} u_{ij}^2)^{1/2}$ is the *Hilbert–Schmidt norm* of the matrix $\mathscr{M}$. The left side of (4.15), of course, is the square of the norm of the operator on E_n mapping v into $\mathscr{M}v$. But we know from elementary linear algebra that if $\lambda_1, \lambda_2, \ldots, \lambda_n$ are the characteristic roots of $\mathscr{M}$ (which must all be real since $\mathscr{M}$ is symmetric) then this operator norm is $\|\mathscr{M}\| = \max\{|\lambda_1|, |\lambda_2|, \ldots, |\lambda_n|\}$ while $|||\mathscr{M}||| = (\lambda_1^2 + \lambda_2^2 + \cdots + \lambda_n^2)^{1/2}$. Since trace

$\mathcal{M} = 0$ we must have, moreover, $\lambda_1 + \lambda_2 + \cdots + \lambda_n = 0$. Thus, if, say, $|\lambda_j| = \max\{|\lambda_1|, |\lambda_2|, \ldots, |\lambda_n|\}$ we have

$$\|\mathcal{M}\|^2 = |\lambda_j|^2 = |-\sum_{i \neq j} \lambda_i|^2 \leqq (n-1) \sum_{i \neq j} \lambda_i^2.$$

Adding $(n-1)|\lambda_j|^2$ to both sides of this inequality we obtain

$$n\|\mathcal{M}\|^2 = n|\lambda_j|^2 \leqq (n-1) \sum_{i=1}^{n} \lambda_i^2 = (n-1)|||\mathcal{M}|||^2$$

and we obtain (4.15) with $\alpha = (n-1)/n$. From Theorem 4.9 we then can conclude that $|F|^p$ is subharmonic for $p \geqq 2 - (1/\alpha) = (n-2)/(n-1)$ whenever F is a solution of the M. Riesz system.

In order to show that, in general, $|F|^p$ is not subharmonic for exponents p less than $(n-2)/(n-1)$ let F be the gradient of the function h whose values are $|x|^{(2-n)}/(2-n)$. Then a simple calculation shows that

$$(4.16) \quad \Delta(|F|^p) = \Delta(|\nabla h|^p) = p(n-1)\{p(n-1) - (n-2)\}|x|^{p(1-n)-2}.$$

Consequently $\Delta(|F|^p) < 0$ if $p < (n-2)/(n-1)$. By theorem (4.4) of Chapter II, therefore, $-|F|^p$ is subharmonic in $E_n - \{0\}$. Thus, $-|F|^p$ satisfies the mean-value inequality (4.2) of that chapter. If $|F|^p$ were subharmonic as well it would also satisfy this mean-value inequality and, therefore, must satisfy the mean-value property. By theorem (1.7) of Chapter II, therefore, $|F|^p$ must be harmonic, but, by equality (4.16), it clearly cannot be harmonic for such a value of p.

One can obtain a large class of systems of conjugate harmonic functions satisfying the M. Riesz equations (4.13) in the following way. Let f be in $L^p(E_n)$, $1 < p < \infty$, and consider the Poisson integral

$$u(x, y) = c_n \int_{E_n} f(t) \frac{y}{(|x-t|^2 + y^2)^{(n+1)/2}}\, dt$$

together with the convolutions

$$v_j(x, y) = c_n \int_{E_n} f(t) \frac{(x_j - t_j)}{(|x-t|^2 + y^2)^{(n+1)/2}}\, dt,$$

$j = 1, 2 \ldots, n$, $x \in E_n$ and $y > 0$. By calculating the various partial derivatives involved we can easily check that $F = (u, V) = (u, v_1, v_2, \ldots, v_n)$ is an M. Riesz system defined in E_{n+1}^+. When $n = 1$ then $v = v_1$ is the function we used originally to define the Hilbert transform $\tilde{f}$ of f (see Section 1). Some of the basic properties of v and $\tilde{f}$, easily deduced from our development of these functions, are the following: $F = u + iv$ is analytic in E_2^+ and $\sup_{y>0} \int_{-\infty}^{\infty} |F(x+iy)|^p\, dx < \infty$; thus, $F \in H^p(E_2^+)$.

Consequently, $f + i\tilde{f}$ is the almost everywhere nontangential limit of F and, also, the limit in the L^p norm of F (i.e.,

$$\lim_{y\to 0} \int_{-\infty}^{\infty} |u(x, y) + iv(x, y) - f(x) - i\tilde{f}(x)|^p \, dx = 0).$$

These properties as well as the other results developed in the first section extend easily to n dimensions and, in this sense, we may consider the M. Riesz transforms to be, among the various singular integral operators, the most natural extensions of the Hilbert transform. We now finish this section by showing how some of these extensions to higher dimensions are obtained.

THEOREM 4.17. *Suppose* $f \in L^p(E_n)$, $1 < p < \infty$, $\tilde{f}_j = R_j f$, $j = 1, 2, \ldots, n$, *are the M. Riesz transforms of* f, $u(x, y)$ *is the Poisson integral of* f *and* $v_j(x, y)$ *is the convolution of* f *with the kernel* $c_n t_j/(|t|^2 + y^2)^{(n+1)/2}$. *Then*

(i) *The function* $F = (u, v_1, \ldots, v_n) \in \mathbf{H}^p(E_{n+1}^+)$;

(ii) $$\int_{E_n} f(x - t) \frac{t_j}{(|t|^2 + y^2)^{(n+1)/2}} \, dt = \int_{E_n} \tilde{f}_j(x - t) \frac{y}{(|t|^2 + y^2)^{(n+1)/2}} \, dt$$

for $j = 1, 2, \ldots, n$;

(iii) $$\lim_{\varepsilon\to 0} \left\{ \int_{E_n} f(x - t) \frac{t_j}{(|t|^2 + \varepsilon^2)^{(n+1)/2}} \, dt - \int_{0<\varepsilon\leqq|t|} f(x - t) \frac{t_j}{|t|^{n+1}} \, dt \right\} = 0$$

at each point of the Lebesgue set of f;

(iv) $(\int_{E_n} \{\sup_{y>0} |v_j(x, y)|\}^p \, dx)^{1/p} \leqq A\|f\|_p$, *where* A *is a constant depending only on the dimension* n *and the index* p.

PROOF. All these facts are either easy consequences of theorems we have already established or can be obtained by modifying in an obvious way the proof of the corresponding one-dimensional result. For example, in order to show (iii) we can use essentially the same argument that we used in order to prove Lemma 1.2. Instead of the function defined in (1.3) we introduce

$$\varphi(t) = \begin{cases} \dfrac{t_j}{(1 + |t|^2)^{(n+1)/2}} - \dfrac{t_j}{|t|^{n+1}} & \text{if } |t| \geqq 1 \\[2ex] \dfrac{t_j}{(1 + |t|^2)^{(n+1)/2}} & \text{if } |t| < 1 \end{cases}$$

and, for $\varepsilon > 0$, $\varphi_\varepsilon(t) = \varepsilon^{-n}\varphi(t/\varepsilon)$. Then the expression within brackets in (iii) is simply $\int_{E_n} f(x-t)\varphi_\varepsilon(t)\,dt$. The fact that

$$\psi(x) = \sup_{|t| \geqq |x|} |\varphi(t)| \leqq \begin{cases} \dfrac{\text{const}}{|x|^n(1+|x|^2)} & \text{if } |x| \geqq 1 \\ \text{const} & \text{if } |x| < 1 \end{cases}$$

implies that $\psi \in L^1(E_n)$ and this allows us to apply Theorem 1.25 of Chapter I to obtain

$$\lim_{\varepsilon \to 0} \int_{E_n} f(x-t)\varphi_\varepsilon(t)\,dt = f(x) \int_{E_n} \varphi(t)\,dt = 0$$

and (iii) is established.

If $\varepsilon > 0$ we claim that

$$(4.18) \qquad \frac{x_j}{(|x|^2 + (y+\varepsilon)^2)^{(n+1)/2}} = c_n \int_{E_n} \frac{y}{(|x-t|^2 + y^2)^{(n+1)/2}} \frac{t_j}{(|t|^2 + \varepsilon^2)^{(n+1)/2}}\,dt$$

for $(x, y) \in E_{n+1}^+$. This is a direct application of Lemma 2.7 of Chapter II, with

$$u(x, y) = \frac{x_j}{(|x|^2 + y^2)^{(n+1)/2}},$$

$y_2 = y$, and $y_1 = \varepsilon$.

It now follows immediately from equality (4.18) that the j-th M. Riesz transform of the function P_y having values

$$P_y(x) = c_n \frac{y}{(|x|^2 + y^2)^{(n+1)/2}}$$

(i.e., the Poisson kernel) is the j-th *Conjugate Poisson kernel*

$$Q_y^{(j)}(x) = c_n \frac{x_j}{(|x|^2 + y^2)^{(n+1)/2}}.$$

Property (ii) now follows easily from this since the Fourier transforms of both sides of the equality must then be the function whose value at $x \in E_n$ is

$$-ie^{-2\pi|x|y}\hat{f}(x)\frac{x_j}{|x|}$$

(see Theorem 1.14 of Chapter I and (2.7) in this chapter). Property (i) is a consequence of (ii), inequality (2.2) of Chapter II and Theorem 2.6. From this last theorem, equality (ii), and Theorems 3.10 and 3.7 of Chapter II we then obtain (iv).

5. Further Results

5.1. In the classical theory of Fourier series the analog of the Hilbert transform is the *conjugate function* of a function f defined on the unit circle $\{z \in \mathbf{C} : |z| = 1\}$. As was the case with the Hilbert transform, the conjugate function $\tilde{f}$ of f can be defined by using methods of the theory of functions of a complex variable or by using pure real variable methods. If we consider f to be a real valued integrable function of the variable $\theta \in [0, 2\pi]$ and let, for $0 < r < 1$,

$$Q(r, \theta) = \frac{1}{\pi} \frac{r \sin \theta}{1 - 2r \cos \theta + r^2}$$

be the *conjugate Poisson kernel*, then the function

$$v(z) = v(re^{i\theta}) = \int_{-\pi}^{\pi} f(\theta - \phi) Q(r, \phi)\, d\phi$$

is a harmonic conjugate to the Poisson integral

$$u(z) = u(re^{i\theta}) = \int_{-\pi}^{\pi} f(\theta - \phi) P(r, \phi)\, d\phi$$

$$= \int_{-\pi}^{\pi} f(\theta - \phi) \frac{1}{2\pi} \frac{1 - r^2}{1 - 2r \cos \phi + r^2}\, d\phi.$$

That is, $F(z) = u(z) + iv(z)$ is analytic in the interior of the unit disk $\{z \in \mathbf{C} : |z| < 1\}$. By making some simple changes in the proof of Lemma 2.6 in Chapter V, we can show that $F(z)$ tends to a limit as z approaches a boundary point $e^{i\theta}$ nontangentially, for almost all θ in $[0, 2\pi]$ (for details see Zygmund [1], Chapter VII). The imaginary part of F, therefore, must also have nontangential boundary values almost everywhere. The function f^c having these values is the conjugate function of f. That is, f^c is defined almost everywhere by the equality

$$f^c(\theta) = \lim_{r \to 1} \int_{-\pi}^{\pi} f(\theta - \phi) Q(r, \phi)\, d\phi.$$

A simple modification of the proof of Lemma 1.2 shows that f^c can also be defined almost everywhere by the equality

$$f^c(\theta) = \frac{1}{2\pi} \lim_{\varepsilon \to 0} \int_{0 < \varepsilon \leqq |\phi| \leqq \pi} f(\theta - \phi) \frac{\sin \phi}{1 - \cos \phi}\, d\phi.$$

If f belongs to $L^2(0, 2\pi)$ and its Fourier series is $\sum_{k=-\infty}^{\infty} c_k e^{ik\theta}$ an argument similar to that used for establishing (2.7) in Chapter V shows that f^c also belongs to $L^2(0, 2\pi)$, its Fourier series is $\sum_{k=-\infty}^{\infty} (-i \operatorname{sgn} k) c_k e^{ik\theta}$, and, hence, $\|f^c\|_2 \leqq \|f\|_2$. We would expect, therefore, that an inequality analogous to (2.11) in Chapter V holds for the conjugate function. In fact, it can be shown that the mapping $f \to f^c$ is of weak type (1, 1) and, thus, with the aid of the Marcinkiewicz interpolation theorem we obtain the inequality $\|f^c\|_p \leqq A_p \|f\|_p$ for $1 < p < \infty$, where A_p depends on p but not on $f \in L^p(0, 2\pi)$. By Theorem 3.15 of Chapter V, however, it suffices to show that this mapping is of restricted weak type (1, 1). This can be done in a very simple way with the following argument:

Let E be a measurable subset of $[0, 2\pi]$ and χ_E its characteristic function. If $u(z)$ and $v(z)$ denote the Poisson integral and the conjugate Poisson integral of χ_E the function $F(z) = u(z) + iv(z)$ is analytic for $|z| < 1$. Thus, for each real number y, $\exp\{yF(z)\}$ is analytic for $|z| < 1$ and, by the mean value theorem,

$$e^{y|E|/2\pi} = \frac{1}{2\pi} \int_0^{2\pi} \exp\{yF(re^{i\theta})\}\, d\theta$$

for $0 \leqq r < 1$. Letting r tend to 1 and applying the dominated convergence theorem of Lebesgue, therefore, we have

$$\frac{1}{2\pi} \int_0^{2\pi} \exp\{y[\chi_E(\theta) + i\chi_E^c(\theta)]\}\, d\theta = e^{y|E|/2\pi}$$

for all real numbers y. Letting $E' = [0, 2\pi] - E$ we may write this equation in the form

$$2\pi e^{y|E|/2\pi} = e^y \int_E \exp\{iy\chi_E^c(\theta)\}\, d\theta + \int_{E'} \exp\{iy\chi_E^c(\theta)\}\, d\theta.$$

By taking complex conjugates of both sides of this last equation with y replaced by $-y$ we obtain

$$2\pi e^{-y|E|/2\pi} = e^{-y} \int_E \exp\{iy\chi_E^c(\theta)\}\, d\theta + \int_{E'} \exp\{iy\chi_E^c(\theta)\}\, d\theta$$

for all real numbers y. Thus, we can solve for $\int_E \exp\{iy\chi_E^c(\theta)\}\, d\theta$ and $\int_{E'} \exp\{iy\chi_E^c(\theta)\}\, d\theta$ and we see that $\int_0^{2\pi} \exp\{iy\chi_E^c(\theta)\}\, d\theta$ depends only on y and $|E|$. If we let $\eta(s) = |\{\theta \in [0, 2\pi] : \chi_E^c(\theta) > s\}|$ and $\nu(s) = |\{\theta \in [0, 2\pi] : \chi_E^c(\theta) < s\}|$ it follows that the integrals

$$\int_{-\infty}^{\infty} e^{iys}\, d\eta(s) \qquad \text{and} \qquad \int_{-\infty}^{\infty} e^{iys}\, d\nu(s)$$

are completely determined by y and $|E|$. Since measures are completely determined by their Fourier transforms, η and ν depend only on $|E|$. Since $\lambda(s) = |\{\theta : |\chi_E^c(\theta)| > s > 0\}| = \eta(s) + \nu(-s)$, the distribution function λ of χ_E^c (see Section 2 of Chapter V) also depends only on $|E|$. If we let χ be the characteristic function of the interval $[0, |E|]$, the conjugate function χ_E^c and its distribution function are particularly simple to calculate and we obtain

$$e^{i\lambda(s)/2} = \frac{\sinh(\pi s) + i \sin(|E|/2)}{\sinh(\pi s) - i \sin(|E|/2)}.$$

From this, we can easily derive the restricted weak type (1, 1) of the conjugate function operator.

5.2. If λ is the distribution function of the Hilbert transform of the characteristic function of the set $E \subset (-\infty, \infty)$ having finite measure it can be shown that

$$\lambda(s) = 2|E|/\sinh(\pi s)$$

for $s > 0$. This, of course, also implies that the Hilbert transform is of restricted weak type (1, 1). For the proof of this fact, as well as another proof of the functional relation at the end of (5.1) see Stein and G. Weiss [2]. For yet another proof of this last formula for $\lambda(s)$ see Calderón [1].

5.3. The equality in (5.2) can be rewritten in terms of the nonincreasing rearrangement of the Hilbert transform $\tilde{\chi}_E$ of χ_E. If we do this we obtain the formula

$$\tilde{\chi}_E^*(t) = \frac{1}{\pi} \sinh^{-1}(2|E|/t).$$

If

$$\tilde{f}(x) = \lim_{\substack{\varepsilon \to 0 \\ R \to \infty}} \int_{R > |t| > \varepsilon} \frac{\Omega(t)}{|t|^n} f(x - t)\, dt$$

is a singular integral operator of the type discussed in Section 2 we can extend the above relation in the following way. Suppose E is a measurable subset of E_n of finite measure, χ_E its characteristic function and $m(t) = (1/t) \int_0^t \tilde{\chi}_E^*(s)\, ds$ (see Section 3 of Chapter V), then

$$tm(t) \leqq \frac{\|\Omega\|}{2} \int_0^t \sinh^{-1}(2|E|/s)\, ds,$$

where $\|\Omega\| = \int_{\Sigma_{n-1}} |\Omega(x')|\, dx'$. (See O'Neill and G. Weiss [1].) This result, together with the interpolation Theorem 3.15 of Chapter V, gives us another proof of the fact that $f \to \tilde{f}$ is a bounded operator on $L^p(E_n)$ (Theorem 2.6).

5.4. We remarked in Section 3 that if the kernel $K(t) = \Omega(t)/|t|^n$ is even we needed to assume more than the integrability of Ω over Σ_{n-1} in order to obtain a bounded singular integral operator on $L^p(E_n)$, $1 < p < \infty$ (though integrability sufficed for the odd kernel case). In fact, we assumed that Ω was square integrable; this enabled us to use the (comparatively) simple argument given in Section 3. It can be shown, however, that Theorem 3.1 is still valid if we replace the square integrability of Ω by the less restrictive condition

$$(*) \qquad \int_{\Sigma_{n-1}} |\Omega(x') \log^+ |\Omega(x')|\,|\, dx' < \infty,$$

where $\log^+ s = \log s$ for $s \geqq 1$ and is 0 for $0 \leqq s < 1$ (see Calderón and Zygmund [4]).

5.5. Theorem 3.1 is less general than Theorem 2.6 in yet another way. While the latter assured us that the singular integral operator in question was defined (a.e.) for all $f \in L^p(E_n)$, in Theorem 3.1 we restricted the domain of the operator to $\mathscr{S}$. When $\varphi \in \mathscr{S}$ the integral

$$\int_{|t| \geqq \varepsilon} \varphi(x - t) \frac{\Omega(t)}{|t|^n}\, dt$$

is well-defined and, thus, the problems described in the footnote following definition (1.8) do not appear here. It is not necessary, however, to restrict ourselves to $\mathscr{S}$ since it can be shown that the integrals

$$\int_{|t| \geqq \varepsilon > 0} f(x - t) \frac{\Omega(t)}{|t|^n}\, dt$$

are absolutely convergent for almost every x, the limits, as $\varepsilon \to 0$, exist in the mean of order p, $1 < p < \infty$, and pointwise almost everywhere whenever $f \in L^p(E_n)$, even when the function Ω satisfies the weaker condition (*) of (5.4) (see Calderón and Zygmund [4]).

5.6. Simple arguments show that the Riesz transforms $R_1, R_2, \ldots, R_n$ satisfy the following properties:

(1) Each R_j commutes with translations;

(2) Each R_j commutes with dilations;

(3) If ρ is a rotation whose matrix is (ρ_{jk}) and T_ρ is the operator on functions f on E_n defined by $(T_\rho f)(x) = f(\rho x)$, then

$$T_{\rho^{-1}} R_j T_\rho = \sum_{k=1}^{n} \rho_{jk} R_k.$$

Moreover, by Theorem 2.6:

(4) each R_j maps $L^2(E_n)$ boundedly into itself.

When $n \geqq 3$, these four properties characterize the Riesz transforms. More precisely, if $R_1, R_2, \ldots, R_n$ are operators on $L^2(E_n)$ satisfying (1), (2), (3), and (4), then they are a fixed constant multiple of the Riesz transforms. We sketch a proof of this fact. By Theorems 3.16 and 3.18 of Chapter I we see that properties (1) and (4) imply that each R_j, when restricted to $\mathscr{S}$, is a convolution operator of the form $R_j\varphi = u_j * \varphi$, $\varphi \in \mathscr{S}$, with u_j a distribution whose Fourier transform $\hat{u}_j = b_j$ is a bounded function. Property (2), together with (1.6) of Chapter I, implies that each b_j is homogeneous of degree 0. Lastly, property (3), together with the fact that the operators T_ρ commute with the Fourier transform (Theorem 1.1 of Chapter IV), implies

$$b_j(\rho^{-1}x) = \sum_{k=1}^{n} \rho_{kj} b_k(x)$$

for all rotations ρ and $x \in E_n$. Fixing $x \in E_n$, $x \neq 0$ and letting ρ range through all the rotations we see that b_j is completely determined by the values $(b_1(x), \ldots, b_n(x)) = b(x)$ (we are, of course, using the fact that b_j is homogeneous of degree 0). If y is any point in E_n such that $|y| = |x|$ let σ be a rotation satisfying $y = \sigma^{-1}x$. Then, for any rotation ρ leaving x fixed we have

$$\sum_{k=1}^{n}\left(\sum_{l=1}^{n} \rho_{lj}\sigma_{kl}\right) b_k(x) = b_j(\sigma^{-1}\rho^{-1}x) = b_j(\sigma^{-1}x) = \sum_{k=1}^{n} \sigma_{kj} b_k(x).$$

That is, the vector $b = b(x)$ has the property that $\sigma\rho b = \sigma b$ for all rotations ρ leaving x fixed. But this means that $b(x)$ must be a multiple of x. This fact, together with equality (2.7), clearly implies that $R_1, R_2, \ldots, R_n$ are constant multiples of the Riesz transforms. When $n = 2$ this argument fails since the fact that $\rho b = b$ for all rotations leaving x fixed does not imply that $b(x)$ is a multiple of x (a rotation in E_2 leaving a point fixed must be the identity). If we replace the rotation group by the orthogonal group, however, we do obtain a characterization of the Riesz transforms which is valid also when $n = 2$.

5.7. Singular integral operators are important tools in the theory of partial differential equations. For example, the Riesz transforms can be used to obtain estimates involving the Laplace operator

$$\Delta = \frac{\partial^2}{\partial x_1^2} + \cdots + \frac{\partial^2}{\partial x_n^2}.$$

Since the Fourier transforms of $\partial^2\varphi/\partial x_j\,\partial x_k$ and $-R_jR_k\Delta\varphi$, for $\varphi \in \mathscr{S}$, must be equal (this is an immediate consequence of (2.7) of this chapter and

Theorem 1.8 of Chapter I) we obtain the relation

$$\frac{\partial^2 \varphi}{\partial x_j\, \partial x_k} = -R_j R_k\, \Delta\varphi.$$

Thus, by Theorem 2.6 we have

$$\left\| \frac{\partial^2 \varphi}{\partial x_j\, \partial x_k} \right\|_p \leqq A_p \|\Delta\varphi\|_p,$$

for $1 < p < \infty$ and A_p is independent of $\varphi \in \mathscr{S}$. For generalizations of these estimates see Stein [3], Chapters III and IV.

5.8. If φ is a testing function then its Riesz transform $R_k\varphi$ is a bounded continuous function. This is the case since we can express $R_k\varphi$, after an integration by parts, as the convolution of the testing function $\partial\varphi/\partial x_k$ with $c_n/(1 - n)|x|^{n-1}$. Thus, if μ is a finite Borel measure on E_n, the integral $\int_{E_n} (R_k\varphi)\, d\mu$ exists and can be used to define the *Riesz transform* ν_k of μ, $k = 1, 2, \ldots, n$, to be the "measure" satisfying

$$\int_{E_n} \varphi\, d\nu_k = -\int_{E_n} (R_k\varphi)\, d\mu$$

for all testing functions φ. This defines the Riesz transform of μ to be a tempered distribution; however, it is not true, in general, that ν_k is also a finite Borel measure for $k = 1, 2, \ldots, n$. In fact, the theory of $\mathbf{H}^p$ spaces developed in Section 4 can be used to show the following generalization of a classical theorem of F. and M. Riesz:

If the Riesz transforms $\nu_1, \ldots, \nu_n$ *of a finite Borel measure* μ *are also finite Borel measures, then* $\mu, \nu_1, \ldots, \nu_n$ *are all absolutely continuous with respect to Lebesgue measure* (see Stein and G. Weiss [1]; an extension of this result can be found in Stein [3], Chapter VII).

5.9. An M. Riesz system (introduced in Section 4) can be realized as a gradient of a harmonic function $\nabla h = F = (u_1, u_2, \ldots, u_n)$ when the domain of F is simply connected. In fact, the second set of Eqs. (4.13) are precisely the condition for the existence of a function h whose gradient is F, while the first equation asserts that the Laplacian of h is 0. By considering the gradient $\nabla u_j = (u_{j1}, u_{j2}, \ldots, u_{jn})$ of each of the components u_j, $j = 1, 2, \ldots, n$, we obtain an n^2-tuple $F^{(2)} = (u_{11}, u_{12}, \ldots, u_{nn})$ *which we shall call the second gradient of* h. We can perform this operation again on each of the components of $F^{(2)}$ and obtain the *third gradient* of h. Continuing in this way we obtain the k-th *gradient* $F^{(k)}$ *of* h. Calderón and Zygmund [6] have shown that $|F^{(k)}|^p$ is subharmonic for $p \geqq (n - 2)/(k + n - 2)$. Thus, by considering $\mathbf{H}^p$ spaces of systems of harmonic functions on E_{n+1}^+ that are k-th gradients we obtain the boundary value results (4.3) and (4.4) for indices $p \geqq (n-1)/(k+n-1)$ (observe that the dimension of the domain is $n+1$). That is, we obtain an $\mathbf{H}^p$ space theory for indices p arbitrarily close to 0.

There exists a class of generalized Cauchy–Riemann equations that is closely connected with the irreducible representations of the orthogonal group $SO(n)$. The k-th gradients can be realized as solutions of GCR systems belonging to this class. For details see Stein and G. Weiss [4].

5.10. The theory of singular integrals treated in this chapter involved kernels of the form $K(x) = \Omega(x)/|x|^n$ with Ω a homogeneous function satisfying certain integrability conditions on the surface Σ_{n-1} of the unit sphere and, in particular, $\int_{\Sigma_{n-1}} \Omega(x')\,dx' = 0$. These were the kernels originally studied by Calderón and Zygmund in [3]. It was pointed out by Cotlar [1] and Hörmander [1] that their development applies, without any major change, to more general kernels K. The most general class consist of distributions K that are locally integrable functions away from the origin satisfying (1) $\hat{K}(x)$ is bounded (where $\hat{K}$ is the Fourier transform of K in the sense of tempered distributions) and (2)

$$\int_{|x| \geqq 2|y|} |K(x-y) - K(x)|\,dx \leqq A < \infty \qquad \text{for all } y \text{ in } E_n$$

(see Stein [3, Chapter II]).

The advantage of this type of approach is that it gives results for $L^1(E_n)$ also, and in particular leads to a generalization of the fact (see Lemma 2.8 of Chapter V) that the Hilbert transform is of weak type (1,1).

Bibliographical Notes

For further discussion on the Hilbert transform see Titchmarsh [2], Zygmund [1] and Loomis [1]. The rotation method discussed in the second section was first developed by Calderón and Zygmund [4]. For further properties of the Riesz transform see Hörvath [1]. The spaces $\mathbf{H}^p(E^+_{n+1})$ were introduced by Stein and G. Weiss [1]. Systems of conjugate harmonic functions have also been studied by Calderón and Zygmund [6], Stein and G. Weiss [4], Küran [1] and Coifman and G. Weiss [2]. The fact that a harmonic solution F of an elliptic system has the property that $|F|^p$ is subharmonic for some $p < 1$ was observed by Calderón.

CHAPTER VII

Multiple Fourier Series

The purpose of this chapter is to give a brief introduction to several parts of the theory of multiple Fourier series. While from the abstract point of view this represents only a particular instance of Fourier analysis on compact abelian groups, our emphasis is on the connection between the analysis on the n-torus and Euclidean n-dimensional space. Thus, in Section 2 we consider the process of "periodization" applied to functions on E_n, which gives us the Poisson summation formula, and in Section 3 we consider the corresponding periodization for multiplier operators. In the fourth and fifth sections we consider another topic, the summability of multiple Fourier series. We study that subject at this stage, instead of earlier in the nonperiodic context (in which very similar results could be proved) mainly for reasons of elegance and appropriateness. The results that are presented are valid only in the case $n > 1$, and are obtained by techniques different from those originally developed for treating similar problems in the more familiar one-dimensional case.

1. Elementary Properties

Let Λ denote the additive group of points in E_n having integral coordinates (the addition being, of course, the one derived from the vector structure of E_n). Λ will be called the *unit lattice*. We consider the coset space E_n/Λ and we identify, in the usual manner, functions defined on E_n/Λ with periodic functions on E_n. More precisely, a function f satisfying $f(x + m) = f(x)$ whenever $m \in \Lambda$ is identified with the function whose value at the coset determined by x is $f(x)$. The elements of Λ are the *periods* of these functions.

There exists a natural identification of E_n/Λ with the *n-torus* $T_n = \{(e^{2\pi i x_1}, e^{2\pi i x_2}, \ldots, e^{2\pi i x_n}) \in \mathbf{C}^n : (x_1, x_2, \ldots, x_n) \in E_n\}$. This identification is given by the mapping $(x_1, x_2, \ldots, x_n) \to (e^{2\pi i x_1}, e^{2\pi i x_2}, \ldots, e^{2\pi i x_n})$ and, from it, we obtain the standard identification of periodic functions on E_n with functions on the n-torus.

A set $D \subset E_n$ will be called a *fundamental domain* if every point of E_n has exactly one translate with respect to Λ in D. A periodic function on E_n is, clearly, uniquely determined by its restriction to a fundamental domain. Obviously, there are infinitely many fundamental domains; one that is both simple and appropriate for our purposes is the *fundamental cube* $Q_n = \{x \in E_n : -1/2 \leqq x_j < 1/2, j = 1, 2, \ldots, n\}$. Integration on T_n can be defined in terms of the Lebesgue integral on Q_n: if f is a function on T_n we write

$$\int_{T_n} f\,dx = \int_{Q_n} f\,dx \tag{1.1}$$

and interpret this equality in the following way. Any function f on T_n, as we have seen, gives rise to a periodic one on E_n, and the restriction of this last function to Q_n is the one appearing in the right side of (1.1). The function f on T_n is defined to be measurable if the corresponding function on Q_n is Lebesgue measurable; if the latter is integrable then the integral on the left side of (1.1) is defined to be equal to the integral on the right side. In the same way, the L^p spaces on T_n are identified with the L^p spaces on Q_n. Similarly, we can identify the class of finite Borel measures on T_n with the class of finite Borel measures concentrated on the fundamental cube Q_n. We shall use the notation $L^p(T_n)$ and $\mathscr{B}(T_n)$ for the L^p spaces on T_n and the Borel measures on T_n, respectively. It should be pointed out, however, that the class of continuous functions on T_n, $C(T_n)$, does *not* correspond to the class of continuous functions on Q_n, but only to those functions which remain continuous on E_n when extended by periodicity. $C(T_n)$ is a Banach subspace of $L^\infty(T_n)$ (that is, we endow the former space with the L^∞ norm). We record here the elementary inclusion relations

$$\mathscr{B}(T_n) \supset L^p(T_n) \supset C(T_n)$$

and remark that, as usual, $L^1(T_n)$ may be considered as identified with the subspace of absolutely continuous measures of $\mathscr{B}(T_n)$.

We shall now discuss Fourier series of measures and functions in T_n. To every element $\mu \in \mathscr{B}(T_n)$ we assign the Fourier series

$$d\mu \sim \sum_{m \in \Lambda} a_m e^{2\pi i m \cdot x}, \tag{1.2}$$

where

$$a_m = \int_{T_n} e^{-2\pi i m \cdot x}\, d\mu(x) \tag{1.3}$$

are the *Fourier–Stieltjes coefficients* of $d\mu$.

To facilitate formal manipulation with Fourier coefficients it is convenient to consider, for each measure μ in $\mathscr{B}(T_n)$, the continuous linear functional

$$L_\mu : f \to \int_{T_n} f\,d\mu, \tag{1.4}$$

which it defines on the space $C(T_n)$. By the Riesz–Markov representation theorem we know that, conversely, every continuous linear functional on $C(T_n)$ is of the form (1.4) for some $\mu \in \mathscr{B}(T_n)$ with $\|L_\mu\| = \|d\mu\|$, where the norm on the left is that of the linear functional L_μ and the norm on the right is the total measure of μ. This identification makes it easy to define the *convolution of two measures* μ_1 and μ_2 in $\mathscr{B}(T_n)$: it is the measure μ which represents (via (1.4)) the linear functional

$$f \to \int_{T_n} f\, d\mu = \int_{T_n}\int_{T_n} f(x + y)\, d\mu_1(x)\, d\mu_2(y). \tag{1.5}$$

From this definition it is clear that $\|d\mu\| \leqq \|d\mu_1\|\,\|d\mu_2\|$ and, with this operation of convolution, $\mathscr{B}(T_n)$ is a commutative Banach algebra. If we put $f(x) = e^{-2\pi i m\cdot x}$ in (1.5) we obtain the *multiplication formula*

$$d\mu_1 * d\mu_2 \sim \sum_{m\in\Lambda} a_m b_m e^{2\pi i m\cdot x}, \tag{1.6}$$

where $\{a_m\}$ and $\{b_m\}$ are the Fourier–Stieltjes coefficients of $d\mu_1$ and $d\mu_2$ respectively.

A cursory examination of (1.5) also leads to an alternative definition of $d\mu_1 * d\mu_2$. It is the measure μ with the property that

$$\mu(E) = \int_{T_n} \mu_1(E - y)\, d\mu_2(y)$$

for any Borel set E. From this it is clear that if μ_1 (or μ_2) is absolutely continuous then μ is also absolutely continuous. The measure μ_1, however, is absolutely continuous if and only if it has a Radon–Nikodym derivative $f \in L^1(T_n)$. In this case the Radon–Nikodym derivative of μ is the function h defined by

$$h(x) = \int_{T_n} f(x - y)\, d\mu_2(y).$$

This integral converges absolutely for almost every x by virtue of Fubini's theorem. If μ_2 is also absolutely continuous, with Radon–Nikodym derivative $g \in L^1(T_n)$, then

$$h(x) = \int_{T_n} f(x - y) g(y)\, dy.$$

Thus, we see that $L^1(T_n)$ inherits the convolution structure from $\mathscr{B}(T_n)$. In particular, the multiplication formula (1.6) holds for L^1 functions as well.

In order to formulate the basic facts concerning completeness of the Fourier expansion, it is convenient to refer to any *finite* sum of the form $\sum a_m e^{2\pi i m\cdot x}$ as a *trigonometric polynomial.*

THEOREM 1.7. (i) *The trigonometric polynomials are dense in* $C(T_n)$ *and* $L^p(T_n)$, $1 \leqq p < \infty$.

(ii) *Suppose that for some* $\mu \in \mathscr{B}(T_n)$, $\int_{T_n} e^{-2\pi i m\cdot x}\, d\mu(x) = 0$ *for all* $m \in \Lambda$; *then* $\mu = 0$.

(iii) *Suppose* $f \in L^2(T_n)$ *and* $f \sim \sum_{m\in\Lambda} a_m e^{2\pi i m\cdot x}$, *then*

$$\sum_{m\in\Lambda} |a_m|^2 = \|f\|_2^2.$$

The correspondence $f \leftrightarrow \{a_m\} = \{\int_{T_n} e^{-2\pi i m\cdot x} f(x)\, dx\}$ *is a unitary mapping of* $L^2(T_n)$ *onto* $l^2(\Lambda)$.

PROOF. Since the trigonometric polynomials form an algebra which separates points of the n-torus, which contains the constants and is closed under complex conjugation, we can apply the Stone–Weierstrass theorem in order to obtain the density of the trigonometric polynomials in $C(T_n)$. This approximation property together with the density of $C(T_n)$ in $L^p(T_n)$, $1 \leqq p < \infty$, implies the corresponding approximation property in $L^p(T_n)$.

In order to show (ii) we observe that the property $\int_{T_n} e^{-2\pi i m\cdot x}\, d\mu(x) = 0$ for all $m \in \Lambda$ is equivalent to

$$\int_{T_n} P(x)\, d\mu(x) = 0$$

for each trigonometric polynomial P. But, by (i), this implies that $\int_{T_n} f(x)\, d\mu(x) = 0$ for all $f \in C(T_n)$. It follows, therefore, that $\mu = 0$.

Given $f \in L^2(T_n)$ then, if N is a positive integer, the infimum of the expression $\|f - \sum_{|m|\leqq N} a_m e^{2\pi i m\cdot x}\|_2$ is attained when the numbers a_m are the *Fourier coefficients* $\int_{T_n} e^{-2\pi i m\cdot x} f(x)\, dx$ of f. This assertion follows, as is well-known, from the fact that the functions $\{e^{2\pi i m\cdot x}\}$ are mutually orthogonal and normalized (in the L^2 norm). Because each $f \in L^2(T_n)$ can be approximated by trigonometric polynomials we know, therefore, that $\lim_{N\to\infty} \|f - \sum_{|m|\leqq N} a_m e^{2\pi i m\cdot x}\|_2 = 0$. Consequently,

$$\|f\|_2 - \Big\| \sum_{|m|\leqq N} a_m e^{2\pi i m\cdot x} \Big\|_2 \to 0$$

as $N \to \infty$ and so $\|f\|_2 = (\sum_{m\in\Lambda} |a_m|^2)^{1/2}$. This shows that the mapping $f \to \{a_m\}$ is an isometry. If it is also onto, then this mapping must be unitary. In fact, let $\{a_m\}$ be given with $\sum_{m\in\Lambda} |a_m|^2 < \infty$ and put $s_N(x) = \sum_{|m|\leqq N} a_m e^{2\pi i m\cdot x}$. Clearly, $\|s_{N_1} - s_{N_2}\|_2 = (\sum_{N_1<|m|\leqq N_2} |a_m|^2)^{1/2}$, if $N_1 < N_2$. Therefore, $\{s_N\}$ is a Cauchy sequence in $L^2(T_n)$ and, thus, it converges in the norm to some function $f \in L^2(T_n)$. Since

$$a_m = \int_{T_n} s_N(x) e^{-2\pi i x \cdot m}\, dx \qquad \text{if } |m| \leqq N,$$

it follows that

$$a_m = \int_{T_n} f(x) e^{-2\pi i x \cdot m}\, dx \qquad \text{for all } m \in \Lambda.$$

This shows (iii) and, therefore, the theorem is proved.

We now derive some useful corollaries of this result.

COROLLARY 1.8. *Suppose $f \in L^1(T_n)$ and $\sum_{m\in\Lambda} |a_m| < \infty$, where $\{a_m\}$ are the Fourier coefficients of f; then f can be modified on a set of measure zero so that it is in $C(T_n)$ and equals $\sum_{m\in\Lambda} a_m e^{2\pi i m\cdot x}$ for all $x \in T_n$.*

PROOF. It follows from the hypotheses that the function whose values are $f(x) - \sum_{m\in\Lambda} a_m e^{2\pi i m\cdot x}$ has zero Fourier coefficients; thus, by (ii) of Theorem 1.7, this function must be 0 almost everywhere.

When k is a positive integer the class $C^{(k)}(Q_n)$ $(= C^{(k)}(T_n))$ consists of the restrictions to Q_n of all periodic functions on E_n belonging to the class $C^{(k)} = C^{(k)}(E_n)$.[1]

COROLLARY 1.9. *Suppose $f \in C^{(k)}(T_n)$ for $k > n/2$, then $\sum_{m\in\Lambda} |a_m| < \infty$, where $\{a_m\}$ are the Fourier coefficients of f.*

PROOF. In the first chapter we defined x^α for $x = (x_1, x_2, \ldots, x_n) \in E_n$ and $\alpha = (\alpha_1, \alpha_2, \ldots, \alpha_n)$ an n-tuple of nonnegative integers to be the number $x_1^{\alpha_1} x_2^{\alpha_2} \cdots x_n^{\alpha_n}$ (we also adopted the convention that $0^0 = 1$). Using this notation and integrating by parts we obtain

$$\int_{T_n} (D^\alpha f)(x) e^{-2\pi i m\cdot x}\, dx = (2\pi i m)^\alpha \int_{T_n} f(x) e^{-2\pi i m\cdot x}\, dx = (2\pi i m)^\alpha a_m,$$

whenever $f \in C^{(k)}(T_n)$ and $\alpha_1 + \alpha_2 + \cdots + \alpha_n \leqq k$. Since $D^\alpha f$ is continuous

[1] That is, those periodic functions f having everywhere defined continuous derivatives $D^\alpha f$ whenever $\alpha = (\alpha_1, \alpha_2, \ldots, \alpha_n)$ is an n-tuple of nonnegative integers satisfying $\alpha_1 + \alpha_2 + \cdots + \alpha_n \leqq k$. Recall that in Chapter I the operator D^α was defined to be

$$\frac{\partial^{\alpha_1}}{\partial x_1^{\alpha_1}} \frac{\partial^{\alpha_2}}{\partial x_2^{\alpha_2}} \cdots \frac{\partial^{\alpha_n}}{\partial x_n^{\alpha_n}}.$$

We warn the reader that, in that chapter, the sum $\alpha_1 + \alpha_2 + \cdots + \alpha_n$ was denoted by $|\alpha|$; however, in the proof of Theorem 1.7 we used the symbol $|m|$ to denote the Euclidean norm of the lattice point m. In order to avoid confusion we will not use here the notation $|\alpha|$ introduced in the first chapter.

it must belong to $L^2(T_n)$; therefore, by Theorem 1.7 (iii)

$$\sum_{\alpha_1+\cdots+\alpha_n=k}\left\{\sum_{m\in\Lambda}|a_m|^2[(2\pi m)^\alpha]^2\right\}<\infty. \tag{1.10}$$

Moreover, a simple calculation shows that there exists a constant $c = c(k, n)$, depending only on the dimension n and on k, such that

$$\sum_{\alpha_1+\cdots+\alpha_n=k}[(2\pi m)^\alpha]^2 \geqq c|m|^{2k}.$$

Thus, by Schwarz's inequality

$$\begin{aligned}\sum_{|m|>0}|a_m| &\leqq \sum_{|m|>0}|a_m|\Big(\sum_{\alpha_1+\cdots+\alpha_n=k}[(2\pi m)^\alpha]^2\Big)^{1/2}c^{-1/2}|m|^{-k}\\ &\leqq \Big(\sum_{|m|>0}|a_m|^2\sum_{\alpha_1+\cdots+\alpha_n=k}[(2\pi m)^\alpha]^2\Big)^{1/2}\Big(\sum_{|m|>0}|m|^{-2k}\Big)^{1/2}c^{-1/2}.\end{aligned}$$

When $k > n/2$ the sum $\sum_{|m|>0}|m|^{-2k}$ is finite; hence, the last expression must also be finite in this case by virtue of inequality (1.10). Thus, the corollary is proved.

The following corollary is the familiar Riemann–Lebesgue theorem in the present context:

COROLLARY 1.11. *If $f\in L^1(T_n)$ and $f\sim\sum_{m\in\Lambda}a_m e^{2\pi i m\cdot x}$ then $a_m\to 0$ as $|m|\to\infty$.*

PROOF. Suppose first that $f\in L^2(T_n)$. Then the conclusion $a_m\to 0$ is an immediate consequence of the fact that $\sum_{m\in\Lambda}|a_m|^2<\infty$ (see Theorem 1.7 part (iii)). For the general $f\in L^1(T_n)$, given $\varepsilon>0$, we can find f_1 and f_2 such that $f=f_1+f_2$ with $f_1\in L^2(T_n)$ and $\|f_2\|_1<\varepsilon$. Then, if $\{a_m^{(j)}\}$ denote the Fourier coefficients of f_j, we have $\lim_{|m|\to\infty} a_m^{(1)}=0$ and $|a_m^{(2)}|\leqq\|f_2\|_1<\varepsilon$. Thus, $\limsup_{|m|\to\infty}|a_m|\leqq\varepsilon$, where $a_m=a_m^{(1)}+a_m^{(2)}$, $m\in\Lambda$, are the Fourier coefficients of f. Since $\varepsilon>0$ is arbitrary, this establishes the corollary.

2. *The Poisson Summation Formula*

Instead of continuing a step-by-step development of the analogy between the properties of Fourier series and those of Fourier integrals, we now turn to the question which is the main concern of this chapter. We can phrase it in the following general way: Suppose we are given an "element" of a function space associated with E_n (e.g., a function on E_n, an operator on functions of E_n, etc.), what is its periodic analog? That is, what object corresponds to it on the n-torus T_n? In addition we shall be interested in understanding how we can deduce the properties of such an object in its periodic form from those we have already established for its non-periodic form.

We consider these matters first in the context of functions defined on E_n. In order to obtain a better idea of the problems involved we proceed formally, disregarding for the moment all questions of convergence.

Let, then, f be some (appropriate) function on E_n. There are at least two ways of obtaining a periodic function from f. The first construction is elementary and does not involve Fourier analysis. We simply write

$$\sum_{m \in \Lambda} f(x + m). \tag{2.1}$$

Since this (formal) sum is taken over all the lattice points of Λ it is clearly periodic (for the passage from x to $x + m'$ merely permutes the terms in (2.1)). We shall refer to the passage from f to the sum (2.1) as the *periodization* of f.

In order to describe the second approach we write

$$f(x) \sim \int_{E_n} \hat{f}(y) e^{2\pi i x \cdot y}\, dy, \tag{2.2}$$

which is simply the Fourier inversion formula, where

$$\hat{f}(y) = \int_{E_n} f(x) e^{-2\pi i x \cdot y}\, dx.$$

Then a periodic analog of the function f (given by (2.2)) is

$$\sum_{m \in \Lambda} \hat{f}(m) e^{2\pi i x \cdot m}. \tag{2.3}$$

The main thrust of the Poisson summation formula is to the effect that the two approaches to a periodic analog of f given by (2.1) and (2.3) are essentially identical. This conclusion can be formulated precisely in many ways. The simplest and most direct is the following.

THEOREM 2.4. *Suppose $f \in L^1(E_n)$. Then the series $\sum_{m \in \Lambda} f(x + m)$ converges in the norm of $L^1(Q_n)$ $(= L^1(T_n))$. The resulting function in $L^1(T_n)$ has the Fourier expansion*

$$\sum_{m \in \Lambda} \hat{f}(m) e^{2\pi i x \cdot m}. \tag{2.5}$$

This means that $\{\hat{f}(m)\}$ are the Fourier coefficients of the L^1 function defined by the series $\sum_{m \in \Lambda} f(x + m)$, where, for any $y \in E_n$,

$$\hat{f}(y) = \int_{E_n} f(x) e^{-2\pi i y \cdot x}\, dx.$$

PROOF. If $Q_n - m$ is the translate of Q_n by the lattice point m we then have

$$\int_{Q_n} \left| \sum_{m \in \Lambda} f(x + m) \right| dx \leqq \sum_{m \in \Lambda} \int_{Q_n} |f(x + m)| \, dx = \sum_{m \in \Lambda} \int_{Q_n - m} |f(x)| \, dx.$$

Since the fundamental cube Q_n is a fundamental domain, the translates $Q_n - m$ are mutually disjoint and their union is E_n. Thus,

$$\sum_{m \in \Lambda} \int_{Q_n - m} |f(x)| \, dx = \int_{E_n} |f(x)| \, dx < \infty.$$

This shows that the series $\sum_{m \in \Lambda} f(x + m)$ converges (absolutely) in the norm of $L^1(Q_n)$. Using the same interchange of summation and integration we can evaluate the Fourier coefficients of $\sum_{m \in \Lambda} f(x + m)$. In fact,

$$\begin{aligned} \int_{Q_n} \left(\sum_{m' \in \Lambda} f(x + m') \right) e^{-2\pi i m \cdot x} \, dx &= \sum_{m' \in \Lambda} \int_{Q_n - m'} f(x) e^{-2\pi i m \cdot x} \, dx \\ &= \int_{E_n} f(x) e^{-2\pi i m \cdot x} \, dx \\ &= \hat{f}(m). \end{aligned}$$

The argument given above for the L^1 convergence also shows that this interchange is justified and, thus, the fact that $\sum_{m' \in \Lambda} f(x + m')$ has the Fourier expansion (2.5) is demonstrated.

The following corollary is a useful special case of Theorem 2.4:

COROLLARY 2.6. *Suppose* $\hat{f}(y) = \int_{E_n} f(x) e^{-2\pi i x \cdot y} \, dx$ *and*

$$f(x) = \int_{E_n} \hat{f}(y) e^{2\pi i x \cdot y} \, dy$$

with $|f(x)| \leqq A(1 + |x|)^{-n-\delta}$ *and* $|\hat{f}(y)| \leqq A(1 + |y|)^{-n-\delta}$, $\delta > 0$ (*thus, both* f *and* $\hat{f}$ *can be assumed to be continuous*). *Then,*

$$\sum_{m \in \Lambda} f(x + m) = \sum_{m \in \Lambda} \hat{f}(m) e^{2\pi i m \cdot x}, \tag{2.7}$$

and, in particular,

$$\sum_{m \in \Lambda} f(m) = \sum_{m \in \Lambda} \hat{f}(m). \tag{2.8}$$

The four series in (2.7) *and* (2.8) *converge absolutely.*[2]

[2] The identity (2.8) is referred to as the *Poisson summation formula*, but we shall also use the same name for (2.7) and, more generally, for Theorem 2.4.

PROOF. Because of our assumption on $\hat{f}$, the Fourier series

$$\sum_{m \in \Lambda} \hat{f}(m) e^{2\pi i m \cdot x}$$

converges absolutely. Therefore, according to Corollary 1.8 and Theorem 2.4, the function $\sum_{m \in \Lambda} f(x + m)$ can be modified on a set of measure zero so as to equal the continuous function $\sum_{m \in \Lambda} \hat{f}(m) e^{2\pi i m \cdot x}$ everywhere. By comparison with $\sum_{m \in \Lambda} (1 + |m|)^{-n-\delta}$, however, we see that $\sum_{m \in \Lambda} f(x + m)$ is a uniformly convergent series whose terms are continuous functions. Thus, its sum is already continuous and we obtain (2.7) for every x.

We shall now give two illustrations of how the Poisson summation formula may be used. In the first example we consider the analog of the Fourier inversion problem considered in Chapter I; that is, we ask to what extent can the series

$$\sum_{m \in \Lambda} a_m e^{2\pi i m \cdot x}$$

be "summed" to $f(x)$ when $a_m = \int_{T_n} f(x) e^{-2\pi i m \cdot x}\, dx$ with $f \in L^1(T_n)$. In view of what we learned in the first chapter it is natural to try to replace the above sum by the limit

$$\lim_{\varepsilon \to 0} \sum_{m \in \Lambda} \Phi(\varepsilon m) a_m e^{2\pi i m \cdot x}, \tag{2.9}$$

where Φ is an appropriate continuous function with $\Phi(0) = 1$. Motivated by Corollary (2.6) we make the following assumptions on Φ:

$$\text{(2.10)} \quad \begin{cases} \text{(i)} \quad \Phi(y) = \hat{\varphi}(y) \qquad \text{with } \int_{E_n} \varphi(x)\, dx = 1 \\ \text{(ii)} \quad |\Phi(y)| \leqq A(1 + |y|)^{-n-\delta},\ |\varphi(x)| \leqq A(1 + |x|)^{-n-\delta}, \end{cases}$$

for some $\delta > 0$.

THEOREM 2.11. *Suppose Φ satisfies the conditions* (2.10) *above, $f \in L^p(T_n)$ and $f(x) \sim \sum_{m \in \Lambda} a_m e^{2\pi i m \cdot x}$. Then*

(a) *If $1 \leqq p < \infty$, the limit* (2.9) *converges to f in the norm of $L^p(T_n)$;*
(b) *If $f \in C(T_n)$, the limit* (2.9) *converges to f uniformly;*
(c) *For any $f \in L^1(T_n)$ the limit* (2.9) *converges to $f(x)$ at each point x of the Lebesgue set of f (and, thus, almost everywhere).*

PROOF. Let us write $\varphi_\varepsilon(x) = \varepsilon^{-n}\varphi(x/\varepsilon)$ and $\Phi^\varepsilon(y) = \Phi(\varepsilon y)$, $\varepsilon > 0$. Then we know that $\Phi^\varepsilon = (\varphi_\varepsilon)\hat{}$. Moreover, because of (2.10) φ_ε and Φ^ε satisfy the conditions of corollary (2.6) and, therefore,

$$\sum_{m \in \Lambda} \Phi(\varepsilon m) e^{2\pi i m \cdot x} = \sum_{m \in \Lambda} \varphi_\varepsilon(x + m). \tag{2.12}$$

Let us denote the right side of (2.12) by $K_\varepsilon(x)$. Then

$$(f * K_\varepsilon)(x) = \sum_{m \in \Lambda} \Phi(\varepsilon m) a_m e^{2\pi i m \cdot x},$$

where the convolution is the one introduced in the first section (see, in particular, (1.6)). Because of our assumptions on Φ, this last series converges absolutely. On the other hand,

$$\int_{T_n} |K_\varepsilon(x)|\, dx \leqq \int_{Q_n} \sum_{m \in \Lambda} |\varphi_\varepsilon(x + m)|\, dx = \int_{E_n} |\varphi_\varepsilon(x)|\, dx = \int_{E_n} |\varphi(x)|\, dx.$$

Since $\|f * K_\varepsilon\|_p \leqq \|f\|_p \|K_\varepsilon\|_1$ (where the norms are those of the spaces $L^p(T_n)$, $1 \leqq p \leqq \infty$), we see that the mappings

$$M_\varepsilon : f \to \sum_{m \in \Lambda} \Phi(\varepsilon m) a_m e^{2\pi i x \cdot m}$$

are uniformly bounded in ε, $\varepsilon > 0$, as operators on $L^p(T_n)$. By the continuity of Φ and the fact that $\int_{E_n} \varphi(x)\, dx = 1$ we have $\lim_{\varepsilon \to 0} \Phi(\varepsilon m) = 1$; therefore, $M_\varepsilon f \to f$ as $\varepsilon \to 0$ (in the L^p norm) whenever f is a trigonometric polynomial. The fact that the trigonometric polynomials are dense in $L^p(T_n)$, for $p < \infty$, and in $C(T_n)$, together with the uniform boundedness we just established, proves parts (a) and (b) of the theorem.

Before proving (c) we should first explain that we say x is a *point of the Lebesgue set of* f provided, after f is continued periodically, x is a point of the Lebesgue set of the continued periodic function on E_n. It is clear that after an appropriate translation we may take our point x to be the origin (which is the center of the fundamental cube Q_n). Let $\tilde{f}(x) = f(x)$ for $x \in Q_n$ and $\tilde{f}(x) = 0$ for $x \in E_n - Q_n$. Since the property of being a member of the Lebesgue set of a function is a local one, 0 is a point of the Lebesgue set of $\tilde{f}$. Now,

$$\begin{aligned}\sum_{m \in \Lambda} \Phi(\varepsilon m) a_m &= \int_{Q_n} f(x) K_\varepsilon(-x)\, dx = \sum_{m \in \Lambda} \int_{Q_n} f(x) \varphi_\varepsilon(-x + m)\, dx \\ &= \int_{Q_n} f(x) \varphi_\varepsilon(-x)\, dx + \sum_{m \neq 0} \int_{Q_n} f(x) \varphi_\varepsilon(-x + m)\, dx.\end{aligned}$$

Because of our assumption on φ,

$$\begin{aligned}|\varphi_\varepsilon(-x + m)| &\leqq A\left(1 + \left|\frac{-x + m}{\varepsilon}\right|\right)^{-n-\delta} \varepsilon^{-n} \\ &= \varepsilon^\delta A(\varepsilon + |-x + m|)^{-(n+\delta)} \\ &\leqq \varepsilon^\delta A' |m|^{-(n+\delta)}, \qquad \text{if } x \in Q_n \text{ and } |m| \geqq 1.\end{aligned}$$

Therefore,

$$\left|\sum_{m\neq 0}\int_{Q_n} f(x)\varphi_\varepsilon(-x+m)\,dx\right| \leqq \varepsilon^\delta A'' \int_{Q_n} |f(x)|\,dx \to 0 \qquad \text{as } \varepsilon \to 0.$$

Also, $\int_{Q_n} f(x)\varphi_\varepsilon(-x)\,dx = \int_{E_n} \tilde{f}(x)\varphi_\varepsilon(-x)\,dx$. But the condition $|\varphi(x)| \leqq A(1+|x|)^{-n-\delta}$ allows us to apply Theorem 1.25 of Chapter I to this last integral and we obtain

$$\lim_{\varepsilon\to 0}\int_{E_n} \tilde{f}(x)\varphi_\varepsilon(-x)\,dx = \tilde{f}(0) = f(0).$$

This proves part (c) and, thus, Theorem 2.11 is completely proved.

Two particular special cases of this theorem are worth noting separately. First consider $\varphi(x) = c_n(1+|x|^2)^{-(n+1)/2}$; then $\Phi(y) = e^{-2\pi|y|}$ (see Theorem 1.14 of Chapter I). In this case the series occurring in (2.9) becomes (if we write t instead of ε)

$$\sum_{m\in\Lambda} e^{-2\pi|m|t} a_m e^{2\pi i m\cdot x}. \tag{2.13}$$

This series, which is absolutely convergent for $t > 0$, is called the *Poisson* (or *Abel–Poisson*) *integral of f*; it is the convolution of f with the *Poisson kernel*

$$P_t(x) = \sum_{m\in\Lambda} e^{-2\pi|m|t} e^{2\pi i m\cdot x}.$$

It is a consequence of (2.12) that $P_t(x) \geqq 0$ for $t > 0$ and $x \in Q_n$. Moreover,

$$\int_{Q_n} P_t(x)\,dx = \int_{Q_n} 1\,dx + \int_{Q_n} \sum_{m\neq 0} e^{-2\pi|m|t} e^{2\pi i m\cdot x}\,dx = 1 + 0 = 1.$$

The second example arises when $\Phi(y) = (1-|y|^2)^\alpha$ if $|y| \leqq 1$ and $\Phi(y) = 0$ if $|y| \geqq 1$. By Theorem 4.15 of Chapter IV we then have

$$\varphi(x) = \hat{\Phi}(x) = \pi^{-\alpha}\Gamma(\alpha+1)|x|^{-(n/2)-\alpha} J_{(n/2)+\alpha}(2\pi|x|).$$

Thus, if $\alpha > (n-1)/2$, it follows from Lemma 3.11 of Chapter IV that Φ (and φ) satisfy hypotheses (2.10). Therefore, in this case we obtain the appropriate convergence for the *Riesz means*[3]

$$\lim_{R\to\infty}\sum_{|m|<R}\left(1-\frac{|m|^2}{R^2}\right)^\alpha a_m e^{2\pi i m\cdot x}, \qquad \alpha > \frac{n-1}{2}. \tag{2.14}$$

[3] The number $(n-1)/2$, the lower bound for the values α for which the functions Φ and $\varphi = \hat{\Phi}$ satisfy (2.10), is called the *critical index* (for the convergence of the Riesz means (2.14)).

We summarize these facts as follows

COROLLARY 2.15. *The conclusions of Theorem* 2.11 *hold, in particular, for the Riesz means* (2.14), *when* α *is larger than the critical index, and for the Abel–Poisson means* (2.13).

The treatment of the Riesz means of order not greater than the critical index is subtler and will be discussed in sections four and five below.

The application of the Poisson summation formula we have just presented concerned periodic analogs of regularization operators arising in the problem of Fourier inversion. The second application is typical of the situation where the Poisson summation formula can be used to give precise estimates involving various elementary (periodic) functions.

In the fourth chapter (see Theorem 4.1) we learned that if α is a complex number satisfying $0 < \operatorname{Re}(\alpha) < n$ and $f(x) = |x|^{\alpha-n}$, then $\hat{f}(x) = \gamma_\alpha |x|^{-\alpha}$ with $\gamma_\alpha = \pi^{-\alpha+n/2}\Gamma(\alpha/2)/\Gamma[(n-\alpha)/2]$. If we were to apply identity (2.7), ignoring all questions of convergence, we would obtain

$$\gamma_\alpha^{-1} \sum_{m\in\Lambda} |x+m|^{\alpha-n} = \sum_{m\in\Lambda} |m|^{-\alpha} e^{2\pi i m\cdot x}. \tag{2.16}$$

This relation, as it stands, cannot be correct because of the infinite term corresponding to $m = 0$ in the right side, and because of the divergence of the series on the left. Nevertheless, the attempt to apply the Poisson summation formula in this context can be made to work after suitable modification. One way of doing this involves the functional equation of the Riemann zeta function and some of its generalizations. This will be described in (6.3) below. Another interpretation of (2.16) is given by the following result:

THEOREM 2.17. *Let* $0 < \operatorname{Re}(\alpha) < n$. *Then the series* $\sum_{|m|>0} |m|^{-\alpha} e^{2\pi i m\cdot x}$ *is the Fourier series of an integrable function on* Q_n *which is of class* C^∞ *on* $Q_n - \{0\}$. *At the origin this function has the same singularity as the function whose values are* $\gamma_\alpha^{-1}|x|^{\alpha-n}$. *That is,*

$$\gamma_\alpha^{-1}|x|^{\alpha-n} + b(x) \sim \sum_{|m|>0} |m|^{-\alpha} e^{2\pi i m\cdot x}, \qquad (x \in Q_n)$$

where $b \in C^\infty(Q_n)$.

PROOF. Let us choose a function η having the following properties: $\eta \in C^\infty(E_n)$, $\eta(x) = 1$ if $|x| \geqq 1$, and η vanishes in a neighborhood of the origin. We then define F by letting $F(x) = \eta(x)|x|^{-\alpha}$ for $x \in E_n$. We claim that there exists a function $f \in L^1(E_n)$ such that

(1) $\hat{f} = F$;

(2) $f(x) = \gamma_\alpha^{-1}|x|^{\alpha-n} + b_1(x)$, where $b_1 \in C^\infty(E_n)$;

(3) For each n-tuple $\beta = (\beta_1, \beta_2, \ldots, \beta_n)$ of nonnegative integers and every positive integer N, $|D^\beta f(x)| = O(|x|^{-N})$ as $|x| \to \infty$.

In fact, write $F(x) = |x|^{-\alpha} + (\eta(x) - 1)|x|^{-\alpha}$ and let f be the inverse Fourier transform of F in the sense of tempered distributions. Then, by (4.1) of Chapter IV $f(x) = \gamma_\alpha^{-1}|x|^{\alpha-n} + b_1(x)$, where b_1 is the inverse Fourier transform of the integrable function with bounded support whose values are $(\eta(x) - 1)|x|^{-\alpha}$. Therefore, $b_1 \in C^\infty(E_n)$ and properties (1) and (2) are established. Since f is the inverse Fourier transform of F, for any n-tuple β of nonnegative integers, $-(2\pi ix)^\beta f(x)$ is the inverse Fourier transform of $D^\beta F$ (see Theorem 1.8 of Chapter I). Note that if the order of the partial derivative $D^\beta F$ is large enough ($\beta_1 + \cdots + \beta_n > n - \mathrm{Re}(\alpha)$), then $D^\beta F \in L^1(E_n)$; consequently, when this is the case, $-(2\pi ix)^\beta f(x)$ is bounded. This shows that $|f(x)| = O(|x|^{-N})$ as $|x| \to \infty$. The rest of property (3) is proved in the same way. The fact that $f \in L^1(E_n)$ follows from (2) and the just established fact that $|f(x)| = O(|x|^{-N})$ as $|x| \to \infty$.

We next apply the Poisson summation formula in the variant given by Theorem 2.4 to the function f we just constructed and $\hat{f} = F$. We obtain

$$\sum_{m\in\Lambda} f(x+m) \sim \sum_{m\in\Lambda} F(m)e^{2\pi ix\cdot m} = \sum_{|m|>0} |m|^{-\alpha}e^{2\pi ix\cdot m}.$$

Since

$$\sum_{m\in\Lambda} f(x+m) = f(x) + \sum_{|m|>0} f(x+m) = \gamma_\alpha^{-1}|x|^{\alpha-n} + b_1(x) + \sum_{|m|>0} f(x+m),$$

this proves the theorem with $b(x) = b_1(x) + \sum_{|m|>0} f(x+m)$.

Various other special series besides $\sum_{|m|>0} |m|^{-\alpha}e^{2\pi im\cdot x}$ can be treated by this method. Some of these are described in (6.1) below.

3. Multiplier Transformations

In Chapter I we studied the class of bounded operators from $L^p(E_n)$ into $L^q(E_n)$ which commute with translations. We found that each such operator T is, in effect, determined by a certain tempered distribution u having the property that $Tf = u * f$ for each $f \in \mathcal{S}$. Taking Fourier transforms we then have $(Tf)\hat{} = \hat{u}\hat{f}$, so that T is equivalent, via the Fourier transform, to the operator which maps each testing function into the product of the Fourier transform of the testing function with the distribution $\hat{u}$. In general, not too much is known about such "multipliers" $\hat{u}$; however, when $p = q$ it can be shown that among other things $\hat{u}$ is a bounded measurable function. Theorem 3.18 of Chapter I asserts that this is the case when $p = 2 = q$. If $p \neq 2$ and $\|u * f\|_p \leqq A\|f\|_p$ for all $f \in \mathcal{S}$, it follows from Theorem 3.20 of Chapter I that $\|u * f\|_{p'} \leqq A\|f\|_{p'}$ for all $f \in \mathcal{S}$, where $(1/p) + (1/p') = 1$. Since 2 must lie between p and p', it then follows from the M. Riesz interpolation theorem (see Theorem 1.3 of Chapter V) that the operator $T: f \to u * f$ is of type (2, 2). Consequently $\hat{u}$ satisfies the hypotheses of Theorem 3.18 and, therefore, must be a bounded measurable function.

It is now our intention to study the analogous multiplier operators in the periodic case and to show that for an important class of these their properties are a priori consequences of their nonperiodic variants.

In the notation of Chapter I when $p < \infty$, we let (L^p, L^q) denote the class of bounded operators T from $L^p(E_n)$ into $L^q(E_n)$ which commute with translations (or, equivalently, (L^p, L^q) is the class of tempered distributions u such that $\|u * \varphi\|_q \leqq A\|\varphi\|_p$ for some $A = A(u)$ and all $\varphi \in \mathcal{S}$). In order to emphasize the fact that these operators act on functions defined on E_n we shall now write $(L^p(E_n), L^q(E_n))$ for this class. Similarly, for the periodic case, we introduce the class $(L^p(T_n), L^q(T_n))$ of all bounded operators $\tilde{T}$ from $L^p(T_n)$ into $L^q(T_n)$ which commute with translations.

In analogy with the nonperiodic case we could identify any $\tilde{T} \in (L^p(T_n), L^q(T_n))$ with convolution (on T_n) by an appropriate tempered distribution. In this case, however, it is much simpler to consider the operators directly.

THEOREM 3.1. *Suppose* $\tilde{T} \in (L^p(T_n), L^q(T_n))$, $1 \leqq p, q \leqq \infty$. *Then there exists a bounded complex valued function* $\lambda: m \to \lambda(m)$ *on the lattice* Λ, *such that*

$$\tilde{T}f \sim \sum_{m \in \Lambda} \lambda(m) a_m e^{2\pi i m \cdot x}, \tag{3.2}$$

whenever $f \sim \sum_{m \in \Lambda} a_m e^{2\pi i m \cdot x}$.

PROOF. Set $\psi_m(x) = (\tilde{T}e_m)(x)$, where $e_m(y) = e^{2\pi i m \cdot y}$, for $m \in \Lambda$. Then $\|\psi_m\|_q \leqq A\|e_m\|_p = A$. If τ_h denotes translation by h (that is, for f periodic $(\tau_h f)(x) = f(x - h)$), the hypothesis that $\tilde{T}\tau_h = \tau_h \tilde{T}$ implies that for each $h \in Q_n$

$$\psi_m(x - h) = e^{-2\pi i m \cdot h} \psi_m(x)$$

for almost every x. By Fubini's theorem, then, this identity holds for almost every $h \in Q_n$ if a certain $x(= x_0)$ is fixed. This means that $\psi_m(h) = e^{2\pi i m \cdot h} e^{-2\pi i m \cdot x_0} \psi_m(x_0)$ for almost every h. Thus,

$$\psi_m(x) = \lambda(m) e^{2\pi i m \cdot x}$$

for almost every x with $\lambda(m) = e^{-2\pi i m \cdot x_0} \psi_m(x_0)$. Moreover, $|\lambda(m)| = |\lambda(m)| \|e_m\|_q = \|\lambda(m) e_m\|_q = \|\psi_m\|_q \leqq A$. Passing to finite linear combinations of the functions $\{e_m\}$ we obtain the representation (3.2) for f a trigonometric polynomial. By a simple limiting argument this extends to all $f \in L^p(T_n)$ and the theorem is established.

Because of Theorem 3.1 we see that it is natural to refer to the transformations $\tilde{T} \in (L^p(T_n), L^q(T_n))$ as *multiplier operators* and the associated sequences $\{\lambda(m)\}$ as *multipliers.*

The following corollary is a consequence of this theorem and part (iii) of (1.7):

COROLLARY 3.3. *$\tilde{T} \in (L^2(T_n), L^2(T_n))$ if and only if $\{\lambda(m)\}$ is a bounded sequence. Moreover, the operator norm $\|\tilde{T}\|$ equals* $\sup_{m\in\Lambda} |\lambda(m)|$.

The L^1 case is similar to the nonperiodic one treated in the first chapter. More precisely, we show the following result:

THEOREM 3.4. *$\tilde{T} \in (L^1(T_n), L^1(T_n))$ if and only if there exists a finite Borel measure μ on T_n, such that*

$$d\mu \sim \sum_{m\in\Lambda} \lambda(m)e^{2\pi i m\cdot x}. \tag{3.5}$$

In this case $\|d\mu\| = \|\tilde{T}\|$.

PROOF. Suppose $\|\tilde{T}f\|_1 \leqq A\|f\|_1$ for all $f \in L^1(T_n)$. For $t > 0$ let $f_t(x) = P_t(x) = \sum_{m\in\Lambda} e^{-2\pi|m|t}e^{2\pi i m\cdot x}$. We showed earlier (see the discussion following (2.13)) that $\|f_t\|_1 = 1$. Thus,

$$\|\tilde{T}f_t\|_1 = \int_{Q_n} \left|\sum_{m\in\Lambda} e^{-2\pi|m|t}e^{2\pi i m\cdot x}\lambda(m)\right| dx \leqq A.$$

Consequently, we can find a sequence $\{t_n\}$ of positive numbers tending to 0 such that $\{\tilde{T}f_{t_n}\}$ tends weakly to a measure μ. That is,

$$\lim_{j\to\infty} \int_{T_n} \tilde{T}f_{t_j}(x)g(x)\, dx = \int_{T_n} g(x)\, d\mu(x)$$

for each $g \in C(T_n)$. It follows that $\|d\mu\| \leqq A$. In particular,

$$\lambda(m) = \lim_{j\to\infty} \int_{T_n} e^{-2\pi i m\cdot x}\tilde{T}f_{t_j}(x)\, dx = \int_{T_n} e^{-2\pi i m\cdot x}\, d\mu(x).$$

Therefore,

$$d\mu \sim \sum_{m\in\Lambda} \lambda(m)e^{2\pi i m\cdot x}.$$

The converse is an immediate consequence of the definition of the convolution operator (see (1.6) and the discussion that follows).

Having established these preliminary results we turn to the main question to be discussed in this section. Suppose T belongs to $(L^p(E_n), L^p(E_n))$. We noticed above that T could be realized in terms of a multiplier

$\hat{u}$, which is a bounded measurable function. In general, $\hat{u}$ is not continuous and so $\hat{u}$ need not be defined on the points of the lattice Λ. Suppose, however, that $\hat{u}$ were continuous at the points of Λ, so that $\lambda(m) = \hat{u}(m)$ is well-defined on Λ. We can then ask whether $\{\lambda(m)\}$ is a multiplier in the class $(L^p(T_n), L^p(T_n))$. If this is the case we shall say that the corresponding operator $\tilde{T}$ on the n-torus arises from the operator T by *periodization.*

Before we formulate the general theorem we examine some of the simpler special cases. When $p = 2$, the conditions that $\hat{u}$ is bounded on E_n and continuous at the points of Λ imply $\sup_{m\in\Lambda} |\hat{u}(m)| = \sup_{m\in\Lambda} |\lambda(m)| < \infty$. By Corollary 3.3, therefore, the periodized operator belongs to $(L^2(T_n), L^2(T_n))$. For the case $p = 1$, we know that $\hat{u}(x) = \hat{\mu}(x)$, where $\hat{\mu}$ is the Fourier transform of a measure $\mu \in \mathscr{B}(E_n)$ (see Theorem 3.19 of Chapter I). The fact that the corresponding periodized operator belongs to $(L^1(T_n), L^1(T_n))$ is then a consequence of Theorem (3.4) and the following observation (which can be considered as still another variant of the Poisson summation formula):

THEOREM 3.6. *Suppose* $\mu \in \mathscr{B}(E_n)$ and $\hat{\mu}$ *is its Fourier transform. Then* $\sum_{m\in\Lambda} \hat{\mu}(m)e^{2\pi i m\cdot x}$ *is the Fourier series of a measure* $\tilde{\mu}$ *on* T_n*; moreover,* $\|d\tilde{\mu}\| \leqq \|d\mu\|$.

PROOF. Consider the linear functional mapping $f \in C(T_n)$ into $\int_{E_n} f(x)\, d\mu(x)$ (we are using the symbol f to denote the periodic extension of f as well). This functional can be realized, by the Riesz representation theorem, in terms of some measure $\tilde{\mu} \in \mathscr{B}(T_n)$. That is,

$$\int_{E_n} f(x)\, d\mu(x) = \int_{T_n} f(x)\, d\tilde{\mu}(x). \tag{3.7}$$

Theorem 3.6 now follows by applying (3.7) to $f(x) = e^{-2\pi i m\cdot x}$ for $m \in \Lambda$.

Observe that it follows from (3.7) that for any Borel set $E \subset Q_n$ $\tilde{\mu}(E) = \sum_{m\in\Lambda} \mu(E + m)$. Thus, the measure $\tilde{\mu}$ arises from μ by a method of periodization analogous to that we introduced for functions (see (2.1) and the discussion following).

The main result of this section can be formulated as follows:

THEOREM 3.8. *Suppose* $1 \leqq p \leqq \infty$ *and* $T \in (L^p(E_n), L^p(E_n))$. *Let* $\hat{u}$ *be the multiplier corresponding to* T *and suppose that* $\hat{u}$ *is continuous at each point of the lattice* Λ. *Set* $\lambda(m) = \hat{u}(m)$ *for* $m \in \Lambda$. *Then there exists a unique periodized operator* $\tilde{T}$ *defined by* (3.2) *such that* $\tilde{T} \in (L^p(T_n), L^p(T_n))$ *and* $\|\tilde{T}\| \leqq \|T\|$.

We begin the proof by considering two lemmas.

LEMMA 3.9. *Suppose f is a continuous periodic function on E_n, then*

$$\lim_{\varepsilon \to 0} \varepsilon^{n/2} \int_{E_n} f(x) e^{-\varepsilon\pi|x|^2}\, dx = \int_{Q_n} f(x)\, dx. \tag{3.10}$$

PROOF. Equality (3.10) is obvious when $f(x) = e^{2\pi i m\cdot x}$ since (see (1.13) of Chapter I)

$$\varepsilon^{n/2} \int_{E_n} e^{2\pi i m\cdot x} e^{-\varepsilon\pi|x|^2}\, dx = e^{-\pi|m|^2/\varepsilon}$$

for $\varepsilon > 0$ and $m \in \Lambda$. Hence, (3.10) holds for any trigonometric polynomial. The lemma now follows by approximating an arbitrary continuous periodic function uniformly on Q_n by such polynomials. The second lemma, which is the principal tool we shall use in the proof of Theorem 3.8, is the following:

LEMMA 3.11. *Suppose P and Q are trigonometric polynomials, T belongs to $(L^p(E_n), L^p(E_n))$, $\tilde{T}$ is defined on the class of trigonometric polynomials by* (3.2) *and $w_\delta(y) = e^{-\pi\delta|y|^2}$ for $\delta > 0$, $y \in E_n$. Then*

$$\lim_{\varepsilon \to 0} \varepsilon^{n/2} \int_{E_n} T(Pw_{\varepsilon\alpha})(x)\overline{Q(x)} w_{\varepsilon\beta}(x)\, dx = \int_{Q_n} (\tilde{T}P)(x)\overline{Q(x)}\, dx, \tag{3.12}$$

whenever $\alpha, \beta > 0$ and $\alpha + \beta = 1$.

PROOF. Since the expressions in (3.12) are linear in P and Q it suffices to prove (3.12) when $P(x) = e^{2\pi i m\cdot x}$ and $Q(x) = e^{2\pi i k\cdot x}$ for $m, k \in \Lambda$. By Plancherel's theorem and the definition of the multiplier $\hat{u}$, the integral in the left side equals $\varepsilon^{n/2} \int_{E_n} \hat{u}(x)\varphi(x)\overline{\psi(x)}\, dx$, when φ and ψ are the Fourier transforms of the functions having values $e^{2\pi i m\cdot x}e^{-\pi\varepsilon\alpha|x|^2}$ and $e^{2\pi i k\cdot x}e^{-\pi\varepsilon\beta|x|^2}$. But, by (1.5) and Theorem 1.13 of Chapter I we have

$$\varphi(x) = e^{-\pi(|x-m|^2/\alpha\varepsilon)}(\alpha\varepsilon)^{-(n/2)} \quad \text{and} \quad \psi(x) = e^{-\pi(|x-k|^2/\beta\varepsilon)}(\beta\varepsilon)^{-(n/2)}.$$

Assume now that $m \neq k$ and, consequently, $|m - k| \geqq 1$. Since $|\hat{u}(x)| \leqq A$ for a suitable constant A, the left side of (3.12) is bounded by

$$\varepsilon^{n/2} A \int_{E_n} e^{-(|x-m|^2/\alpha\varepsilon)\pi}(\alpha\varepsilon)^{-(n/2)} e^{-(|x-k|^2/\beta\varepsilon)\pi}(\beta\varepsilon)^{-(n/2)}\, dx$$

$$\leqq \varepsilon^{n/2} A \left[\int_{|x-m| \geqq 1/2} + \int_{|x-k| \geqq 1/2} \right].$$

In the integral extended over $\{x \in E_n : |x - m| \geqq \tfrac{1}{2}\}$, the factor $\varepsilon^{n/2} e^{-(|x-m|^2/\alpha\varepsilon)\pi}(\alpha\varepsilon)^{-(n/2)}$ tends uniformly to 0 as ε tends to 0, while the factor $e^{-(|x-k|^2/\beta\varepsilon)\pi}(\beta\varepsilon)^{-(n/2)}$ has total integral 1 when extended over E_n. It follows, therefore, that $\varepsilon^{(n/2)}\int_{|x-m|\geqq 1/2}$ tends to 0 with ε. The same argument, with the roles of m and k interchanged shows that $\lim_{\varepsilon\to 0}\varepsilon^{n/2}\int_{|x-k|\geqq 1/2} = 0$. Since, for $m \neq k$,

$$\int_{T_n}(\tilde{T}P)(x)\overline{Q(x)}\,dx = \int_{Q_n}\lambda(m)e^{2\pi im\cdot x}e^{-2\pi ik\cdot x}\,dx = 0,$$

equality (3.12) is established for this case.

In case $m = k$, the left side of (3.12) equals

$$\lim_{\varepsilon\to 0}(\varepsilon\alpha\beta)^{-(n/2)}\int_{E_n}\hat{u}(x)e^{-\pi(|x-m|^2/\varepsilon)(1/\alpha+1/\beta)}\,dx. \tag{3.13}$$

Since $1/\alpha + 1/\beta = 1/\alpha\beta$, (3.13) is the limit, as $\varepsilon\to 0$, of the Gauss-Weierstrass integral of $\hat{u}$. By Theorem 1.25 of Chapter I, therefore, this limit is $\hat{u}(m)$ provided m belongs to the Lebesgue set of $\hat{u}$. But this is the case since $\hat{u}$ is assumed to be continuous at m. This proves equality (3.12) when $P(x) = e^{2\pi im\cdot x} = Q(x)$ (since, in this case, $\int_{T_n}(\tilde{T}P)(x)\overline{Q(x)}\,dx = \lambda(m)$) and the lemma is established.

We now pass to the proof of Theorem 3.8. In order to avoid certain irrelevant technicalities we shall temporarily assume that $1 < p < \infty$. Let q be the conjugate exponent to p; then $1/p + 1/q = 1$ and $1 < q < \infty$. We first prove that there exists $A \leqq \|T\|$ such that

$$\left(\int_{Q_n}|(\tilde{T}P)(x)|^p\,dx\right)^{1/p} \leqq A\left(\int_{Q_n}|P(x)|^p\,dx\right)^{1/p} \tag{3.14}$$

for all trigonometric polynomials P. If Q is also a trigonometric polynomial then

$$\left|\int_{E_n}(T(Pw_{\varepsilon\alpha}))(x)\overline{Q(x)}w_{\varepsilon\beta}(x)\,dx\right| \leqq \|T\|\,\|Pw_{\varepsilon\alpha}\|_p\|Qw_{\varepsilon\beta}\|_q, \tag{3.15}$$

where the norms are taken with respect to E_n and w_δ, for $\delta > 0$, is the function introduced in Lemma 3.11. Set $\alpha = (1/p)$, $\beta = (1/q)$ multiply both sides by $\varepsilon^{n/2}$ and let $\varepsilon\to 0$. By Lemma 3.11 the left side converges to $\int_{Q_n}(\tilde{T}P)(x)\overline{Q(x)}\,dx$. On the other hand, by Lemma 3.9

$$\lim_{\varepsilon\to 0}\varepsilon^{n/2}\|Pw_{\varepsilon/p}\|_p\|Qw_{\varepsilon/q}\|_q$$

$$= \lim_{\varepsilon\to 0}\left[\varepsilon^{n/2}\int_{E_n}|P(x)|^p e^{-\varepsilon\pi|x|^2}\,dx\right]^{1/p}\left[\varepsilon^{n/2}\int_{E_n}|Q(x)|^q e^{-\varepsilon\pi|x|^2}\,dx\right]^{1/q}$$

$$= \left[\int_{Q_n}|P(x)|^p\,dx\right]^{1/p}\left[\int_{Q_n}|Q(x)|^q\,dx\right]^{1/q}.$$

Combining this with (3.15) we have

$$\left|\int_{Q_n} (\tilde{T}P)(x)\overline{Q(x)}\,dx\right| \leqq \|T\| \left(\int_{Q_n} |P(x)|^p\,dx\right)^{1/p} \left(\int_{Q_n} |Q(x)|^q\,dx\right)^{1/q}.$$

Finally, taking the supremum over all polynomials Q satisfying $\int_{Q_n} |Q(x)|^q\,dx \leqq 1$, we obtain (3.14). This shows that the restriction of $\tilde{T}$ to trigonometric polynomials is a bounded operator on $L^p(T_n)$ with bound not exceeding $\|T\|$. This restriction then has a unique bounded extension to all of $L^p(T_n)$, and it is this extension that satisfies the conclusions of Theorem 3.8.

We now pass to the cases $p = 1$ and $p = \infty$. These extreme cases are, in fact, much simpler than the general one we just considered. The result for $p = 1$ is (as we have already pointed out) an immediate consequence of Theorem 3.19 of Chapter I and Theorems 3.4 and 3.6 of this chapter. When $p = \infty$ we argue as follows: By Theorem 3.20 of Chapter I, if $T \in (L^\infty(E_n), L^\infty(E_n))$, then $T \in (L^1(E_n), L^1(E_n))$; moreover, the proof of Theorem 3.20 shows that the norm of T, as an operator on $L^1(E_n)$, is no larger than its norm as an operator on $L^\infty(E_n)$. Thus from the just established case $p = 1$, $\tilde{T} \in (L^1(T_n), L^1(T_n))$ and, by (3.4), there exists a finite Borel measure μ on T_n such that $\|d\mu\| = \|\tilde{T}\|$ and $\tilde{T}f = f * d\mu$. For $f \in L^\infty(T_n)$, however, $\|f * d\mu\|_\infty \leqq \|f\|_\infty \|d\mu\|$ and the theorem is also proved in this case.

COROLLARY 3.16. *The conclusions of Theorem* 3.8 *are still valid if, instead of assuming $\hat{u}$ is continuous at the points of Λ, we assume that*

$$\lim_{\varepsilon \to 0} \varepsilon^{-n} \int_{|t| \leqq \varepsilon} [\hat{u}(m - t) - \hat{u}(m)]\,dt = 0 \tag{3.17}$$

for each $m \in \Lambda$.

In fact, the assumption (3.17) suffices to show that the limit (3.13) exists and has value $\hat{u}(m)$ (see (4.10) in Chapter I).

An important class of these multiplier transforms arises from the singular integral operators that were studied in the last chapter. There we showed (see also Theorem 4.7, Corollary 4.12, and (5.10) of Chapter IV) that the multiplier corresponding to a singular integral operator T is a homogeneous function of degree 0, Ω_0, which is continuous on the surface of the unit sphere (hence, continuous everywhere but the origin) and, moveover, has the property $\int_{\Sigma_{n-1}} \Omega_0(x')\,dx' = 0$. If we set $\hat{u}(x) = \Omega_0(x)$, for $x \neq 0$, and $\hat{u}(0) = 0$, then condition (3.17) of Corollary 3.16 is satisfied; consequently, the operator $\tilde{T}$ defined by

$$(\tilde{T}f)(x) \sim \sum_{m \neq 0} \Omega_0(m) a_m e^{2\pi i m \cdot x}$$

whenever

$$f(x) \sim \sum_{m \in \Lambda} a_m e^{2\pi i m \cdot x}$$

belongs to $(L^p(T_n), L^p(T_n))$ for $1 < p < \infty$. A particular example arises when $\Omega_0(x) = P^{(k)}(x)/|x|^k$, where $P^{(k)}$ is a harmonic polynomial on E_n that is homogeneous of degree $k \geqq 1$ (see Theorem 4.5 of Chapter IV). The special cases $\Omega_0(x) = -ix_j/|x|, j = 1, 2, \ldots, n$, are known as the *periodic Riesz transforms* (see (2.7) of Chapter VI).

We shall now show that Theorem 3.8 has a converse. Suppose λ is a continuous function on E_n and the family $\{\lambda(m)\}$, $m \in \Lambda$, consists of the multipliers of an operator in $(L^p(T_n), L^p(T_n))$. Can we conclude that λ is a multiplier of an operator in $(L^p(E_n), L^p(E_n))$? It is easy to see that this cannot be the case since the assumption we are making on λ bears only on its values at the points of the lattice Λ, while the conclusion would involve its behavior on all of E_n. To formulate a meaningful converse we observe that if λ were a multiplier of an operator in $(L^p(E_n), L^p(E_n))$ then, for each $\varepsilon > 0$, the function $\lambda(\varepsilon x)$ would also be a multiplier of an operator T of type $(L^p(E_n), L^p(E_n))$ whose norm depends only on λ and not on ε. In fact, we have $(Tf)\hat{}(x) = \lambda(x)\hat{f}(x)$ for each $f \in \mathcal{S}$. Thus, if we define T_ε to be the operator $\delta_\varepsilon^{-1} T \, \delta_\varepsilon$ (δ_ε being the dilation operator introduced immediately before (1.6) of Chapter I) we obtain again a multiplier operator whose multiplier is the function having the values $\lambda(\varepsilon x)$. Moreover, since $\|\delta_\varepsilon f\|_p = \varepsilon^{-n/p}\|f\|_p$ and $\|\delta_\varepsilon^{-1} f\|_p = \varepsilon^{n/p}\|f\|_p$, we see that $\|T_\varepsilon\| = \|T\|$. In view of this observation, it is not surprising that the following is a converse of (3.8):

THEOREM 3.18. *Let λ be a continuous function on E_n. Suppose that for each $\varepsilon > 0$ there exists an operator $\tilde{T}_\varepsilon \in (L^p(T_n), L^p(T_n))$ given by*

$$(3.19) \qquad (\tilde{T}_\varepsilon f)(x) \sim \sum_{m \in \Lambda} \lambda(\varepsilon m) a_m e^{2\pi i m \cdot x}$$

whenever $\{a_m\}$ are the Fourier coefficients of $f \in L^p(T_n)$. We assume that the norms $\|\tilde{T}_\varepsilon\|$ of the operators $\tilde{T}_\varepsilon$ are uniformly bounded. Then λ is a multiplier of type $(L^p(E_n), L^p(E_n))$; moreover, if T is the corresponding operator, its operator norm $\|T\|$ does not exceed $\sup_{\varepsilon > 0} \|\tilde{T}_\varepsilon\|$.

PROOF. We dispose first of the case $p = \infty$ by showing that it can be reduced to the case $p = 1$. In fact, (3.19) implies immediately the duality identity

$$\int_{T_n} (\tilde{T}_\varepsilon f)(x) g(-x)\,dx = \int_{T_n} (\tilde{T}_\varepsilon g)(x) f(-x)\,dx \tag{3.20}$$

whenever f and g are (say) trigonometric polynomials. From this it follows (see the analogous argument in the proof of Theorem 3.20 of Chapter I) that $\|\tilde{T}_\varepsilon\|_1 \leqq \|\tilde{T}_\varepsilon\|_\infty$ (the subscripts are used here to denote the norms of $\tilde{T}_\varepsilon$ as an operator on $L^1(T_n)$ and on $L^\infty(T_n)$, respectively). Thus, if the case $p = 1$ of this theorem were established, we would obtain an operator $T \in (L^1(E_n), L^1(E_n))$, having multiplier λ, satisfying $\|T\|_1 \leqq \sup_{\varepsilon>0} \|\tilde{T}_\varepsilon\|_1 \leqq \sup_{\varepsilon>0} \|\tilde{T}_\varepsilon\|_\infty$. Finally, again appealing to Theorem 3.20 of Chapter I, we would obtain the fact that $T \in (L^\infty(E_n), L^\infty(E_n))$ and $\|T\|_\infty \leqq \sup_{\varepsilon>0} \|\tilde{T}_\varepsilon\|_\infty$.

We shall, therefore, assume that $1 \leqq p < \infty$ and make use of the following appropriate partition of unity:

LEMMA 3.21. *There exists a nonnegative continuous function η having compact support in E_n and satisfying*

(a) $\eta(0) = 1$;
(b) $\sum_{m\in\Lambda} [\eta(x + m)]^p \equiv 1$.

To prove this lemma choose any nonnegative continuous function η_1 with compact support in E_n such that $\eta_1(0) = 1$, $\eta_1(m) = 0$ if $m \in \Lambda - \{0\}$ and $\eta_1(x) > 0$ for $x \in \bar{Q}_n$. Let $\eta_2(x) = \eta_1(x)/\sum_{m\in\Lambda} \eta_1(x + m)$. Then, clearly, $\eta_2(0) = 1$ and $\sum_{m\in\Lambda} \eta_2(x + m) \equiv 1$. We now need only take $\eta = \eta_2^{1/p}$.

Let us now return to the proof of the theorem. We assume, for simplicity, that $\|\tilde{T}_\varepsilon\|_p \leqq 1$, $\varepsilon > 0$. Then $|\lambda(\varepsilon m)| \leqq 1$ for $m \in \Lambda$ and $\varepsilon > 0$ (see the proof of Theorem 3.1). Since the set $\{\varepsilon m : \varepsilon > 0, m \in \Lambda\}$ is dense in E_n it follows that λ is bounded. When $f \in L^2(E_n)$, therefore, $\lambda\hat{f}$ also belongs to $L^2(E_n)$; hence, the latter function is the Fourier transform of a square integrable function. In particular, this allows us to define Tf, for $f \in \mathscr{D}$ $(\subset \mathscr{S})$, to be the function whose Fourier transform is $\lambda\hat{f}$; that is, $(Tf)\hat{}(x) = \lambda(x)\hat{f}(x)$. We shall show that

$$\|Tf\|_p \leqq \|f\|_p. \tag{3.22}$$

In order to do so, define $\tilde{f}_\varepsilon$, for $\varepsilon > 0$, to be the dilated and periodized version of f, viz.,

$$\tilde{f}_\varepsilon(x) = \varepsilon^{-n} \sum_{m\in\Lambda} f\left(\frac{x + m}{\varepsilon}\right).$$

Then, using the Poisson summation formula (2.7), we obtain

(3.23) $$\tilde{f}_\varepsilon(x) = \sum_{m\in\Lambda} \hat{f}(\varepsilon m)e^{2\pi i m\cdot x}.$$

We now claim that, for each $x \in E_n$,

(3.24) $$\lim_{\varepsilon\to 0} \varepsilon^n[\tilde{T}_\varepsilon \tilde{f}_\varepsilon](\varepsilon x) = [Tf](x).$$

From (3.19) and (3.23) we obtain

(3.25) $$\varepsilon^n[\tilde{T}_\varepsilon \tilde{f}_\varepsilon](\varepsilon x) = \varepsilon^n \sum_{m\in\Lambda} \lambda(\varepsilon m)\hat{f}(\varepsilon m)e^{2\pi i\varepsilon m\cdot x}.$$

Now, λ is bounded, $\hat{f}$ is rapidly decreasing at ∞ (i.e., $\lim_{|x|\to\infty} |x|^k|\hat{f}(x)| = 0$ for all positive integers k) and both λ and $\hat{f}$ are continuous. Consequently, by the definition of the Riemann integral, the right side of (3.25) tends to

$$\int_{E_n} \lambda(t)\hat{f}(t)e^{2\pi i x\cdot t}\, dt = (Tf)(x)$$

as ε tends to 0 (see Corollary 1.21 of Chapter I). Therefore, equality (3.24) is established. Moreover, since η is continuous and $\eta(0) = 1$, we also have

(3.26) $$\lim_{\varepsilon\to 0} \varepsilon^n[\tilde{T}_\varepsilon \tilde{f}_\varepsilon](\varepsilon x)\eta(\varepsilon x) = (Tf)(x)$$

for each $x \in E_n$.

Because of the periodicity of $\tilde{T}_\varepsilon \tilde{f}_\varepsilon$ we have

$$\begin{aligned}\varepsilon^{np} \int_{E_n} |(\tilde{T}_\varepsilon \tilde{f}_\varepsilon)(\varepsilon x)\eta(\varepsilon x)|^p\, dx &= \varepsilon^{np-n} \int_{E_n} |(\tilde{T}_\varepsilon \tilde{f}_\varepsilon)(x)|^p[\eta(x)]^p\, dx \\ &= \varepsilon^{np-n} \sum_{m\in\Lambda} \int_{Q_n} |(\tilde{T}_\varepsilon \tilde{f}_\varepsilon)(x)|^p[\eta(x+m)]^p\, dx.\end{aligned}$$

Thus, by Lemma 3.21 and the assumption $\|\tilde{T}_\varepsilon\|_p \leqq 1$,

(3.27) $$\int_{E_n} |\varepsilon^n(\tilde{T}_\varepsilon \tilde{f}_\varepsilon)(\varepsilon x)\eta(\varepsilon x)|^p\, dx \leqq \varepsilon^{np-n} \int_{Q_n} |\tilde{f}_\varepsilon(x)|^p\, dx.$$

For ε sufficiently small the support of $\varepsilon^{-n}f(x/\varepsilon)$ lies entirely in Q_n and, in this case, $\varepsilon^{-n}f(x/\varepsilon) = \tilde{f}_\varepsilon(x)$ for $x \in Q_n$. Thus, for small ε the right side of (3.27) equals

$$\varepsilon^{np-n} \int_{Q_n} |\varepsilon^{-n}f(x/\varepsilon)|^p\, dx = \varepsilon^{np-n} \int_{E_n} |\varepsilon^{-n}f(x/\varepsilon)|^p\, dx = \int_{E_n} |f(x)|^p\, dx.$$

Now, applying (3.26) and Fatou's lemma to the left side of (3.27) we obtain

$$\int_{E_n} |(Tf)(x)|^p \, dx \leqq \int_{E_n} |f(x)|^p \, dx,$$

which is the desired inequality (3.22). Since the class $\mathscr{D}$ is dense in $L^p(E_n)$, $1 \leqq p < \infty$, this proves the theorem.

Theorem 3.18, the remarks immediately preceding it, and Theorem 3.8 have the following obvious corollary:

COROLLARY 3.28. $\sup_{\varepsilon>0} \|\tilde{T}_\varepsilon\| = \|T\|$, *where* $\|\tilde{T}_\varepsilon\|$ *and* $\|T\|$ *denote the operator norms of the operators* $\tilde{T}_\varepsilon$ *and* T *occurring in Theorem* 3.18.

4. Summability Below the Critical Index (Negative Results)

Let $\sum_{m\in\Lambda} a_m e^{2\pi i m\cdot x}$ be the Fourier series of an integrable function f. COROLLARY (2.15) asserts that the equality

$$\lim_{R\to\infty} \sum_{|m|<R} \left(1 - \frac{|m|^2}{R^2}\right)^\alpha a_m e^{2\pi i m\cdot x} = f(x) \tag{4.1}$$

is valid (almost everywhere and also if the limit is taken in the L^1 norm), provided α is larger than the critical index $(n - 1)/2$. It is natural to ask what happens when $\alpha \leqq (n - 1)/2$. In this section we shall be concerned mainly with the two cases $\alpha = (n - 1)/2$ and $\alpha = 0$.

In order to gain some perspective about this question we begin by reviewing the classical results when $n = 1$. In this case summability at the critical index ($\alpha = 0$) and ordinary convergence coincide. Kolmogoroff showed that there exists an L^1 function for which the limit (4.1) fails to exist for every x when $\alpha = 0$. Nevertheless, there are localization results for L^1 functions which assert that (4.1) holds for a given x as long as f is sufficiently regular in an arbitrarily small neighborhood of x. For a function f in $L^p(T_1)$, $p > 1$, there is a result of Carleson and Hunt which shows that the Fourier series of f converges to $f(x)$ almost everywhere. There is also an older result of M. Riesz which shows that convergence in the norm holds when $1 < p < \infty$.

We shall see now that if $n > 1$, the situation has a rather different character. This is clear from the following three assertions:

THEOREM 4.2. *There exists an* $f \in L^1(T_n)$, $n > 1$, *such that*

$$\limsup_{R\to\infty} \left| \sum_{|m|<R} \left(1 - \frac{|m|^2}{R^2}\right)^{(n-1)/2} a_m e^{2\pi i m\cdot x} \right| = \infty$$

for almost every x. *Such a function* f *can be constructed so that it is supported in an arbitrarily small given neighborhood of the origin.*[4]

[4] This theorem shows at once that neither almost everywhere summability nor localization can hold at the critical index.

THEOREM 4.3. *The trigonometric series*

$$\sum_{m \neq 0} |m|^{-(n/2)+1/2} e^{2\pi i m \cdot x} \tag{4.4}$$

diverges almost everywhere. More particularly,

$$\limsup_{R \to \infty} \left| \sum_{0 < |m| < R} |m|^{-(n/2)+1/2} e^{2\pi i m \cdot x} \right| = \infty$$

for almost every $x \in Q_n$.

We know by Theorem 2.17 that the series (4.4) is the Fourier series of the function whose values are $\gamma_{(n-1)/2}^{-1} |x|^{-(n+1)/2} + b(x)$, where $b \in C^\infty(Q_n)$. Thus, we also have the following corollary:

COROLLARY 4.5. *There exists a function which belongs to* $L^p(T_n)$, $p < 2n/(n+1)$, *whose Fourier series diverges almost everywhere.*

Therefore, for the general function of $L^p(T_n)$, $p < 2$, convergence almost everywhere fails if the dimension n is sufficiently large. It is an open problem to determine what happens for $p = 2$.

These three results are based on the following lemma:

LEMMA 4.6. *If* $n > 1$ *then*

$$\limsup_{R \to \infty} \left| \sum_{|m| < R} \left(1 - \frac{|m|^2}{R^2}\right)^{(n-1)/2} e^{2\pi i m \cdot x} \right| = \infty \tag{4.7}$$

for almost every $x \in Q_n$.

We break up the proof of (4.6) into several steps. In doing so we set

$$K_R^\alpha(x) = \sum_{|m| < R} \left(1 - \frac{|m|^2}{R^2}\right)^\alpha e^{2\pi i m \cdot x}$$

and shall often write K_R for $K_R^{(n-1)/2}$.

LEMMA 4.8. *If* x^0 *is a point for which*

$$\limsup_{R \to \infty} \left| \sum_{|m| < R} \left(1 - \frac{|m|^2}{R^2}\right)^{(n-1)/2} e^{2\pi i m \cdot x^0} \right| < \infty$$

then

$$\sup_{\alpha > (n-1)/2} \sup_{R > 0} \left| \sum_{|m| < R} \left(1 - \frac{|m|^2}{R^2}\right)^\alpha e^{2\pi i m \cdot x^0} \right| < \infty. \tag{4.9}$$

This simple lemma indicates that it might be useful to reduce our consideration to the case $\alpha > (n-1)/2$ where the Poisson summation formula is directly applicable.

To prove the lemma we begin by observing that the identity [5]

$$t^{\delta+\beta} = \frac{\Gamma(\delta+\beta+1)}{\Gamma(\delta+1)\Gamma(\beta)} \int_0^t (t-s)^{\beta-1} s^{\delta}\, ds$$

and the change of variables $s = r^2 - |m|^2$ implies

$$\int_{|m|}^{R} (R^2 - r^2)^{\beta-1}\left(1 - \frac{|m|^2}{r^2}\right)^{(n-1)/2} r^n\, dr$$

$$= \int_{|m|}^{R} (R^2 - r^2)^{\beta-1}(r^2 - |m|^2)^{(n-1)/2} r\, dr$$

$$= \frac{R^{2\beta+n-1}\Gamma[(n+1)/2]\Gamma(\beta)}{2\Gamma[(n+1)/2+\beta]}\left(1 - \frac{|m|^2}{R^2}\right)^{\beta+(n-1)/2}$$

Putting

$$\beta = \alpha - \frac{n-1}{2} > 0, \qquad c_\alpha = \frac{2\Gamma(\alpha+1)}{\Gamma[(n+1)/2]\Gamma(\beta)}$$

and interchanging the order of summation and integration we therefore obtain

$$K_R^\alpha(x) = \sum_{|m|<R} \left(1 - \frac{|m|^2}{R^2}\right)^{\alpha} e^{2\pi i m\cdot x}$$

$$= \sum_{|m|<R} \left\{ c_\alpha R^{-2\alpha} \int_{|m|}^{R} (R^2 - r^2)^{\beta-1}\left(1 - \frac{|m|^2}{r^2}\right)^{(n-1)/2} r^n\, dr \right\} e^{2\pi i m\cdot x}$$

$$= c_\alpha R^{-2\alpha} \int_0^R \left\{ \sum_{|m|<r} \left(1 - \frac{|m|^2}{r^2}\right)^{(n-1)/2} e^{2\pi i m\cdot x} \right\} (R^2 - r^2)^{\beta-1} r^n\, dr;$$

that is,

(4.10) $$K_R^\alpha(x) = c_\alpha R^{-2\alpha} \int_0^R K_r(x)(R^2 - r^2)^{\beta-1} r^n\, dr.$$

Now, if $\limsup_{R\to\infty} |K_R(x^0)| < \infty$ then $\sup_{R>0} |K_R(x^0)| = A < \infty$. By (4.10) then

$$|K_R^\alpha(x^0)| \leqq A\left\{ c_\alpha R^{-2\alpha} \int_0^R (R^2 - r^2)^{\beta-1} r^n\, dr \right\}$$

[5] This equality follows immediately, via a change of variables, from the well-known relationship between the beta and gamma functions: $\int_0^1 t^{x-1}(1-t)^{y-1}\, dt = \Gamma(x)\Gamma(y)/\Gamma(x+y)$.

This last expression within brackets, however, is identically equal to one (this can be easily seen from (4.10) by considering the constant term). Thus,

$$\sup_{\alpha>(n-1)/2}\ \sup_{R>0} |K_R^\alpha(x^0)| \leqq \sup_{R>0} |K_R(x^0)|$$

and Lemma 4.8 is proved.

We now describe the set S of points for which (we claim) that (4.7) takes place. Let S be the set of points x where the denumerable collection of real numbers $\{|x - m| : m \in \Lambda\}$ is linearly independent over the rationals. In one dimension the set S is obviously empty; however, we will show that the complement of S has measure 0 when $n > 1$.

LEMMA 4.11. *If $x^0 \in S$, then*

$$\limsup_{R\to\infty} \left| \sum_{|m|<R} \left(1 - \frac{|m|^2}{R^2}\right)^{(n-1)/2} e^{2\pi i m\cdot x^0} \right| = \infty.$$

PROOF. If $\alpha > (n - 1)/2$, in view of the remarks made before (2.14), we can apply the Poisson summation formula and obtain

(4.12)
$$K_R^\alpha(x) = \pi^{-\alpha}\Gamma(\alpha + 1)R^{(n/2)-\alpha} \sum_{m\in\Lambda} J_{(n/2)+\alpha}(2\pi R|x - m|)/|x - m|^{(n/2)+\alpha}.$$

Using the asymptotic estimates for the Bessel functions in lemma 3.11 of Chapter IV (see also the accompanying footnote) we obtain for $x^0 \notin \Lambda$

$$K_R^\alpha(x^0) = c_\alpha R^{(n-1)/2-\alpha} \sum \frac{\cos(2\pi R\gamma_m + \delta_\alpha)}{\gamma_m^{\alpha+(n+1)/2}} + E(R, \alpha) \tag{4.13}$$

where $\gamma_m = |x^0 - m|$, and $\sup_{\alpha>(n-1)/2}\ \sup_{R>1} R|E(R, \alpha)| < \infty$.

The identity (4.13) implies that

$$\lim_{T\to\infty} \frac{1}{T}\int_1^T K_R(x^0)e^{2\pi i\lambda R}\, dR \text{ exists } (=a(\lambda)), \text{ if } \lambda \geq 0, \tag{4.14}$$

with $a(\lambda) = c\gamma_m^{-n}$, if $\lambda = \gamma_m$, and $a(\lambda) = 0$ otherwise. (To prove this integrate (4.13) with respect to R; then let $\alpha \to (n - 1)/2$, and finally let $T \to \infty$). The following lemma shows that (4.14) is incompatible with $\limsup_{R\to\infty} |K_R(x^0)| < \infty$, if $x^0 \in S$.

LEMMA 4.15 *Suppose K is a bounded real function on $[1, \infty)$, and $\lim_{T\to\infty} \frac{1}{T}\int_1^T K(t)e^{2\pi i\lambda t}\, dt = a(\lambda)$ exists for every $\lambda \geq 0$. Let $\{\lambda_j\}$ be the set of λ's for which $a(\lambda) \neq 0$, and assume this set is linearly independent over the* *rationals. Then*

(4.16) $$\Sigma|a(\lambda_j)| < \infty .$$

PROOF Let $\lambda_1, \ldots \lambda_n$ be any finite sub-collection of $\{\lambda_j\}$. Write $a(\lambda_j) = |a(\lambda_j)|e^{i\mu_j}$, and $A(t) = \prod_1^N (1 + \cos(2\pi\lambda_j t - \mu_j))$. By the linear independence of $\{\lambda_j\}$, $\lim_{T\to\infty} \frac{1}{T}\int_1^T A(t)\,dt = 1$; moreover

$$\lim_{T\to\infty} \frac{1}{T}\int_1^T K(t)A(t)\,dt = \sum_1^N |a(\lambda_j)|.$$

From this it follows that $\sum_1^N |a(\lambda_j)| \leqq \sup_{t\geq 1} |K(t)|$; hence (4.16) is proved, and so is lemma 4.15. Since

$$\sum_m \gamma_m^{-n} \approx \sum_{m\neq 0} |m|^{-n} = \infty,$$

we see that $\limsup_{R\to\infty} |K_R(x^0)| = \infty$, if $x^0 \in S$. Therefore, Lemma 4.11 is established and to complete the proof of our main lemma (Lemma 4.6) we need only show that the following result is true:

LEMMA 4.17. *If $n > 1$, the complement of S has measure* 0.

PROOF. Suppose $a_{m_1}, a_{m_2}, \ldots, a_{m_k}$ are nonzero rational numbers (associated with the lattice points $m_1, \ldots, m_k$). Let

$$\Phi(x) = \sum_{j=1}^{k} a_{m_j}|x - m_j|.$$

Then $\Phi(x) \not\equiv 0$ since Φ has singularities at $m_j, j = 1, 2, \ldots, k$. Moreover, Φ is real-analytic on the connected set $E_n - \Lambda$. But a nonzero real-analytic function can vanish only on a set of Lebesgue measure 0. Thus, the relation

(4.18) $$\Phi(x) = \sum_{j=1}^{k} a_{m_j}|x - m_j| = 0$$

can hold only in a set of measure 0. Since there are only a countable number of such point sets (each associated with a choice of nonzero rationals $a_{m_1}, \ldots, a_{m_k}$), their union also has measure 0. But this union is clearly $E_n - S$.[6] This proves Lemma 4.17.

[6] The reader should note that this argument breaks down for $n = 1$ because Φ consists of distinct "pieces" of linear functions in this case; or to put it another way, $E_n - \Lambda$ is not connected, when $n = 1$.

Let us see what we have accomplished. Let μ_0 be the Dirac measure. Then

$$d\mu_0 \sim \sum_{m\in\Lambda} e^{2\pi i m\cdot x}.$$

Lemma 4.6, therefore, shows that the Fourier–Stieltjes series of $d\mu_0$ is not summable by the Riesz means at the critical index for almost every $x \in Q_n$. To complete the proof of Theorem 4.2 we shall replace μ_0 by "sharply peaked" L^1 functions.

Let ψ be a nonnegative C^∞ function in E_n with support in the unit ball $|x| \leqq 1$ which satisfies $\int_{E_n} \psi(x)\,dx = 1$. Let $\Phi(y) = \int_{E_n} \psi(x)e^{-2\pi i x\cdot y}\,dx$ and $\varphi_\varepsilon(x) = \varepsilon^{-n}\sum_{m\in\Lambda}\psi((x-m)/\varepsilon)$. Then, by the Poisson summation formula,

$$\varphi_\varepsilon(x) \sim \sum_{m\in\Lambda} \Phi(\varepsilon m)e^{2\pi i m\cdot x}. \tag{4.19}$$

We shall show that we can find two null sequences, $\{\varepsilon_k\}$ and $\{\delta_k\}$, of positive numbers so that the function f defined by

$$f(x) = \sum_{k=1}^{\infty} 2^{-k}(\varphi_{\varepsilon_k}(x) - \varphi_{\delta_k}(x)) \tag{4.20}$$

belongs to $L^1(T_n)$ and satisfies the conclusions of Theorem 4.2.

Let S_R be the operator defined by

$$(S_R f)(x) = \sum_{|m|<R}\left(1 - \frac{|m|^2}{R^2}\right)^{(n-1)/2} a_m e^{2\pi i m\cdot x} \qquad \text{when } f \sim \sum_{m\in\Lambda} a_m e^{2\pi i m\cdot x}.$$

We first observe that

$$\sup_{0<R} |(S_R\varphi_\varepsilon)(x)| \leqq A\varepsilon^{-n}. \tag{4.21}$$

In fact,

$$\sup_{0<R} |(S_R\varphi_\varepsilon)(x)| \leqq \sum_{m\in\Lambda} |\Phi(\varepsilon m)| = \sum_{|m|<1/\varepsilon} |\Phi(\varepsilon m)| + \sum_{|m|\geqq 1/\varepsilon} |\Phi(\varepsilon m)|.$$

Since the functions with values $|\Phi(x)|$ and $|x|^{n+1}|\Phi(x)|$ are bounded, we have

$$\text{(i)} \qquad \sum_{|m|<1/\varepsilon} |\Phi(\varepsilon m)| \leqq \|\Phi\|_\infty \sum_{|m|<1/\varepsilon} 1 = \|\Phi\|_\infty N_\varepsilon,$$

where N_ε ($\leqq$ constant times ε^{-n}) is the number of lattice points satisfying $|m| < 1/\varepsilon$, and

$$\text{(ii)} \qquad \sum_{|m|\geqq 1/\varepsilon} |\Phi(\varepsilon m)| \leqq \Big\{\sup_{x\in E_n} |x|^{n+1}|\Phi(x)|\Big\} \sum_{|m|\geqq 1/\varepsilon} |\varepsilon m|^{-(n+1)}$$

$$= \Big\{\sup_{x\in E_n} |x|^{n+1}|\Phi(x)|\Big\}\varepsilon^{-n-1} \sum_{|m|\geqq 1/\varepsilon} |m|^{-(n+1)} \leqq B\varepsilon^{-n}.$$

These two inequalities imply (4.21).

We shall also construct subsets $\mathscr{E}_k \subset Q_n$ with measures $|\mathscr{E}_k| \geqq 1 - 1/k$ and an increasing sequence of positive numbers $\{R_k\}$ such that

$$\sup_{R \leqq R_k} |(S_R f)(x)| \geqq k, \qquad \text{if } x \in \mathscr{E}_k. \tag{4.22}$$

Once this is done, it is easy to see that we have established Theorem 4.2. It is clear that the first conclusion of this theorem follows immediately from (4.22). The second is a consequence of the fact that, in Q_n, φ_ε has its support inside the solid sphere of radius ε about the origin; therefore, by subtracting from f partial sums of the series (4.20), we obtain functions satisfying both the first and second parts of (4.2).

Suppose now that ε_j, δ_j, R_j, and $\mathscr{E}_j$ have been determined when $1 \leqq j \leqq k - 1$; we shall describe how to choose ε_k, δ_k, R_k, and $\mathscr{E}_k$. We always select $\varepsilon_k \leqq \delta_k$ and choose δ_k so small that

$$\sup_{R \leqq R_{k-1}} |S_R(\varphi_{\varepsilon_k} - \varphi_{\delta_k})| \leqq 1. \tag{4.23}$$

If this can be done for each k, we have, of course,

$$\sup_{R \leqq R_k} |S_R(\varphi_{\varepsilon_{k'}} - \varphi_{\delta_{k'}})| \leqq 1 \qquad \text{if } k' > k. \tag{4.23'}$$

Since

$$\begin{aligned}\sup_{R \leqq R_{k-1}} |S_R(\varphi_{\varepsilon_k} - \varphi_{\delta_k})| &\leqq \sum_{|m| < R_{k-1}} |\Phi(\varepsilon_k m) - \Phi(\delta_k m)| \\ &\leqq A(\delta_k - \varepsilon_k) \sum_{|m| < R_{k-1}} |m| \leqq A'\delta_k (R_{k-1})^{n+1},\end{aligned}$$

we see that (4.23) can be realized if δ_k is small enough.

Now, let A_k be a positive number such that

$$A_k \geqq \sup_{0 < R < \infty} \left\{ \left| S_R\left(\sum_{j<k} 2^{-j}(\varphi_{\varepsilon_j} - \varphi_{\delta_j})(x) - 2^{-k}\varphi_{\delta_k}(x)\right) \right| \right\}. \tag{4.24}$$

Because of (4.21) such a finite constant A_k can be found.

So far A_k and δ_k have been fixed and ε_k is restricted only by the condition $\varepsilon_k \leqq \delta_k$. We now impose on ε_k a condition which, in essence, implies that φ_{ε_k} is sufficiently close to $d\mu_0$. That is, we know that if R_k is large enough then, in view of the almost everywhere divergence in Lemma 4.6,

$$\sup_{R \leqq R_k} 2^{-k} |(S_R\, d\mu_0)(x)| > A_k + k + 2 \tag{4.25}$$

on a set $\mathscr{E}_k \subset Q_n$ having measure $|\mathscr{E}_k| \geqq 1 - 1/k$. The Fourier series of φ_ε, however, converges termwise to that of $d\mu_0$ as ε tends to 0. Therefore, by choosing ε_k sufficiently small we have

$$\sup_{R \leqq R_k} 2^{-k} |(S_R \varphi_{\varepsilon_k})(x)| > A_k + k + 1 \qquad \text{for } x \in \mathscr{E}_k. \tag{4.26}$$

Thus, (4.25) and (4.26) determine our choice of ε_k, R_k, and $\mathscr{E}_k$.

We now turn to the proof of (4.22). Write

$$f = \Big\{\sum_{j<k} 2^{-j}(\varphi_{\varepsilon_j} - \varphi_{\delta_j}) - 2^{-k}\varphi_{\delta_k}\Big\} + 2^{-k}\varphi_{\varepsilon_k} + \Big\{\sum_{j>k} 2^{-j}(\varphi_{\varepsilon_j} - \varphi_{\delta_j})\Big\}.$$

Consider $\sup_{R \leqq R_k} |(S_R f)(x)|$ for $x \in \mathscr{E}_k$. In view of (4.24) the contribution from the terms within the first bracket is at most A_k. By (4.26) the middle term contributes at least $A_k + k + 1$. Finally, the last term contributes at most 1 (by (4.23)). This proves (4.22) and Theorem 4.2 is completely proved.

We shall now prove Theorem 4.3. We claim first that, if for some x^0

$$\sup_{0<R<\infty} \Big| \sum_{0<|m|<R} |m|^{-(n/2)+1/2} e^{2\pi i m \cdot x^0} \Big| < \infty,$$

then

$$\sup_{0<R<\infty} R^{-(n/2)+1/2} \Big| \sum_{0<|m|<R} e^{2\pi i m \cdot x^0} \Big| < \infty. \tag{4.27}$$

In fact, if we put $\sigma_R = \sum_{0<|m|<R} |m|^{-(n/2)+1/2} e^{2\pi i m \cdot x^0}$, then

$$\sum_{0<|m|<R} e^{2\pi i m \cdot x^0} = \int_1^R t^{(n/2)-1/2}\, d\sigma_t = R^{(n/2)-1/2}\sigma_R - (n/2 - 1/2) \int_1^R \sigma_t t^{(n/2)-3/2}\, dt.$$

From this it is clear that the assumption $\sup_{0<R<\infty} |\sigma_R| < \infty$ implies (4.27).

We next show that

$$\sup_{0<R<\infty} R^{-(n/2)+1/2} \Big| \sum_{0<|m|<R} e^{2\pi i m \cdot x^0} \Big| = \infty$$

for each x^0 satisfying

$$\sup_{0<R<\infty} \Big| \sum_{|m|<R} \Big(1 - \frac{|m|^2}{R^2}\Big)^{(n-1)/2} e^{2\pi i m \cdot x^0} \Big| = \infty$$

(by Lemma 4.6, almost all $x^0 \in Q_n$ satisfy this last relation). That is, we will show that

$$\sup_{0<R<\infty} R^{-(n/2)+1/2} |K_R^0(x^0) - 1| = \infty$$

for each x^0 satisfying

$$\sup_{0<R<\infty} |K_R(x^0)| = \sup_{0<R<\infty} |K_R^{(n-1)/2}(x^0)| = \infty.$$

It is easier to do this for n odd and, in fact, the simplest case is the one occurring when $n = 3$. Let us consider this case; thus, the critical index is $(n-1)/2 = 1$. For x^0 fixed write

$$F_\alpha(t) = t^\alpha K_{\sqrt{t}}^\alpha(x^0) = \sum_{|m|^2<t} (t - |m|^2)^\alpha e^{2\pi i m \cdot x^0}.$$

Obviously $(d/dt)F_\alpha(t) = \alpha F_{\alpha-1}(t)$ for $\alpha \geqq 1$. We claim that

$$|F_2(t)| \leqq At^{3/2} \qquad \text{for } 1 \leqq t < \infty. \tag{4.28}$$

In fact, when $\alpha > (n-1)/2$ the Poisson summation formula can be applied to $K_R^\alpha(x^0)$ in order to obtain the expression (4.12). This, together with the estimate $|J_{(n/2)+\alpha}(2\pi R)| \leqq AR^{-(1/2)}$, $R \geqq 1$, (see Lemma 3.11 in Chapter IV) shows that

$$|K_R^\alpha(x^0)| \leqq AR^{(n-1)/2-\alpha} \qquad \text{for } 1 \leqq R < \infty$$

which, by the definition of F_2, is equivalent to (4.28).

If $|R^{-(n/2)+1/2}K_R^0(x^0)| \leqq A$ then (since $n = 3$)

$$\frac{1}{2}\left|\frac{d^2F_2(t)}{dt^2}\right| = |F_0(t)| \leqq At^{1/2}, \qquad 1 \leqq t < \infty. \tag{4.29}$$

Using the Taylor development of F_2 up to the second derivative, we see that

$$|F_2(t+h) - F_2(t) - hF_2'(t)| \leqq \frac{h^2}{2} \sup_{t \leqq t' \leqq t+h} |F_2''(t')|.$$

If we put $h = t^{1/2}$, (4.28) and (4.29) imply $|F_2'(t)| \leqq Bt$, $1 \leqq t < \infty$. Thus, $|F_1(t)| \leqq Bt/2$, and, therefore, $\sup_{1 \leqq R < \infty} |K_R(x^0)| \leqq B$. This shows that

$$\sup_{R>0} R^{-(n/2)+1/2}|K_R^0(x^0) - 1| = \infty,$$

whenever $\sup_{R>0} |K_R(x^0)| = \infty$. This completes the proof of Thorem 4.3 in case $n = 3$. The other cases $n > 1$ are similar but somewhat more complicated. The needed machinery is presented in (6.10) below.

5. Summability Below the Critical Index

Suppose $\sum_{m\in\Lambda} a_m e^{2\pi i m\cdot x}$ is the Fourier series of a function f and

$$S_R^\alpha(f)(x) = \sum_{|m|<R} \left(1 - \frac{|m|^2}{R^2}\right)^\alpha a_m e^{2\pi i m\cdot x}$$

are the corresponding Riesz means of order α. We shall denote by $S_*^\alpha(f)(x)$ the associated "maximal function"; that is,

$$S_*^\alpha(f)(x) = \sup_{0<R<\infty} |S_R^\alpha(f)(x)|.$$

In this section we shall see that if $f \in L^p(T_n)$, $1 < p < \infty$, then there are positive results for summability below the critical index. The theorem that is proved does not, in view of the counterexamples of Section 4, completely dispose of the problems of summability of multiple Fourier series.

Nevertheless, it does represent the furthest extent of our present knowledge of this subject.[7]

THEOREM 5.1. *Suppose* $1 < p < \infty, n > 1$ *and* $\alpha > (n - 1)|\frac{1}{2} - 1/p|$. *Then*

(a) $\|S_*^\alpha(f)\|_p \leqq A_{p,\alpha}\|f\|_p$;
(b) $\lim_{R\to\infty} S_R^\alpha(f)(x) = f(x)$ *for almost every* x;
(c) $\|S_R^\alpha(f) - f\|_p \to 0$ *as* $R \to \infty$.

This theorem will be deduced from two lemmas; the first will be a reformulation of the results obtained in the second section when α was greater than the critical index, the second will be an L^2 result for α near zero and, finally, a combination of these two estimates obtained by using the complex convexity of the norms of linear operators (Theorem 4.1 in Chapter V). The first lemma is as follows:

LEMMA 5.2. *When* $\alpha > (n - 1)/2$ *conclusion* (a) *of Theorem* 5.1 *is true.*

We remind the reader that when $\alpha > (n - 1)/2$, conclusions (b) and (c) of Theorem 5.1 are contained in Theorem 2.11, and in particular its Corollary 2.15 above.

PROOF. Let $f \in L^p(T_n)$. We also use the symbol f to denote the periodic extension to E_n. By Corollary 2.15 (see also the comments preceding it) we know that

$$S_R^\alpha(f)(x) = \int_{Q_n} \sum_{m\in\Lambda} \varphi_\varepsilon(y + m)f(x - y)\, dy = \int_{E_n} \varphi_\varepsilon(y)f(x - y)\, dy$$

for an appropriate φ, where $\varphi_\varepsilon(x) = \varepsilon^{-n}\varphi(x/\varepsilon)$ and $\varepsilon = 1/R$.

Observe that in proving estimate (a) for $S_*^\alpha(f) = \sup_{0<R<\infty} |S_R^\alpha(f)|$, it suffices to restrict our attention to $R \geqq 1$, (i.e., $\varepsilon \leqq 1$) since when $R < 1$ then $S_R^\alpha(f) = a_0$, and $|a_0| \leqq \|f\|_1 \leqq \|f\|_p$. Now let $\tilde{f}(x) = 0$ if $|x| > 1$, and $\tilde{f}(x) = f(x)$, if $|x| \leqq 1$. Then, clearly,

$$\left(\int_{E_n} |\tilde{f}(x)|^p\, dx\right)^{1/p} \leqq A\left(\int_{Q_n} |f(x)|^p\, dx\right)^{1/p}.$$

Also

$$(f * \varphi_\varepsilon)(x) = \int_{|y|\leqq 1} \varphi_\varepsilon(x - y)f(y)\, dy + \int_{|y|>1} \varphi_\varepsilon(x - y)f(y)\, dy.$$

[7] See, however, the recent result of Fefferman [1].

We have already observed (see the paragraph preceding (2.14)) that, in this case,

$$\varphi(x) = \pi^{-\alpha}\Gamma(\alpha + 1)|x|^{-(\alpha+n/2)}J_{\alpha+n/2}(2\pi|x|).$$

Thus, by (3.12) of Chapter IV, $|\varphi(x)| \leqq A(1 + |x|)^{-n-\delta}$ with $\delta = \alpha - (n-1)/2$. Consequently, as was shown in the proof of Theorem (2.11),

$$\left|\int_{|y|>1} \varphi_\varepsilon(x-y)f(y)\,dy\right| \leqq A'\varepsilon^\delta\left(\sum_{m\neq 0}|m|^{-n-\delta}\right)\int_{Q_n}|f(x)|\,dx.$$

But

$$\int_{|y|\leqq 1} \varphi_\varepsilon(x-y)f(y)\,dy = \int \varphi_\varepsilon(x-y)\tilde{f}(y)\,dy$$

and, therefore, by (3.9) of Chapter II,

$$\|S_*^\alpha(f)\|_p \leqq A_{p,\alpha}\|f\|_p$$

if $p > 1$ and $\alpha > (n-1)/2$.

Let us make some general remarks concerning the Riesz means S_R^α. When we introduced them we made the implicit assumption that the index α is nonnegative. Nevertheless, nothing in the definition we gave impedes our considering these means when α is negative or even complex. In fact, the behavior of S_R^α as an analytic function of α will be one of the main tools we shall use. In the following L^2 result the means of negative order are already involved.

LEMMA 5.3. *Let*

$$M^\alpha(f)(x) = \sup_{0<R<\infty}\left(\frac{1}{R}\int_0^R |S_t^\alpha(f)(x)|^2\,dt\right)^{1/2}.$$

Then

$$\|M^\alpha(f)\|_2 \leqq A_\alpha\|f\|_2 \tag{5.4}$$

if $f \in L^2(T_n)$ *and* $\alpha > -\frac{1}{2}$.[8]

PROOF. We introduce the auxiliary function $G^\alpha(f)$ defined by

$$G^\alpha(f)(x) = \left(\int_0^\infty |S_R^{\alpha+1}(f)(x) - S_R^\alpha(f)(x)|^2\,\frac{dR}{R}\right)^{1/2}.$$

We claim that

$$\|G^\alpha(f)\|_2 \leqq A_\alpha\|f\|_2 \tag{5.5}$$

[8] Notice that the quantities $M^\alpha(f)$ represent averages of the type of expressions we want eventually to dominate.

if $f \in L^2(T_n)$ and $\alpha > -\frac{1}{2}$. In fact, by Parseval's relation and Fubini's theorem

$$\|G^\alpha(f)\|_2^2 = \int_0^\infty \left(\sum_{|m|<R} \left| \left(1 - \frac{|m|^2}{R^2}\right)^{(\alpha+1)} - \left(1 - \frac{|m|^2}{R^2}\right)^\alpha \right|^2 |a_m|^2 \right) \frac{dR}{R}$$

$$= \sum_{m \neq 0} |a_m|^2 \int_{|m|}^\infty \frac{|m|^4}{R^4} \left[1 - \frac{|m|^2}{R^2}\right]^{2\alpha} \frac{dR}{R} = C_\alpha \sum_{m \neq 0} |a_m|^2.$$

The third equality is a consequence of the fact that

$$\int_{|m|}^\infty \frac{|m|^4}{R^4} \left[1 - \frac{|m|^2}{R^2}\right]^{2\alpha} \frac{dR}{R} = \int_1^\infty s^{-5}(1 - s^{-2})^{2\alpha}\, ds = C_\alpha$$

$$= [2(2\alpha + 1)(2\alpha + 2)]^{-1}$$

and the last integral converges if $\alpha > -\frac{1}{2}$. Therefore,

$$\|G^\alpha(f)\|_2^2 = C_\alpha\{\|f\|_2^2 - |a_0|^2\},$$

which is, of course, a more precise form of (5.5). From (5.5) the lemma follows easily. To begin with, notice that by the definition of $G^\alpha(f)$

$$\sup_{0<R<\infty} \frac{1}{R} \int_0^R |S_t^{\alpha+1}(f)(x) - S_t^\alpha(f)(x)|^2\, dt \leqq [G^\alpha(f)(x)]^2,$$

and, thus

$$M^\alpha(f)(x) \leqq M^{\alpha+1}(f)(x) + G^\alpha(f)(x). \tag{5.6}$$

Consequently, applying (5.6) repeatedly,

$$M^\alpha(f) \leqq M^{\alpha+k}(f) + G^\alpha(f) + G^{\alpha+1}(f) + \cdots + G^{\alpha+k-1}(f). \tag{5.7}$$

Now, if we choose $k > n/2$ then $\alpha + k > (n-1)/2$. Since $M^{\alpha+k}(f)$ is clearly less than or equal to $S_*^{\alpha+k}(f)$, we can apply Lemma 5.2 (with $p = 2$) and obtain $\|M^{\alpha+k}(f)\|_2 \leqq A\|f\|_2$. This inequality, together with (5.5) (applied to $G^\alpha(f), G^{\alpha+1}(f), \ldots, G^{\alpha+k-1}(f)$) gives us (5.4) and the lemma is proved.

We can use Lemma 5.3 once we have obtained the formula which expresses the Riesz means of a given order as averages of the Riesz means of lower orders. The required calculation was already carried out in the proof of Lemma 4.8. In fact, the identity

$$\left(1 - \frac{|m|^2}{R^2}\right)^{\beta+\delta} = C_{\beta,\delta} R^{-2\beta-2\delta} \int_{|m|}^R (R^2 - t^2)^{\beta-1} t^{2\delta+1} \left(1 - \frac{|m|^2}{t^2}\right)^\delta dt,$$

with $C_{\beta,\delta} = 2\Gamma(\delta + \beta + 1)/\Gamma(\delta + 1)\Gamma(\beta)$, gives us immediately the desired expression; namely,

$$(5.8) \qquad S_R^{\beta+\delta} = C_{\beta,\delta} R^{-2\beta-2\delta} \int_0^R (R^2 - t^2)^{\beta-1} t^{2\delta+1} S_t^\delta \, dt.$$

It then follows that

$$|S_R^{\beta+\delta}(f)(x)| \leqq C_{\beta,\delta} R^{-2\beta-2\delta} \left(\int_0^R |(R^2 - t^2)^{\beta-1} t^{2\delta+1}|^2 \, dt \right)^{1/2} \times R^{1/2} \cdot R^{-1/2} \left(\int_0^R |S_t^\delta(f)(x)|^2 \, dt \right)^{1/2};$$

therefore, upon taking the supremum over all $R > 0$, we obtain

$$(5.9) \qquad S_*^{\beta+\delta}(f)(x) \leqq C'_{\beta,\delta} M^\delta(f)(x) \qquad \text{if } \beta > \tfrac{1}{2}.$$[9]

Finally, given $\alpha > 0$, choose β and δ so that $\beta + \delta = \alpha$, with $\delta > -\frac{1}{2}$ and $\beta > \frac{1}{2}$. Then, combining (5.9) and Lemma 5.3, we obtain the following basic inequality in the L^2 case:

LEMMA 5.10. *If $\alpha > 0$ then $\|S_*^\alpha(f)\|_2 \leqq A_\alpha \|f\|_2$ for all f in $L^2(T_n)$.*

Let us now pause momentarily in our efforts toward proving Theorem 5.1 in order to see what has already been accomplished. The principal part of that theorem is the maximal inequality (part (a)); once it is established, the rest is routine. When p is near 1 or ∞, part (a) is in effect already contained in Lemma 5.2. Moreover, Lemma 5.10 gives us the desired inequality in the special case $p = 2$. The general case can be deduced from these special cases by applying the interpolation theorem that was established in the fourth section of Chapter V. Although the ideas we use in doing this are simple, the details required are somewhat complicated. What is needed first is a strengthening of the two special cases just mentioned which allows orders α that are complex. This strengthening is a consequence of a very simple general principle which asserts that if we have an estimate for $S_*^\alpha(f)$ with $\alpha \geqq 0$, then there exists a corresponding estimate for $S_*^{\alpha'}(f)$ with α' complex and satisfying $\mathrm{Re}\,(\alpha') > \alpha$. That this is so follows from formula (5.8) with $\delta = \alpha, \beta + \delta = \alpha'$ (and, thus

[9] Notice that, if $\beta > \frac{1}{2}$, then

$$R^{1-4\beta-4\delta} \int_0^R |(R^2 - t^2)^{\beta-1} t^{2\delta+1}|^2 \, dt = \int_0^1 |(1 - t^2)^{\beta-1} t^{2\delta+1}|^2 \, dt < \infty.$$

Re $(\beta) > 0$). In fact, we have

$$S_*^{\alpha'}(f) \leqq S_*^{\alpha}(f) C_{\beta,\delta} \int_0^1 (1 - s^2)^{\mathrm{Re}(\beta)-1} s^{2\alpha+1}\, ds$$
$$= \frac{|\Gamma(\alpha'+1)|\Gamma(\mathrm{Re}\,(\beta))}{\Gamma(\mathrm{Re}\,(\alpha')+1)|\Gamma(\beta)|} S_*^{\alpha}(f).$$

Since $\Gamma(\alpha'+1) \leqq |\Gamma(\mathrm{Re}\,(\alpha')+1)|$ this gives us

(5.11) $$S_*^{\alpha'}(f) \leqq D_\beta S_*^{\alpha}(f)$$

with $D_\beta = \Gamma(\mathrm{Re}\,(\beta))/|\Gamma(\beta)|$.

Since the operators $f \to S_*^{\alpha}(f)$ are not linear, and we wish to apply an interpolation theorem for linear operators, we need to introduce another technical device. In order to do this we let $\mathscr{C}$ denote the class of nonnegative measurable functions on T_n having only a finite number of distinct values. Suppose $R \in \mathscr{C}$, then

(5.12) $$|S_{R(x)}^{\alpha}(f)(x)| \leqq S_*^{\alpha}(f)(x).$$

We assert, however, that the following converse of this inequality holds:

(5.13) $$\sup_{R \in \mathscr{C}} \|S_{R(x)}^{\alpha}(f)(x)\|_p = \|S_*^{\alpha}(f)(x)\|_p.$$

This follows from the obvious fact that we can find a sequence $R_1(x)$, $R_2(x), \ldots, R_j(x), \ldots$ in $\mathscr{C}$ such that

$$\lim_{j\to\infty} |S_{R_j(x)}^{\alpha}(f)(x)| = \sup_{0<R<\infty} |S_R^{\alpha}(f)(x)| = S_*^{\alpha}(f)(x) \qquad \text{for all } x \in T_n.$$

We now fix an element $R \in \mathscr{C}$ and consider the *linear* operators

(5.14) $$f \to S_{R(x)}^{\alpha}(f)(x)$$

and their analytic dependence on the parameter α. Suppose f is integrable and $f(x) \sim \sum_{m\in\Lambda} a_m e^{2\pi i m\cdot x}$. Since $R(x)$ takes on only a finite number of values (in particular, it is bounded) it is easy to see that

$$S_{R(x)}^{\alpha}(f)(x) = \sum_{|m| \leqq R(x)} \left(1 - \frac{|m|^2}{R^2(x)}\right)^{\alpha} a_m e^{2\pi i m\cdot x}$$

is analytic in α and is of admissible growth in any vertical strip of the complex α plane, in the sense of Theorem 4.1 in Chapter V.

Let $\alpha = \mu + i\nu$, $\mu_0 > 0$ and $\mu_1 > (n-1)/2$. Then, by (5.12), (5.11), and (5.10) with $\alpha' = \mu_0 + i\nu$, $\alpha = \mu_0/2$, $\beta = (\mu_0/2) + i\nu$, we have

(5.15) $$\|S_{R(x)}^{\mu_0+i\nu}(f)(x)\|_2 \leqq A_0(\nu)\|f\|_2$$

for all simple functions f, where

$$A_0(\nu) \leqq A_0 \frac{\Gamma(\mu_0/2)}{|\Gamma((\mu_0/2) + i\nu)|} \leqq A_0' e^{\pi|\nu|}.$$ [10]

What is important to observe is that the bound for $A_0(\nu)$ does not depend on the particular choice of $R \in \mathscr{C}$. Similarly, by (5.12), (5.11), and (5.2) with $\alpha' = \mu_1 + i\nu$, $\alpha = \frac{1}{2}[\mu_1 + (n-1)/2]$, $\beta = \frac{1}{2}[\mu_1 - (n-1)/2] + i\nu$, we have

$$\|S_{R(x)}^{\mu_1 + i\nu}(f)(x)\|_{p_1} \leqq A_{1,p_1}(\nu)\|f\|_{p_1} \tag{5.16}$$

for all simple f, $1 < p_1 < \infty$, $[\mu_1 > (n-1)/2]$, where $A_{1,p_1}(\nu) \leqq A'_{1,p_1} e^{\pi|\nu|}$. Here again the constants are independent of $R(x)$.

We now invoke the interpolation Theorem 4.1 of Chapter V. In keeping with the notation set down there we let $0 < t < 1$, $p_0 = 2$ and p_1 be the exponent arising in (5.16). Then, if $\mu = \mu_0(1 - t) + \mu_1 t$ and $1/p = [(1-t)/p_0] + [t/p_1] = (1-t)/2 + t/p_1$, we obtain

$$\|S_{R(x)}^{\mu}(f)(x)\|_p \leqq A_p\|f\|_p. \tag{5.17}$$

Again A_p is independent of the $R(x)$ chosen from $\mathscr{C}$. But (5.17) and (5.13) give us the inequality:

$$\|S_*^{\mu}(f)\|_p \leqq A_p\|f\|_p. \tag{5.18}$$

We claim that the restrictions on p and μ in (5.18) are precisely those of Theorem 5.1; that is, $1 < p < \infty$ and $\mu > (n-1)|\frac{1}{2} - 1/p|$. Recall that $\mu_0 > 0$, $\mu_1 > (n-1)/2$ and $1 < p_1 < \infty$ are otherwise arbitrary, while $\mu = \mu_0(1-t) + t\mu_1$ and $1/p = (1-t)/2 + t/p_1$. Assume first that $p \leqq 2$. If we set $p_1 = 1$, $\mu_0 = 0$, $\mu_1 = (n-1)/2$, then an obvious calculation gives us $\mu = (n-1)[1/p - \frac{1}{2}]$. It is also clear that μ is a continuous function of p_1, μ_0, and μ_1 (we see this by expressing t in terms of these parameters). Thus, by continuity, we can always realize any μ satisfying $\mu > (n-1)(1/p - \frac{1}{2})$ by choosing $p_1 > 1$, $\mu_0 > 0$ and $\mu_1 > (n-1)/2$. If $p \geqq 2$ the argument is similar once we begin by setting $p_1 = \infty$. This concludes the proof of part (a) of Theorem 5.1.

Parts (b) and (c) follow from this by the general principles concerning convergence of families of operators developed in the second chapter (see, in particular, Theorem 3.12) when we observe that, if we choose f from the class of trigonometric polynomials (which is dense in $L^p(T_n)$ when $p < \infty$), then $\lim_{R\to\infty} S_R^{\alpha}(f)(x)$ converges uniformly to f, for each $\alpha \geqq 0$.

[10] More precise estimates for $A_0(\nu)$ can be made by using the asymptotic formula $|\Gamma((\mu/2) + i\nu)| \sim \sqrt{2\pi}|\nu|^{(\mu-1)/2}e^{-\pi|\nu|/2}$, as $\nu \to \infty$, but they are not needed in order to apply the interpolation Theorem 4.1 of Chapter V.

6. Further Results

6.1. An application of reasoning similar to that used in the proof of Theorem 2.17 shows that each of the following are Fourier series of functions that are continuously differentiable and periodic on the fundamental cube except at the origin; moreover, they have the indicated behavior:

(a) $\sum_{m \neq 0} |m|^{-n} e^{2\pi i m \cdot x}$ is asymptotic to $c \cdot \log(1/|x|)$ as $|x| \to 0$;

(b) If P_k is a homogeneous harmonic polynomial of degree $k \geqq 1$, then $\sum_{m \neq 0} |m|^{-n-k} P_k(m) e^{2\pi i m \cdot x}$ is a bounded function;

(c) $\sum_{m \neq 0} |m|^{-\alpha-k} b(|m|) P_k(m) e^{2\pi i m \cdot x}$ is asymptotic to $c|x|^{-(n-\alpha)-k} \times b(1/|x|) P_k(x)$ as $|x| \to 0$, when P_k is a homogeneous harmonic polynomial of degree $k \geqq 0$, $0 < \alpha < n$ and $b(x)$ is an appropriate function slowly varying near ∞; (examples of allowable b's are functions whose values for $|x|$ large are $(\log |x|)^a$, $(\log(\log |x|))^c$, $\exp(\log |x|)^{1/2}$, etc., and also products and quotients of such functions). For the one-dimensional theory see Zygmund [1] and for the n-dimensional one see Wainger [1].

6.2. Some examples of a different nature may also be obtained by the aid of the Poisson summation formula:

(a) $\sum_{m \neq 0} e^{ci|m|\log|m|} |m|^{-\varepsilon-n/2} e^{2\pi i m \cdot x}$ is the Fourier series of a continuous periodic function, if $c \neq 0$ and $\varepsilon > 0$;

(b) If the dimension n is even then $\sum_{|m|>1} |m|^{-n} (\log |m|)^{-1} \times e^{ci|m|\log(|m|)^a} e^{2\pi i m \cdot x}$, with $c \neq 0$ and $0 < a < 2/n$, is the Fourier series of a function of class $C^{n/2}$, but is not absolutely convergent.

In the one-dimensional case example (a) is due to Hardy and Littlewood. For both (a) and (b) see Wainger [1], where other examples are also studied. Example (b), incidentally, shows that the simple Corollary 1.9 cannot be improved very much.

6.3. The following considerations give an interpretation of the heuristic formula (2.16). Let $P_k(x)$ be a homogeneous harmonic polynomial of degree k. Fix x and consider the following two functions of α:

$$\sum_{m \neq 0} \frac{P_k(m)}{|m|^{k+\alpha}} e^{2\pi i m \cdot x}, \qquad \text{for } \mathrm{Re}(\alpha) > n,$$

and

$$\gamma_{\alpha,k}^{-1} \sum_{m \neq 0} \frac{P_k(x+m)}{|x+m|^{k+n-\alpha}}, \qquad \text{for } \mathrm{Re}(\alpha) < 0,$$

where $\gamma_{\alpha,k} = i^{-k} \pi^{n/2-\alpha} \Gamma((k+\alpha)/2)/\Gamma((n+k-\alpha)/2)$. These two functions of α are equal in the sense that they are analytic continuations of each other (see Hecke [1] and Bochner [6]).

6.4. Let $S \subset E_n$ be convex, symmetric, open and bounded. Suppose $|S| > 2^n$. Then S contains at least one lattice point different from the origin. This classical theorem of Minkowski can be proved by making use of the Poisson summation formula in the following way. Let $S_{1/2} = \{x: 2x \in S\}$ and let φ denote the characteristic function of $S_{1/2}$. We set $f = \varphi * \varphi$; thus, $\hat{f} = |\hat{\varphi}|^2 \geqq 0$. It can be shown that if the origin is the only lattice point in S then $f(m) = 0$ when $m \neq 0$. Moreover, the Poisson summation formula $\sum f(m) = \sum \hat{f}(m)$ holds. If 0 is the only lattice point within S, therefore, $f(0) = \sum \hat{f}(m) \geqq \hat{f}(0)$. But this contradicts the fact that $f(0) = \int \varphi(x)\, dx = |S_{1/2}|$ and $\hat{f}(0) = \int f(x)\, dx = |S_{1/2}|^2$. The argument sketched goes back to Siegel [1].

6.5. Suppose $\sum_{m \in \Lambda} a_m e^{2\pi i m \cdot x}$ is a given formal trigonometric series. The n series

$$-i \sum_{m \neq 0} a_m \frac{m_k}{|m|} e^{2\pi i m \cdot x}, \qquad k = 1, \ldots, n,$$

are the Riesz transforms of the given series. Let

$$u_0(x, t) = \sum_{m \in \Lambda} a_m e^{-2\pi |m| t} e^{2\pi i m \cdot x}$$

and

$$u_k(x, t) = -i \sum_{m \neq 0} a_m \frac{m_k}{|m|} e^{-2\pi |m| t} e^{2\pi i m \cdot x} \qquad \text{for } t > 0$$

(assuming that the series defining u_k, $k = 0, \ldots, n$, converges). We observe that the $(n+1)$-tuple $F = (u_0, u_1, \ldots, u_n)$ satisfies the generalized Cauchy–Riemann equations (4.13) of Chapter VI. In analogy with the theory described there we say that $F \in \mathbf{H}^p(T_n)$ if $\sup_{t>0} \int_{T_n} |F(x, t)|^p\, dx < \infty$.

(a) $F \in \mathbf{H}^p(T_n)$, $1 < p < \infty$, if and only if $\sum_{m \in \Lambda} a_m e^{2\pi i m \cdot x}$ is the Fourier series of a function in $L^p(T_n)$. This can be proved by using the results developed in Section 3.

(b) Suppose $(n-1)/n < p < \infty$ and $F \in \mathbf{H}^p(T_n)$, then

$$\int_{T_n} \sup_{t>0} |F(x, t)|^p\, dt < \infty$$

and $\lim_{t \to 0} F(x, t)$ exists almost everywhere and in the metric of $L^p(T_n)$.

(c) The special case $p = 1$ implies the following extension of the classical F. and M. Riesz theorem (see also (5.8) in Chapter VI). Suppose $\sum_{m \in \Lambda} a_m e^{2\pi i m \cdot x}$ and $\sum_{m \neq 0} a_m (m_k/|m|) e^{2\pi i m \cdot x}$, $k = 1, 2, \ldots, n$. are Fourier–Stieltjes series of finite measures, then these measures are absolutely continuous.

Both (b) and (c) can be established by arguments that are analogous to those presented in Chapter VI. See also Shapiro [1].

(d) A result that complements the last result (which has no analog when $n = 1$) is as follows. Suppose that $\sum_{m\in\Lambda} a_m e^{2\pi i m\cdot x}$ and $\sum_{m\neq 0} a_m(m_k/|m|)e^{2\pi i m\cdot x}$, $k = 1, 2, \ldots, n$, are Fourier–Stieltjes series of finite measures, then for each homogeneous (of degree r) harmonic polynomial $P_r(x)$, $r = 0, 1, 2, \ldots$, the series $\sum_{m\neq 0} (P_r(m)/|m|^r)a_m e^{2\pi i m\cdot x}$ is the Fourier series of a function in $L^1(T_n)$. This is a consequence of techniques presented in Stein [3], Chapter VII.

6.6. Theorem 2.11 and Lemma 5.2 have the following more general formulation. Suppose $\varphi \in L^1(E_n)$ with $\int_{E_n} \varphi(x)\,dx = 0$. Let $\varphi_\varepsilon(x) = \varepsilon^{-n}\varphi(x/\varepsilon)$ and $K_\varepsilon(x) = \sum_{m\in\Lambda} \varphi_\varepsilon(x + m)$. Consider $(f * K_\varepsilon)(x) = \int_{T_n} f(x - y)K_\varepsilon(y)\,dy$. Then

(a) If $f \in L^p(T_n)$, $1 \leqq p < \infty$, then $f * K_\varepsilon \to f$ in the $L^p(T_n)$ norm as $\varepsilon \to 0$;

(b) Suppose, in addition, that $\psi(x) = \sup_{|x|\leqq|x'|} |\varphi(x')|$ is integrable on E_n. Then $(f * K_\varepsilon)(x) \to f(x)$ at each point of the Lebesgue set of f. Moreover, if $\tilde{f}(x) = f(x)$ for $|x| \leqq 1$ and $\tilde{f}(x) = 0$ if $|x| > 1$ then

$$\sup_{\varepsilon>0} |(f * K_\varepsilon)(x)| \leqq A m_{\tilde{f}}(x).$$

6.7. Let λ be a complex-valued function defined on E_n and of class C^n except at the origin. Suppose that there exists a constant A such that $|D^\alpha\lambda(x)| \leqq A/|x|^{|\alpha|}$ whenever $0 \leqq |\alpha| \leqq n$. Then the sequence $\{\lambda(m)\}$, $m \in \Lambda$, $(\lambda(0) = 0)$ is a multiplier sequence of type $(L^p(T_n), L^p(T_n))$ for $1 < p < \infty$. This theorem goes back to Marcinkiewicz [1], where a somewhat stronger assertion is proved. Under the same conditions, λ is also a multiplier in the nonperiodic case (i.e., $\lambda \in (L^p(E_n), L^p(E_n))$, for $1 < p < \infty$), but for historical reasons this type of result appeared only much later (see Mihlin [1] and Hörmander [1]). The nonperiodic version could also be deduced directly from the periodic version by making use of Theorem 3.18. See also Stein [3], Chapter IV.

6.8. Theorem 5.1 shows that $\|S_*^{(n-1)/2}(f)\|_p \leqq A_p\|f\|_p$, $1 < p < \infty$, if $f \in L^p(T_n)$. This result for the critical index may be extended by considering the growth of A_p as $p \to 1$ or $p \to \infty$. In fact, a refinement of the proof of Theorem 5.1 shows that $A_p \leqq A(p - 1)^{-2}$ for $1 < p < 2$ and $A_p \leqq Ap$ for $2 \leqq p < \infty$.

(a) In view of the estimate for A_p for p near 1 we obtain the inequality $\|S_*^{(n-1)/2}(f)\|_1 \leqq A\int_{T_n} |f|(\log^+|f|)^2 dx + B$. Consequently, if $|f|(\log^+|f|)^2$ is integrable, $\lim_{R\to\infty} S_R^{(n-1)/2}(f)(x) = f(x)$ for almost every x.

(b) From the estimate for A_p for large p it follows that if f is bounded then there exists $a > 0$ such that $\int_{T_n} \exp\{aS_*^{(n-1)/2}(f)(x)\}\,dx < \infty$.

For part (a) see Stein [4]; the estimate for A_p needed to establish (b) is implicitly contained there.

6.9. We have observed that localization does not hold for arbitrary functions in $L^1(T_n)$ at the critical index (see Theorem 4.2 and the accompanying footnote). Localization, however, does hold under stronger assumptions. If we suppose $\int_{Q_n} |f| \log^+ |f|\,dx < \infty$, for example, localization does hold. In particular, if at a given point x_0, f satisfies the Dini condition $\int_{Q_n} |f(x_0 - t) - f(x_0)|\,|t|^{-n}\,dt < \infty$ then $S_R^{(n-1)/2}(f)(x_0) \to f(x_0)$ as $R \to \infty$. This is a consequence of the following result:

$$\sup_{R>0} \left(\int_{Q_n} |\Delta_R(x)|^p\,dx \right)^{1/p} \leqq Ap,\ 1 \leqq p < \infty,$$

where

$$\Delta_R(x) = K_R^{(n-1)/2}(x) - \pi^{(n-1)/2}\Gamma\left(\frac{n+1}{2}\right) R^{1/2}|x|^{-n+1/2} J_{n-1/2}(2\pi R|x|).$$

For details see Stein [7].

6.10. The following convexity theorem for Riesz means of numerical series can be used in the proof of Theorem 4.3. Consider a numerical series $\sum_{k\geqq 0} c_k$ and its Riesz means $\sigma_R^\alpha = \sum_{0\leqq k<R} (1 - k/R)^\alpha c_k$, $\alpha \geqq 0$. Suppose that $\sigma_R^{\alpha_j} = O(R^{a_j})$ as $R \to \infty$, for $j = 0, 1$. Then, if $0 < \theta < 1$, $\alpha = \alpha_0(1 - \theta) + \alpha_1\theta$, and $a = a_0(1 - \theta) + a_1\theta$, we can conclude that $\sigma_R^\alpha = O(R^a)$ as $R \to \infty$. See Riesz [1], and Chandrasekharan and Minukshandaran [1]. To apply this result in the proof of Theorem 4.3 write

$$\sigma_R^\alpha = K_{R^{1/2}}^\alpha(x^0) = \sum_{|m|^2<R} \left(1 - \frac{|m|^2}{R}\right)^\alpha e^{2\pi i m\cdot x^0}$$

for

$$x_0 \notin \Lambda \text{ (where } c_k = \sum_{|m|^2=k} e^{2\pi i m\cdot x^0}).$$

We know (see the inequality following (4.28)) that $\sigma_R^\alpha = O(R^{(1/2)[(n-1)/2-\alpha]})$ if $\alpha > (n-1)/2$. If it were true that $\sigma_R^0 = O(R^{(1/2)[(n-1)/2]})$ then the convexity theorem just quoted would imply that $\sigma_R^{(n-1)/2} = O(1)$, which contradicts Lemma 4.6, for almost every x^0.

Bibliographical Notes

For general references dealing with the elementary aspects of multiple Fourier series and the Poisson summation formula, see the classical book of

Bochner [3] and also his later tract [6]. The significance of the critical index (see Corollary 2.15) goes back to Bochner's paper [7]. The n-dimensional form of Theorem 2.17 may be found in Wainger's memoir [1].

The main results of the third section, Theorems 3.8 and Corollary 3.16 are due to de Leeuw [1]. The general ideas of the argument leading to Theorem 3.18 are classical, but the present sharp form seems to be new. See also Igari [1].

The failure of localization at the critical index for $L^1(T_n)$ was proved in Bochner's paper [7] that was alluded to above. The existence of an $L^1(T_n)$ function ($n > 1$) whose Fourier series is almost nowhere summable by the Riesz means at the critical index is contained in Stein [5]. The present stronger form of Theorem 4.2, however, is new; the same is true of Theorem 4.3. All of these results have as one of their starting points the technique introduced by Bochner, which involves the set S of points x where $\{|x - m|\}$ are linearly independent over the rationals. For Theorem 5.1 see Stein [4]; also [2], and [7] which are closely related. The reader may also consult the Survey paper of Shapiro [1], which covers several topics in multiple Fourier series.

Bibliography

Banach, S.

[1] "Sur la convergence presque partout des fonctionnelles linéaires," *Bull. Sci. Math. 50* (1926), 27–32, 36–43.

Bari, N.

[1] *A Treatise on Trigonometric Series*, Pergamon Press, New York, 1964.

Benedek, A., and Panzone, R.

[1] "The spaces L^p with mixed norm," *Duke Math. J. 28* (1961), 301–324.

Besicovitch, A.

[1] "Sur la nature des fonctions à carré sommable mesurables," *Fund Math. 4* (1923), 172–195.

[2] "A general form of the covering principle and relative differentiation of additive functions," *Proc. of the Cambridge Phil. Soc. 41* (1945), 103–110.

[3] "A general form of the covering principle and relative differentiation of additive functions II," *ibidem 42* (1946), 1–10.

Blumenson, L. E.

[1] "A derivation of *n*-dimensional spherical coordinates," *Am. Math. Monthly 67* (1960), 63–66.

Boas, R. P., and Bochner, S.

[1] "On a theorem of M. Riesz for Fourier Series," *Jour. London Math. Soc. 14* (1939), 62–73.

Bochner, Salomon

[1] "Group invariance of Cauchy's formula in several variables," *Ann. of Math. 45* (1944), 686–707.

[2] "Boundary values of analytic functions of several variables and of almost periodic functions," *Ann. of Math. 45* (1944), 708–722.

[3] *Vorlesungen über Fouriersche Integrale*, Leipzig, 1932.

[4] *Lectures on Fourier Integrals*, Ann. of Math. Studies, No. 42, Princeton University Press, 1959.

[5] "Classes of holomorphic functions of several variables in circular domains," *Proc. Nat. Acad. Sci. U.S.A. 46* (1960), 721–723.

[6] *Harmonic Analysis and the Theory of Probability*, University of California Press, Berkeley, 1955.

[7] "Summation of multiple Fourier series by spherical means," *Trans. Amer. Math. Soc. 40* (1936), 175–207.

[8] "Bounded analytic functions in several variables and multiple Laplace integrals," *Amer. Jour. Math. 59* (1937), 732–738.

[9] "Über Faktorfolgen für Fouriersche Reihen," *Acta Szeged 4* (1929), 125–129.

Bochner, S., and Chandrasekharan, K.

[1] *Fourier Transforms*, Ann. of Math. Studies, No. 19, Princeton University Press, 1949.

Bochner, S., and Martin, W. T.

[1] *Several Complex Variables*, Princeton University Press, 1948.

Boerner, Herman

[1] *Representations of Groups*, North-Holland Publishing Company, Amsterdam, 1970.

Brelot, M.

[1] *Eléments de la Théorie Classique du Potentiel*, Centre de documentation universitaire, Paris, 1965.

Calderón, A. P.

[1] "Singular integrals," *Bull. A.M.S. 72* (1966), 427–465.

[2] "Intermediate spaces and interpolation, the complex method," *Studia Math. 24* (1964), 113–190.

[3] *Integrales Singulares y sus Aplicaciones a Ecuaciones Diferenciales Hiperbolicas*, Cursos y seminarios de Matematica, Universidad de Buenos Aires, Fasc. 3.

[4] "On the behavior of harmonic functions at the boundary," *Trans. A.M.S. 68* (1950), 47–54.

Calderón, A. P., and Zygmund, A.

[1] "Note on the boundary values of functions of several complex variables," *Contributions to Fourier Analysis*, Ann. of Math. Studies, No. 25, Princeton University Press, 1950.

[2] "On the theorem of Hausdorff-Young and its extensions," *Contributions to Fourier Analysis*, Ann. of Math. Studies, No. 25, Princeton University Press, 1950.

[3] "On the existence of certain singular integrals," *Acta Math. 88* (1952), 85–139.

[4] "On singular integrals," *Amer. J. Math. 18* (1956), 289–309.

[5] "Singular integral operators and differential equations," *Amer. J. Math. 79* (1959), 901–921.

[6] "On higher gradients of harmonic functions," *Studia Math. 24* (1964), 211–226.

Carleson, L.

[1] "On the existence of boundary values of harmonic functions of several variables," *Ark. Mat. 4* (1962), 393–399.

Cartwright, M. L.

[1] "On the relation between the different types of Abel summation," *Proc. London Math. Soc. 31* (1930), 81–96.

Chandrasekharan, K., and Minakshisundaran, S.

[1] *Typical Means*, Oxford University Press, 1952.

Coifman, R. R., and Weiss, G.

[1] "Representations of compact groups and spherical harmonics," *L'Ens. Math. 14* (1968), 121–173.

[2] "On subharmonicity inequalities involving solutions of generalized Cauchy–Riemann equations," *Studia Math. 36* (1970), 77–83.

Cotlar, M.

[1] *Condiciones de Continuidad de Operadores Potenciales y de Hilbert*, Cursos y seminarios de Matematica, Universidad de Buenos Aires, Fasc. 2.

de Leeuw, K.

[1] "On L^p multipliers," *Ann. of Math. 91* (1965), 364–379.

Ehrenpreis, L.

[1] *Fourier Analysis in Several Complex Variables*, Wiley-Interscience, New York, 1970.

Erdelyi, A. (director)

[1] *Higher Transcendental Functions*, Vols. I–III, Bateman Manuscript Project, McGraw-Hill, New York, 1955.

Fefferman, Charles

[1] "The multiplier problem for the ball," *Annals of Math.* (to appear in 1971).

Fleming, W. H.

[1] *Functions of Several Variables*, Addison-Wesley, Reading, Massachusetts, 1965.

Gagliardo, Emilio

[1] "Una struttura unitaria in diverse famiglie di spazi funzionali," *Ricerche Mat. 10* (1961), 244–281.

Gelfand, I. M., and Shilov, G. E.

[1] *Generalized Functions*, Vol. I, Academic Press, New York, 1964.

Goldberg, R. R.

[1] *Fourier Transforms*, Cambridge Tracts in Math. and Math. Physics, No. 52, Cambridge, 1965.

de Guzman, M.

[1] "A covering lemma with applications to differentiability of measures and singular integral operators," *Studia Math. XXXIV* (1970), 299–317.

Hardy, G. H., and Littlewood, J. E.

[1] "A maximal theorem with function-theoric applications," *Acta Math. 54* (1930), 81–116.

Hardy, G. H., and Wright, E. M.

[1] An Introduction to the Theory of Numbers, 3rd edition, Oxford, 1956.

Hecke, E.

[1] *Matematische Werke*, Göttingen, 1959.

Helgason, S.

[1] *Differential Geometry and Symmetric Spaces*, Academic Press, New York, 1962.

Herz, C. S.

[1] "Bessel functions of matrix argument," *Ann. of Math. 61* (1955), 474–523.

Hewitt, E., and Ross, K. A.

[1] *Abstract Harmonic Analysis I*, Springer, Berlin, 1963.

Hille, E., and Phillips, R. S.

[1] *Functional Analysis and Semi-groups*, Am. Math. Soc. Coll. Publication XXXI, Providence, 1957.

Hirschman, I. I., jr.

[1] "A convexity theorem for certain groups of transformations," *J. d'Analyse Math. 2* (1953), 209–218.

Hörmander, L.

[1] "Estimates for translation invariant operators on L^p spaces," *Acta Math. 104* (1960), 93–139.

[2] *Linear Partial Differential Operators*, Springer, Berlin, 1963.

Horváth, J.

[1] "Sur les fonctions conjuguées à plusieurs variables," *Indag. Math. 15* (1953), 17–29.

Hua, L. K.

[1] *Harmonic Analysis of Functions of Several Complex Variables in the Classical Domains*, Vol. 6, Translations of Math. Monographs, Am. Math. Soc., Providence, 1963.

Hunt, R. A.

[1] "On $L(p, q)$ spaces," *L'Ens. Math.*, *12* (1966), 249–275.

Hunt, R. A., and Wheeden, R. L.

[1] "On the boundary values of harmonic functions," *Trans. Amer. Math. Soc. 132* (1968), 307–322.

Igari, S.

[1] "Fourier analysis," notes to a course given at the University of Wisconsin, 1968.

Jessen, B., Marcinkiewicz, J., and Zygmund, A.

[1] "Note on the differentiability of multiple integrals," *Fund. Math. 25* (1935), 217–234.

Kellog, O. D.

[1] *Foundations of Potential Theory*, Ungar Publ. Co., New York, 1929.

Koecher, M.

[1] "Positivitäts bereiche im R^m," *Amer. J. Math. 79* (1957), 575–597.

Kolmogorov, A. N.

[1] "Sur les fonctions harmoniques conjuguées et les séries de Fourier," *Fund. Math. 7* (1925), 23–28.

Korányi, A.

[1] "Harmonic functions on hermitian hyperbolic space," *Trans. Amer. Math. Soc. 135* (1969), 507–516.

[2] "The Poisson integral for generalized half planes and bounded symmetric domains," *Ann. of Math. 82* (1965), 332–350.

Korányi, A., and Wolf, J.

[1] "Realization of hermitian symmetric spaces as generalized half planes," *Ann. of Math. 81* (1965), 265–288.

Krein, S. G., and Petunin, J. J.

[1] "Scales of Banach spaces," *Uspehi Math. Nauk*, *21* (1966), No. 2 (128), 89–168.

Krein, S. G., and Semenov, E. M.
[1] "On a space scale," *Soviet Math. Dokl. 2* (1961), 706–710.

Küran, Ü.
[1] "On subharmonicity of nonnegative functions," *J. London Math. Soc. 40* (1965), 41–46.

Lions, J. L.
[1] "Théorèmes de traces et d'interpolation I, II," *Ann. Scuola Norm. Sup. Pisa, 13* (1959), 389–403; *15* (1960), 317–331; III: *J. Math. Pures Appl. 42* (1963), 195–203.

Lions, J. L., and Peetre, J.
[1] "Sur une classe d'espaces d'interpolation," *Inst. Hautes Etudes Sci. Publ. Math. 19* (1964), 5–68.

Loomis, L.
[1] "A note on Hilbert's transform," *Bull. Amer. Math. Soc. 52* (1946), 1082–1086.

Lorentz, G. G.
[1] "Some new functional spaces," *Ann. of Math. 51* (1950), 37–55.

Magenes, E.
[1] "Spazi d'interpolazione ed equazioni a derivate parziali," *Atti. Cong. Un. Mat. Ital.*, Genoa, 1963, 134–197.

Marcinkiewicz, J.
[1] "Sur les multiplicateurs dés series de Fourier," *Studia Math. 8* (1939), 78–91.
[2] "Sur l'interpolation d'operations," *C. R. Acad. des Sciences, Paris 208* (1939), 1272–1273.

Marcinkiewicz, J., and Zygmund, A.
[1] "On the summability of double Fourier series," *Fund. Math. 32* (1939), 112–132.

Mihlin, S. G.
[1] *Multidimensional Singular Integrals and Integral Equations*, Internat. Series of Monographs in Pure and Applied Math., Vol. 83, Pergamon Press, 1965.

Naimark, M. A.
[1] *Normed Rings*, P. Noordhoff, Ltd., Groningen, 1964.

Oklander, E. T.
[1] "L_{pq} interpolators and the theorem of Marcinkiewicz," *Bull. of the A.M.S. 72* (1966), 49–53.
[2] "On interpolation on Banach spaces," Ph.D. thesis, University of Chicago, 1964.

O'Neil, R., and Weiss, Guido
[1] "The Hilbert transform and rearrangement of functions," *Studia Math. 23* (1963), 189–198.

Plancherel, M., and Polya, G.
[1] "Fonctions entières et intégrales de Fourier multiples," *Comm. Math. Helv. 9* (1937), 224–248.

Plessner, A.

[1] "Über die Verhalten analytischer funktionen am Rande ihres Definitions-bereiches," *J. fur reine und Angewandte Math. 159* (1927), 219–227.

Pyatetskii-Shapiro, I. I.

[1] *Géométrie des Domaines Classiques et Théorie des Fonctions Automorphes*, Dunod, Paris, 1969.

Rado, T.

[1] *Subharmonic Functions*, Chelsea Publ. Co., New York, 1949.

Riesz, M.

[1] "Sur un théorème de la moyenne et ses applications," *Acta Sz. 1* (1923), 114–126.

[2] "Sur les maxima des formes bilinéaires et sur les fonctionnelles linéaires," *Acta Math. 49* (1926), 465–497.

Rockafellar, R. T.

[1] *Convex Analysis*, Princeton University Press, 1970.

Rothaus, O. S.

[1] "Domains of positivity," *Abh. Math. Sem. Hamburg 24* (1960), 189–235.

Royden, H. L.

[1] *Real Analysis*, Macmillan, New York, 1963.

Rudin, W.

[1] *Fourier Analysis on Groups*, Interscience Publ., New York, 1962.

Saks, S.

[1] *Theory of the Integral*, Hafner Publ. Co., New York, 1938.

Salem, R., and Zygmund, A.

[1] "A convexity theorem," *Proc. Nat. Acad. U.S.A. 34* (1948), 443–447.

Schwartz, L.

[1] *Théorie des Distributions*, Hermann, Paris, 1957.

Shapiro, V.

[1] "Fourier series in several variables," *Bull. Amer. Math. Soc. 70* (1964), 48–93.

Siegel, C. L.

[1] "Über Gitterpunkte in convexen Körpern und ein damit zusammenhängendes Extremal problem," *Acta Math. 65* (1935), 307–323.

Stein, E. M.

[1] "Functions of exponential type," *Ann. of Math. 65* (1957), 582–592.

[2] "Interpolation of linear operators," *Trans. Amer. Math. Soc. 83* (1956), 482–492.

[3] *Singular Integrals and Differentiability Properties of Functions*, Princeton University Press, 1970.

[4] "Localization and summability of multiple Fourier series," *Acta Math.* 100 (1958), 93–147.

[5] "On limits of sequences of operators," *Ann. of Math. 74* (1961), 140–170.

[6] "Note on the boundary values of holomorphic functions," *Ann. of Math. 82* (1965), 351–353.

[7] "On certain exponential sums arising in multiple Fourier series," *Ann. of Math. 73* (1961), 87–109.

Stein, E. M., and Weiss, Guido
[1] "On the theory of harmonic functions of several variables," *Acta Math. 103* (1960), 26–62.
[2] "An extension of a theorem of Marcinkiewicz and some of its applications," *J. Math. Mech. 8* (1959).
[3] "On the interpolation of analytic families of operators on H^p spaces," *Tohoku Math. J. 9* (1957), 318–339.
[4] "Generalizations of the Cauchy–Riemann equations and representations of the rotation group," *Amer. J. Math. 90* (1968), 163–196.
[5] "Interpolation of operators with change of measures," *Trans. Amer. Math. Soc. 87* (1958), 159–172.

Stein, E. M., Weiss, Guido, and Weiss, M.
[1] "H^p classes of holomorphic functions in tube domains," *Proc. Nat. Acad. Sci. U.S.A. 52* (1964), 1035–1039.

Stein, E. M., and Weiss, N. J.
[1] "On the convergence of Poisson integrals," *Trans. A.M.S. 140* (1969), 34–54.

Streater, R. F., and Wightman, A. S.
[1] *PCT, Spin and Statistics, and All That*, Benjamin, New York, 1964.

Taibleson, M. H.
[1] "Translation invariant operators, duality, and interpolation, II," *J. Math. Mech. 14* (1965), 821–840.

Tamarkin, J. D., and Zygmund, A.
[1] "Proof of a theorem of Thorin," *Bull. A.M.S. 50* (1944), 279–282.

Thorin, G. O.
[1] "An extension of a convexity theorem due to M. Riesz," *Kungl. Fysiografiska Saellskapet i Lund Forhaendlinger, 8* (1939), No. 14.

Titchmarsh, E. C.
[1] "Additional note on conjugate functions," *J. London Math. Soc. 4* (1929), 204–206.
[2] *Introduction to the Theory of Fourier Integrals*, Clarendon Press, Oxford, 1962.

Valentine, F. A.
[1] *Convex Sets*, McGraw-Hill, New York, 1964.

Vilenkin, N. J.
[1] *Special Functions and the Theory of Group Representations*, Vol. 22, Translations of Math. Monographs, Am. Math. Soc., Providence, 1968.

Vinberg, E. B.
[1] "Homogeneous cones," *Dokl. Russ. Ac. Sci. 133* (1960) (in Russian).

Wainger, S.
[1] "Special trigonometric series in k dimensions," *Mem. Amer. Math. Soc. 59* (1965).

Watson, G. N.
[1] *A Treatise on the Theory of Bessel Functions*, Cambridge University Press, Cambridge, 1922.

Weil, A.

[1] *L'intégration dans les Groupes Topologiques et ses Applications*, Hermann, Paris, 1965.

Weiss, Guido

[1] "An interpolation theorem for sublinear operators on H^p spaces," *Proc. Amer. Math. Soc. 8* (1957), 92–99.

[2] *Analisis Armonico en Verias Variables. Teoria de los Espacios H^p*. Cursos y seminarios de Matematica, Universidad de Buenos Aires, fasc. 9.

[3] *Harmonic Analysis*, M.A.A. Studies in Mathematics, Vol. 3, I.I. Hirschman, Jr., ed., Prentice-Hall, 1965, 124–178.

Weyl, H.

[1] *The Classical Groups*, Princeton University Press, 1946.

Wiener, N.

[1] *The Fourier Integral and Certain of its Applications*, Cambridge University Press, Cambridge, 1935.

[2] "The ergodic theorem," *Duke Math. J. 5* (1939), 1–18.

Yosida, K.

[1] *Functional Analysis*, Springer, Berlin, 1968.

Zygmund, A.

[1] *Trigonometric Series*, 2nd ed., Cambridge University Press, Cambridge, 1968.

[2] "On a theorem of Marcinkiewicz concerning interpolation of operations," *Journal de Math. 35* (1956), 223–248.

[3] "On the boundary values of functions of several complex variables," *Fund. Math. 36* (1949), 207–235.

Index